Electrooptics

PHENOMENA, MATERIALS
AND APPLICATIONS

Electrooptics

PHENOMENA, MATERIALS AND APPLICATIONS

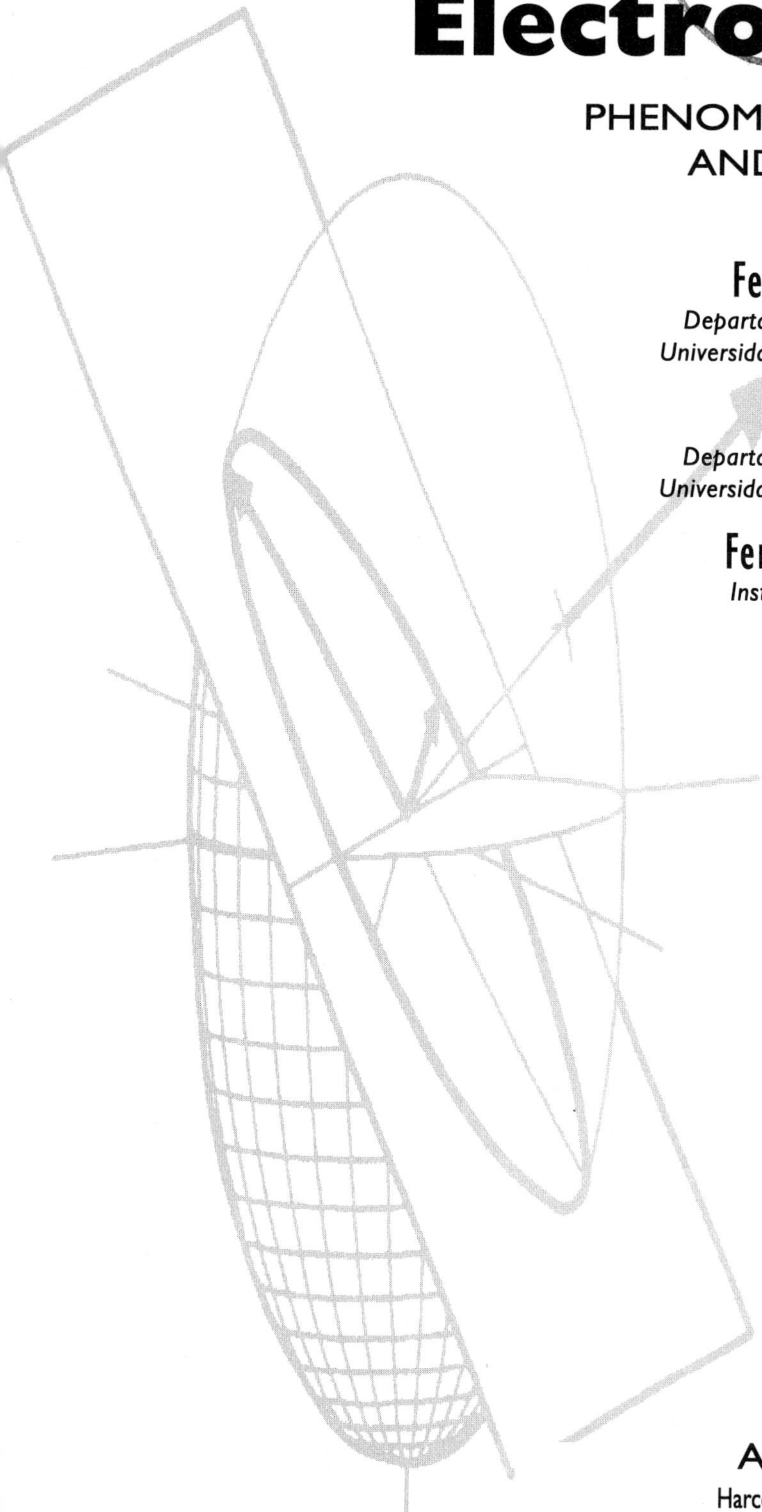

Fernando Agulló-López
Departamento de Física de Materiales
Universidad Autónoma de Madrid, Spain

José Manuel Cabrera
Departamento de Física de Materiales
Universidad Autónoma de Madrid, Spain

Fernando Agulló-Rueda
Instituto de Ciencia de Materiales
de Madrid (CSIC), Spain

ACADEMIC PRESS
Harcourt Brace & Company, Publishers
London San Diego New York
Boston Sydney Tokyo Toronto

ACADEMIC PRESS LIMITED
24/28 Oval Road
London NW1 7DX

United States Edition published by
ACADEMIC PRESS INC.
San Diego, CA 92101

This book is printed on acid-free paper

A catalogue record for this book is available from the British Library.

ISBN 0-12-044512-3

Typeset by Alden Multimedia
and printed in Great Britain by The University Printing House, Cambridge

Contents

6. Organic Electrooptic Materials

Series Foreword

Lasers and Optical Engineering

In the last three or four decades, we have seen the return of optics to its former position of scientific and engineering prominence. It was largely the development of the laser which proved to be the catalyst for this re-birth, and its consequent spawning of dependant technologies such as fibre optics, optical communications, optical computing, laser materials processing and holography. The impact of this "new-wave" in optics has been so overwhelming that it is now possible to perceive of it as a distinct academic discipline in its own right, the roots of which lie in classical optics, materials science, electromagnetism, electronics and computing. The practitioners of this discipline may be physicists, chemists, botanists, biologists, mechanical, civil and electrical engineers, and the applicability of their skills stretches across the entire spectrum of science and technology.

It is to reflect the systems and applied nature of modern optics and to affirm our belief that we are witnessing the evolution of a new breed of engineer, the optical engineer, that we have introduced this series of books on Lasers and Optical Engineering. Our overall objective is to project the multi-disciplinary and wide-ranging coverage of optical engineering and underpin this with a strong commitment to fundamental principles. The series is aimed at the advanced student and the practising optical engineer, whether in industry or academia. Our chosen authors will represent a balance between the established practitioner and the younger scientist.

The first book in the series concentrates on an area which lies at the very heart of optical engineering, namely electrooptics itself. We can make very little progress in optical engineering without an understanding of the materials, the principles and the application of electrooptic phenomena. This book, by Fernando Agulló-López and his colleagues is, therefore, a highly appropriate opener to the series and sets the scene to our cross-disciplinary philosophy. Future planned volumes will cover Optical Thin Films, Non-linear Optics, Lasers in Medicine, Optical Fibre Sensing, Optical Storage Technologies, Optical Image and Information Processing, Imaging Science, Laser Design & Development, Semiconductor Laser Technology, Adaptive Optic Systems, Lens and Optical System Design, Robotics and Machine Vision, and Laser Safety.

Dr John Watson
University of Aberdeen
April 1994

Preface

The interrelation between electrical and optical phenomena is the subject of an applied science called Optoelectronics, which results from the synergy of various fields of Science and Engineering: Optics, Electronics and Materials. It covers diverse topics, such as electrochromism, electroluminescence, displays, laser physics, solar cells and light detectors, which, aside from their scientific relevance, are the basis for advanced technologies in Communications, Defence, Energy and Computers. A particular, but fundamental aspect of that vast area is traditionally known as Electrooptics and refers specifically to the local changes induced by an applied electric field on the optical parameters of materials (refractive index and absorption coefficient), namely the Kerr and Pockels effects.

This is a book on Electrooptics in the above restricted sense. There are several classical books and monographs dealing mainly with particular aspects of Electrooptics, e.g. fundamentals, experimental techniques, materials, or applications. Much progress has recently been made on efficient electrooptic materials, specially semiconductor nanostructures and organics. This has produced new promising applications, such as the self-electrooptic-effect device (SEED). Therefore, it appears opportune to produce a book that encompasses all these aspects and offers a coherent, balanced and up-to-date presentation of the subject. An effort has been made to provide a sound and rigorous physics background, while keeping a tutorial level and avoiding too complex subtleties. The book, which has arisen from several courses and seminars given by the authors at the Autonomous University of Madrid, is suitable for postgraduate courses in Advanced Optics, Optoelectronics, Electrooptics, Laser Physics and Applications, etc. In addition to providing the scientific background, it is intended to take the reader to the frontiers of research, so that they may dig, without excessive effort, into the more specialized literature.

The book is organized as follows. Chapter 1 provides a general introduction to electrooptics and a succinct up-to-date survey of related phenomena not covered in the other chapters, such as electroluminescence, diode lasers and electrochromism. Chapters 2 to 4 describe the formalism and basic concepts used throughout the book, such as light propagation in anisotropic media, the Pockels and Kerr effects, measuring techniques and nonlinear susceptibilities. Nonlinear optics provides the appropriate framework for an in-depth description of electrooptic phenomena. Chapters 5 and 6 deal with electrooptic materials, both inorganic and organic, whereas Chapter 7 reviews their main bulk applications. Chapter 8 and 9 refer to waveguides and semiconductor nanostructures, such as quantum wells and superlattices, and related devices. These structures offer ample possibilities of design and of integration in optoelectronic integrated circuits. Finally, Chapter 10 and 11 are devoted to the photorefractive effect and its potential applications. This effect derives directly from the electrooptic behaviour and offers a very good example of charge-transport-induced nonlinearities.

A number of colleagues have contributed to many parts of the book with their comments, suggestions, criticisms, or research collaborations. The following deserve special mention: Dr J. Etchebarría, Dr P. Durán, Dr M. Carrascosa, Dr A. García-Cabañes, Dr L. Arizmendi, Dr C. Zaldo, Dr E. E. Mendez, Dr H. T. Grahn, L. Viña and Dr F. Meseguer, and our research students R. Müller, J. Olivares, M. A. Díaz and A. Alcázar, E. Serano and M. Aguilar. Our appreciation to all of them. Thanks are also due to Yayo for the typing of one chapter, and to Rocío, Carmen, and Javier for their careful draftsmanship.

Last but not least, we deeply appreciate the patience, stimulus and understanding of our wives Carmen, Josefina and Esther.

Chapter One

The Vast World of Electrooptics

1.1. Introduction

Electrooptics is a vast world. The interrelation between electric fields and optical phenomena presents a large variety of facets and offers an immense diversity of applications. A general optoelectronic system consists of three main functional stages as illustrated by the block diagram in Figure 1.1. The first block involves the generation of the optical signal and may include a laser, a light-emitting diode (LED), or a conventional light source. The second block stands for the control and manipulation of the optical signal and includes such devices as switches, modulators, deflectors, multiplexers, frequency-doublers, and so on. Finally, the third block refers to the detection and display of the signal. All the three blocks rely heavily on a variety of *electrooptic phenomena.*

Diode lasers or LEDs convert an electrical signal into coherent or incoherent light emission, respectively. The corresponding electrooptic mechanisms are then called *active.* Control and manipulation of light depend on *passive* electrooptic phenomena, not implying light generation. They include *local* mechanisms as the *Pockels* and *Kerr* effects as well as those involving charge-transport processes as *electrochromic* and *photorefractive* effects. Finally, light detection requires the conversion of an optical signal into an electrical signal and involves mostly *photoconductive*, *photoelectric* or *photovoltaic* effects. They are, in a way, the inverse of those operating in stage 1. The display of the transmitted information involves the conversion of electrical into optical signals. Here, again, electrooptic phenomena are required, as in stage 1, to carry out this task. In summary, electrooptics plays a key role in all stages of an optoelectronic system.

The treatment of all this variety of phenomena constitutes the objective of a vast science called *optoelectronics*, *optical electronics* or *photonics*, although many authors use the name *electrooptics.* In fact, excellent books on the subject, offering similar or strongly overlapping contents (although quite different levels) use a diversity of titles including: Optical Electronics [1], Quantum Electronics [2,3], Optoelectronics [4], Laser Electronics [5] and Photonics [6]. To some extent, the title is a matter of personal choice, although Photonics may imply a broader coverage. Related, but better differentiated, subjects include Quantum Optics, Statistical Optics, Coherent Optics, and Fourier Optics.

This book focuses on the passive phenomena acting in stages 2 and 3 of Figure 1.1 and so electrooptics is used here in a more restricted and historical sense. It mainly concerns local effects induced by an electric field on the optical parameters of a material, and more specifically the Pockels (linear) and Kerr (quadratic) effects on the refractive index. This restriction permits a more coherent and in-depth presentation of the subject. However, in order to provide a wider perspective of the whole range of phenomena and to

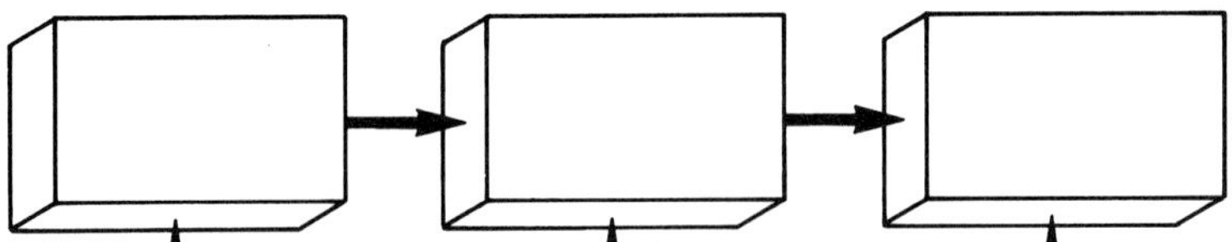

Figure 1.1 *Functional stages in an optoelectronic system.*

permit a comparative evaluation of the field, the other passive (electrochromism) as well as active (electroluminescence) electrooptic effects will be sketchy discussed in this first chapter.

1.2. Historical notes

The first discovery of electrooptic phenomena dates back to the experiments of John Kerr reported in 1875 [7]. He studied the birefringence induced by an electric field in some liquids, e.g. CS_2, and in glass (*Kerr effect*). It was a reversible effect, whose magnitude was proportional to the square of the electric field. Curiously, the first-order electrooptic effect was discovered in later experiments carried out simultaneously but independently by Roentgen [8] and Kundt [9] in 1883 on quartz (lacking inversion symmetry). Pockels [10] proposed an adequate explanation for these observations, which became known as the *Pockels effect.*

Many more electrooptic materials have since then been discovered and investigated. One may mention first reports on KDP and DKDP [11,12], $LiNbO_3$ [13,14], $BaTiO_3$ [15,16], etc. The Pockels and Kerr effects entered into the era of optoelectronic technologies early in the 1950s when the use of KDP and ADP for light modulation and switching was proposed and demonstrated. In modern devices $LiNbO_3$ is the material of choice, where modulation frequencies of 10–100 GHz have been already reached.

The field of transparent electrooptic ceramics, as an alternative to single crystals, starts with the first report [17] on PLZT (lead lanthanum zirconate titanate) in 1971. This material continues to keep a leading position among transparent ceramics in present-day technology and a number of devices are already in the market.

Liquid crystals [18], intermediate phases between the liquid and solid state, were discovered in 1888. The field reached maturity with the classical book by de Gennes [19] and has continued to develop at a high rate in both the scientific and the technological aspects. Their remarkable electrooptical properties have led to a large diversity of products in the consumer electronics market. However, they still keep the handicap of a relatively slow response time (a few ms), hindering their use in faster technologies. An important breakthrough was the discovery of ferroelectric liquid crystals [20] in 1975. They operate at a much higher speed (a few μs), and so offer good opportunities for application in such fields as complex displays and spatial light modulators.

In 1961, second-harmonic generation in quartz was discovered by Franken and coworkers [21]. This was the onset of a new field, *Nonlinear Optics* [22,23], which has reached a high level of maturity and has set the foundation for the application of optics to communication and computing technologies. On the other hand, its formulation in terms of susceptibilities provides the natural framework to describe the classical Pockels and Kerr electrooptic effects as nonlinear optical phenomena. Consequently, the fields of Electrooptics and Nonlinear Optics are now intimately connected and the search for efficient electrooptic materials runs parallel to that for materials presenting high nonlinear responses.

Recently, this search has focused on organic compounds [24], stimulated by the capabilities and flexibility of organic molecular engineering. This constitutes a relevant aspect of a broad emerging field, *molecular electronics*, which aims to develop electronic devices at a molecular level. In particular, active electrooptic (nonlinear) molecules incorporated in polymer hosts may provide the adequate chemical and mechanical performance needed for industrial utilization.

In the mid-1930s, another electrooptic effect, *electroluminescence*, was discovered by Destriau [25] in ZnS powders doped with Cu. It consists of the nonthermal emission of light induced by the application of an electric field. Although the effect has not yet met the initial expectancies, it is presently used in electroluminescent displays. Prior to this bulk effect, injection electroluminescence was discovered in 1923 by Losev [26] on accidental *p–n* junctions in SiC, and more thoroughly investigated and understood by Haynes and others in the 1950s and 1960s [27,28]. The rapid progress in the field has led to a number of useful *light-emitting diodes* (LED) using GaAs, GaN, ZnS and others.

First reports of electric field-induced changes in the colour of a material, *electrochromism*, dates back to 1951 [29] on Na_xWO_3. Another significant electrochromic effect was the electrolytic coloration of alkali halides [30]. All these phenomena involve electron (and even ion) transport and should be distinguished from the changes in the absorption spectra associated with the Pockels and Kerr electrooptic effects, which do not involve any transfer of charge and generally require much higher fields. The first electrochromic display device was reported in 1969 [31].

The main historical breakthrough that has fostered the study and application of all the above electrooptic phenomena is the advent of lasers in 1959 [32]. It set the origin for a tremendous rise in electrooptic developments and the surge of a new scientific field, *optoelectronics* or *photonics*. In fact, electrically pumped lasers are genuine electrooptic devices that convert an electrical current into a coherent optical beam. Of particular relevance in industrial and economic terms are *semiconductor diode lasers* [33–35], which constitute a major fraction of the laser market, as illustrated in Figure 1.2. In the last few years we have witnessed an ever growing improvement of performance characteristics and miniaturization of diode lasers and optoelectronic components to meet the demands of the communication and information technologies. This effort has led to the rise of the field of *integrated optics* [37] as the optical counterpart to microelectronics. A very important step in that direction has been the fabrication of *multiple quantum well* (MQW) lasers [38,39] by molecular beam epitaxy (MBE) or metal-organic chemical vapour deposition (MOCVD). These techniques (particularly MBE) permit precise control of the multilayer nanostructures required for efficient laser emission. Very recent developments in the field include: diode lasers operating in the blue region of the visible spectrum, the demonstration of organic LEDs, and the discovery of efficient visible emission from porous silicon. This latter effect may lead to silicon-based optoelectronics.

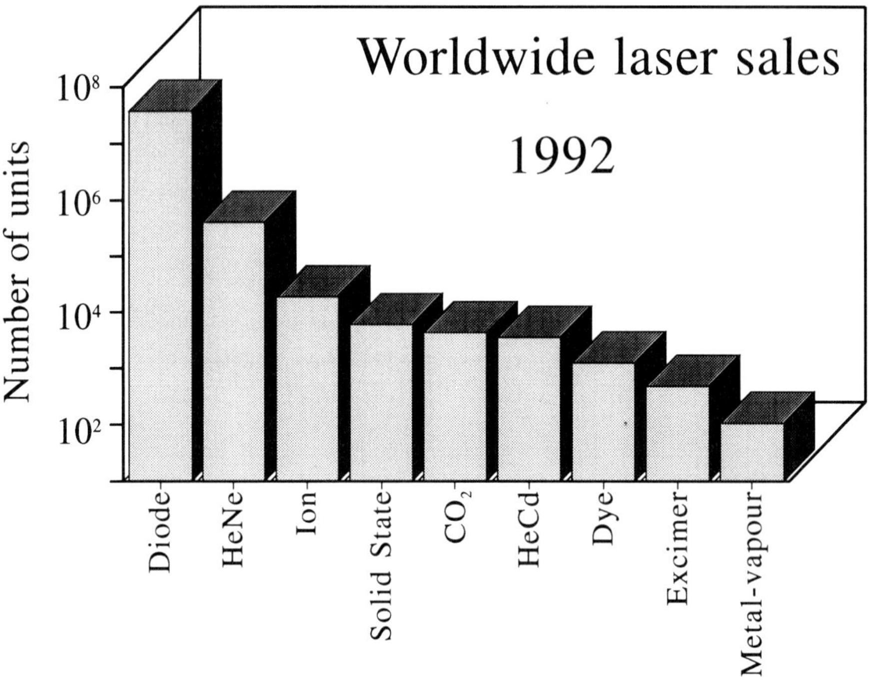

Figure 1.2 *Laser market share in logarithmic scale. [36].*

1.3. Injection electroluminescence: LEDs

This term refers to the light radiated when a forward bias voltage is applied to a *p–n* junction [40]. In the absence of an applied electric field, the Fermi levels on both sides of the junction are matched. On applying a forward bias, minority carriers are injected, electrons into the *p* side and holes in the *n* material, so that a steady current density

$$j = j_e + j_h \tag{1.1}$$

is established. The current–voltage relationship is

$$j = j_0(e^{eV/\beta kT} - 1). \tag{1.2}$$

For $V < 0$ (*reverse bias*) almost no current flows. For $V > 0$ (*forward bias*), the current density increases exponentially with V (Figure 1.3).

The injected minority carriers recombine with the majority carriers and yield photons with energy equal to the semiconductor bandgap E_g. The resulting electroluminescent device is called a *light-emitting diode* (LED). In order to be a useful LED, the semiconductor should have a *direct gap*, i.e. the maximum of the valence band and the minimum of the conduction band should occur at the same **k** value (generally $\mathbf{k} = 0$). Radiative transitions in indirect gap materials require the assistance of phonons and so have emission yields many orders of magnitude lower than for direct gap semiconductors. Consequently they are much less efficient for LED fabrication. However, introduction of suitable dopants assists the emission process and leads to efficient operation. There are already a number of commercial LED materials whose main characteristics are summarized in Table 1.1.

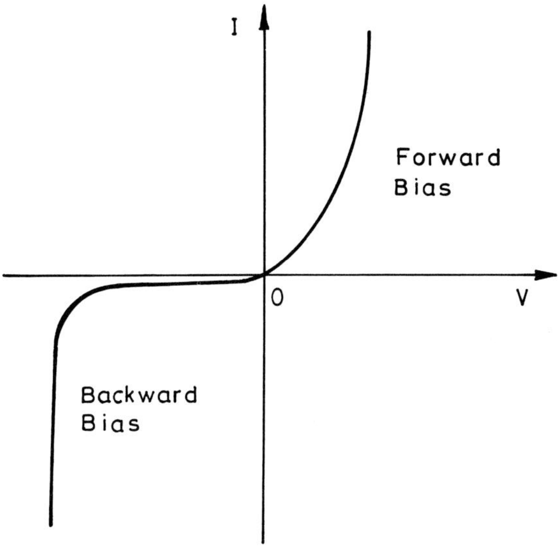

Figure 1.3 *I–V curve for a p–n junction.*

Table 1.1 *Some inorganic semiconductors used for light-emitting diodes (LEDs).*

Infrared	Red	Orange	Yellow	Green	Blue	Violet	Ultraviolet
GaAs	GaAsP	GaAsP	GaAsP	GaP	βSiC	BeTe	GaN
GaInAsP	CdSe	Cu_2O	GaSe	ZnTe	ZnSe	CuBr	ZnS
InP	AlGaAs	GaInP	$ZnSiAs_2$	$CuGaS_2$	In_2O_3	Ga_2S_3	αSiC
CdTe	GaInAsP	GaAlInP		CdP_2	$CuAlSe_2$		SnO_2
$CdGeP_2$	$AgInS_2$			$MgSiP_2$	CdS		$CuAlS_2$
$CuGaSe_2$					GaS		ZnO
$CuInS_2$							GaS_2

The power emitted by a LED is proportional to the injection current at low currents and experiences saturation at high currents. The spectral emission full width at half maximum intensity (FWHM) is proportional to the square of the wavelength of maximum intensity, and varies from about 1200 Å at 1.3 μm (infrared) to 200 Å at 0.55 μm (green) [41]. The response time is limited primarily by the recombination time of the injected minority carriers. A typical value is 10 ns, leading to an operating bandwidth of 100 MHz.

Both surface-emitting and edge-emitting LEDs are available. In the first case one has an approximately Lambertian angular distribution of the emitted light. In the second case the angular profile becomes narrower.

LEDs are extensively used in alphanumeric displays, moving-message panels, car brake lights, indicators of electronic equipment, etc. In the last decade they have been displaced by less power consuming LCDs (see section 6.8) in displays of battery operated devices like watches and calculators.

1.3.1. Polymer LEDs

Recent progress in the processing of conjugated polymers, such as PPV (poly(phenylenevinylene)), has permitted very high luminescence yields (50%) from exciton decay [42]. Therefore, it has become possible to fabricate electroluminescent devices or LEDs that operate through the injection of electrons and holes in a polymer film from suitable electrodes [43–45]. The electron–hole pairs recombine and give rise to light emission. Voltages as low as 10 V across a 100 nm thick film have been successful. Blue organic-polymer LEDs have been recently obtained [46].

1.3.2. Porous silicon

A great deal of interest has followed the discovery of visible photoluminescence from porous silicon, prepared by electrochemical etching of a silicon wafer [47,48]. The structure of the luminescent material can be thought of crystalline silicon wires with diameters of a few nanometres. Quantum confinement of electrons and holes in those quantum wires may account for the luminescence behaviour, although other explanations have also been advanced. LEDs of porous silicon emitting different colours have been recently demonstrated [49,50]. This achievement could provide monolithic integration of lasers and other optoelectronic devices into present silicon integrated circuits and will stimulate an intense research effort in the coming years.

1.4. Bulk electroluminescence

Pure injection electroluminescence considered in the previous section appears in a thin layer along the interface between the *n* and *p* sides of a semiconductor single crystal. Electroluminescence in polycrystalline semiconductors considered in this section, on the contrary, is emitted throughout the entire volume of the material, and therefore we will call it bulk electroluminescence.

Bulk electroluminescent materials (*phosphors*) are polycrystalline wide-gap II–VI semiconductors (formed by atomic elements of the columns II and VI of the periodic table) doped with some ion, which acts as the luminescent centre [51,52]. Therefore the characteristics of the emission depend very much on the doping (see Table 1.2).

Table 1.2 ***Phosphors employed for electroluminescent displays in the different regions of the visible spectrum.***

Colour		Reference
Red	ZnS:Sm,F	[53]
	ZnS:Sm,Cl	[53]
	ZnS:Sm,P	[54]
	CaS:Eu	[53]
	CaS:Eu,F,Cu,Br	[54]
Green	ZnS:Tb	[53]
	ZnS:TbF3	[53]
	ZnS:Tb,F	[54]
Blue	ZnS:Cu,Cl	[53]
	ZnS:Tm	[53]
	SrS:Ce	[53]
	SrS:Ce,F	[54]

Electrons injected into the phosphor are accelerated by the electric field. When these electrons collide they excite the luminescent centres, by mechanisms not well understood yet. Luminescence is produced when these centres recombine radiatively [53].

Bulk electroluminescence is very attractive for large-area applications in illumination and in flat-panel displays [55], because single-crystal LEDs are too expensive. Although much effort was devoted to its development in the 1950s, practical applications were restricted by the rapid degradation of the devices. Since the 1970s new improvements have made possible the fabrication of commercial devices. Two main types can be distinguished: powder and thin film devices.

1.4.1. Powder electroluminescence devices

Bulk electroluminescence was discovered in powders of ZnS:Cu [25], but the luminescence degraded very fast due to migration of Cu ions through S vacancies. Renewed interest in the 1970s was possible after the improvements made in ZnS:Mn phosphors by Vecht [56]. Powder electroluminescence devices are very suitable for large area applications because they do not require expensive vacuum deposition techniques. They are used, for example, as backlighting elements in LCD displays.

Typical configurations of powder electroluminescence displays for d.c. and a.c. voltage operation are shown in Figure 1.4(a), (b) [52]. The different layers are deposited on top of a glass substrate, which serves as the front surface of the device. The active phosphor layer consists of small powder particles (close to 1 μm in diameter). For the d.c. device, ZnS:Cu,Mn particles are coated with Cu_xS, whereas for the a.c. device, ZnS:Cu,Mn,Cl particles are embedded in a dielectric matrix. The phosphor layer is sandwiched between two contact layers to apply the voltage, which is of the order of 100 V. The back electrode is a metal such as aluminium. The front electrode must be of a transparent conducting material such as indium-tin oxide (ITO) or SnO_2. Between this electrode and the phosphor there are ZnS:Mn and insulator layers for the d.c. and a.c. devices, respectively.

1.4.2. Thin-film electroluminescence devices (TFEL)

TFEL devices are very promising for flat-panel colour displays like television and computer monitors. The main advantages are their high brightness, high resolution with reduced cross-talk between individual pixels, large viewing angle, broad range of operation temperatures, stability, long life, and intrinsic memory. The main disadvantage is the high operating voltage (of the order of 100 V), which makes the control electronics more expensive than for LCD displays. The most developed are devices operating with an a.c. voltage, which are called ACTFELs. A typical device [57] (Figure 1.4(c)) consists of a phosphor semiconductor thin film (0.5–1.0 μm thick) sandwiched between two insulator layers (e.g. SiO_xN_y, Y_2O_3 or $PbTiO_3$, 0.3 μm). An Al back electrode and a transparent front electrode (e.g. indium-tin oxide or SnO_2) serve to apply the voltage. The different layers are deposited on top of a glass substrate. Light is emitted when the applied voltage exceeds the threshold voltage, which corresponds to a field of about 1 MV/cm in the phosphor film [54]. The emission reaches saturation at high voltages. The behaviour has some hysteresis and emission continues until the voltage decreases a few tens of volts below the threshold voltage. The different colours are obtained by using appropriate phosphors (see Table 1.2).

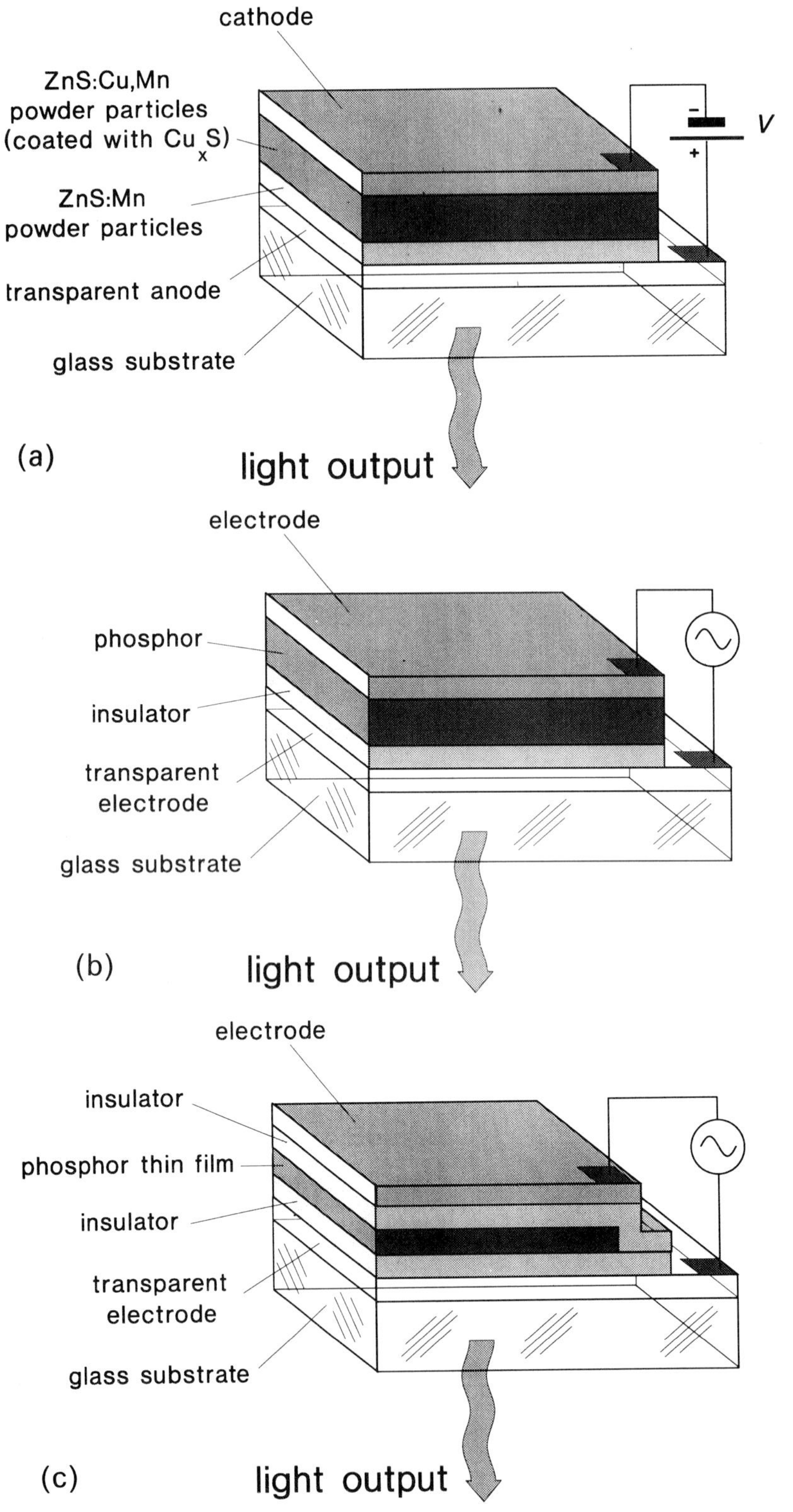
cathode
ZnS:Cu,Mn
powder particles
(coated with Cu_xS)
ZnS:Mn
powder particles
transparent anode
glass substrate
V
(a)
light output
electrode
phosphor
insulator
transparent
electrode
glass substrate
(b)
light output
electrode
insulator
phosphor thin film
insulator
transparent
electrode
glass substrate
(c)
light output

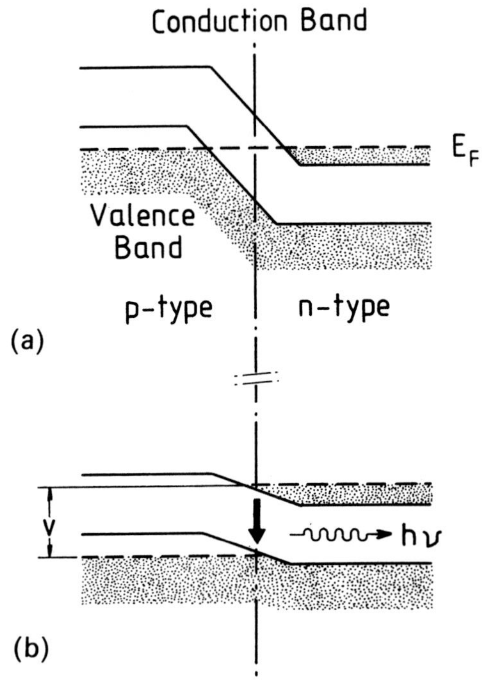

Figure 1.5 *Degenerated* p–n *junction (a) without and (b) with an applied electric field.*

1.5. Diode lasers

The basic physical principle of diode lasers [58] is the same as that of LEDs. However, in order to obtain *gain* by stimulated emission, one has to achieve a population inversion of carriers. This situation is reached in heavily doped (*degenerated*) junctions on application of a forward bias voltage V, as illustrated in Figure 1.5. Moreover, to convert the light amplifier into an oscillator (*laser*), a feedback mechanism has to be provided. This can be accomplished by using the reflectivity at the crystal faces or by means of external mirrors or gratings. The active or light-emitting region at the junction constitutes an optical waveguide since its refractive index is higher than in the bulk of the material due to the excess concentration of carriers. However, in *homojunction* lasers, involving a *p–n* junction of the same semiconductor, the *threshold current*, required for lasing action is very high at room temperature. Therefore, these lasers have to be operated at low temperature in the continuous-wave (c.w.) mode. This behaviour is related to the poor electronic and optical confinement at the junction region, leading to high absorption losses. In fact, injected electrons and holes diffuse away from the junction plane and reach distances of the order of the diffusion lengths before experiencing recombination. This main drawback can be overcome with the single (SH) and double (DH) [59] heterostructure lasers, illustrated in Figure 1.6. The active layer is, then, bounded by two other semiconductors with larger bandgaps. The advantage of these devices is that they achieve much better electronic and

Figure 1.4 *Electroluminescence (EL) display devices: (a) d.c. EL powder display; (b) a.c. EL powder display; (c) thin-film electroluminescence display (TFEL).*

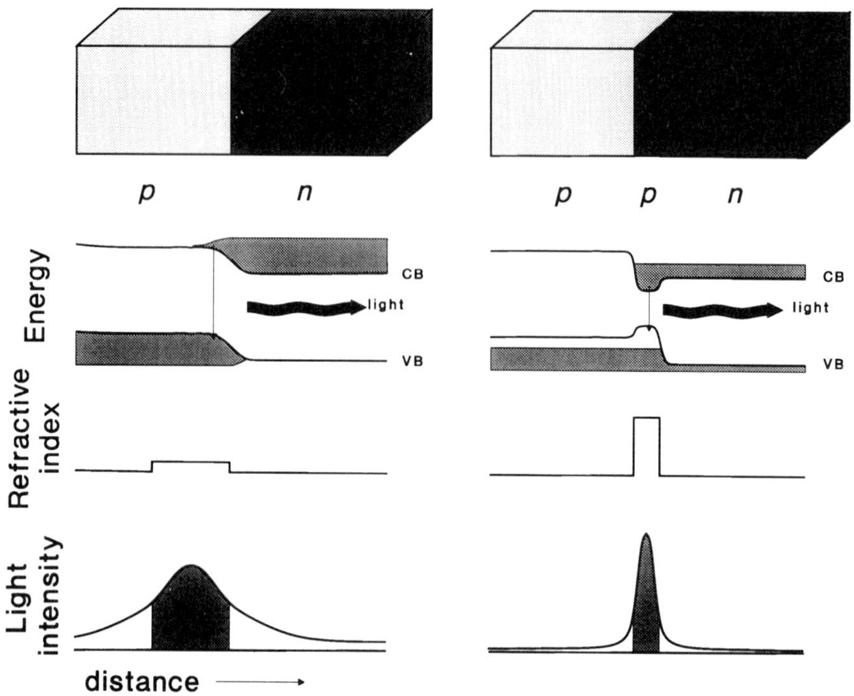

Figure 1.6 *Single (SH) (left) and double heterostructure (DH) (right) diode lasers.*

optical confinement. Considering, as an example, a DH laser, carrier recombination is confined to the active layer, since the diffusion of electrons and holes into the adjacent *p* and *n* regions, respectively, is hindered by an energy barrier. On the other hand, the superior optical confinement derives from the abrupt index steps (due to the change in bandgap) at

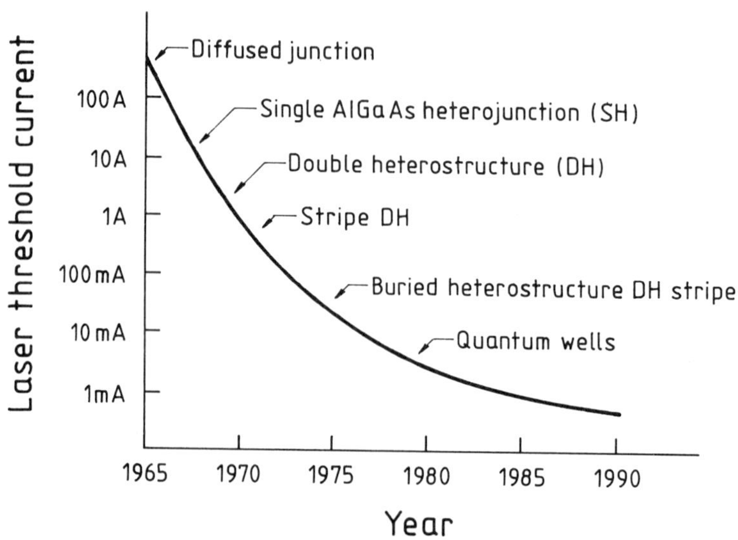

Figure 1.7 *Threshold current of diode lasers vs year. (Adapted from [36].)*

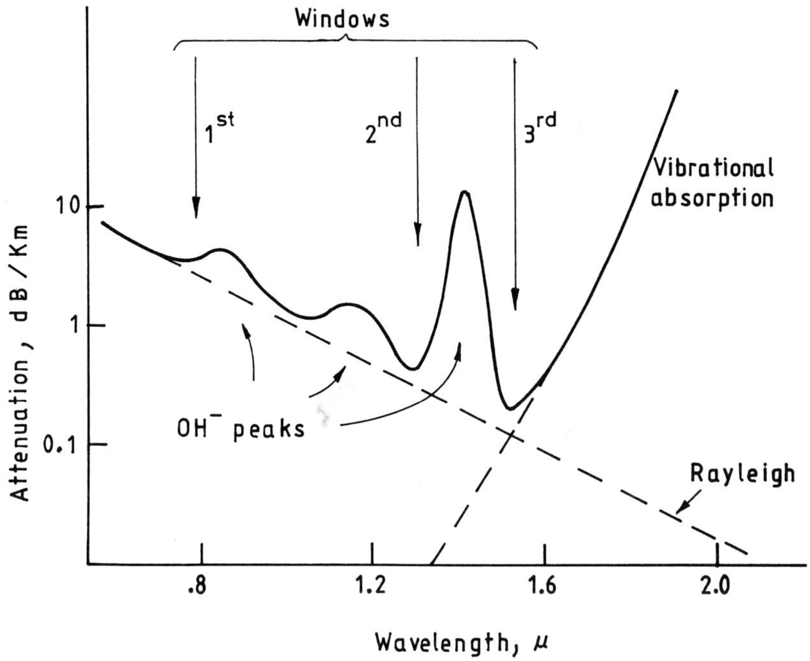

Figure 1.8 *Attenuation of a silica fibre as a function of wavelength. The Rayleigh, vibrational and OH– impurity contributions are shown. The three windows for optical communications are indicated by arrows.*

the interfaces limiting the active layer. In fact, DH lasers show a much reduced threshold current density (Figure 1.8) and can be operated at room temperature in the c.w. mode.

Due to the mechanism of light emission, diode lasers emit at a wavelength close to the bandgap of the active semiconductor. GaAs has been so far the most widely used material. It emits at 840 nm, which corresponds to the first window for optical silica fibre communications (Figure 1.9). More recently various materials have been developed for other infrared wavelengths, for example the quaternary InGaAsP emitting at 1.3 μm (second window) and at 1.55 μm (third window).

An important feature of diode lasers is their capability of high frequency modulation. In fact, they act as electrooptic modulators responding to variations of the pumping current. Bit modulation rates higher than 1 Gb/s have already been achieved. However, it is often preferable to use a hybrid philosophy, combining the laser with an external modulator (e.g. an $LiNbO_3$-based one). Higher frequencies are then attainable.

Shortcomings of the diode lasers that may be relevant in some applications are a broad spectral range and a large beam divergence. The former property is associated to the interband optical transitions responsible for the emission. It can be used to our advantage if tunability is looked for (see below). On the other hand, the high beam divergence derives from the small thickness of the emitting region.

1.5.1. Narrow-line and tunable lasers

In some applications, as in coherent optical transmission systems, a very narrow spectral width of the laser emission is essential. In order to achieve such a goal [60], wavelength-sensitive elements have to be incorporated to the laser. They may consist of

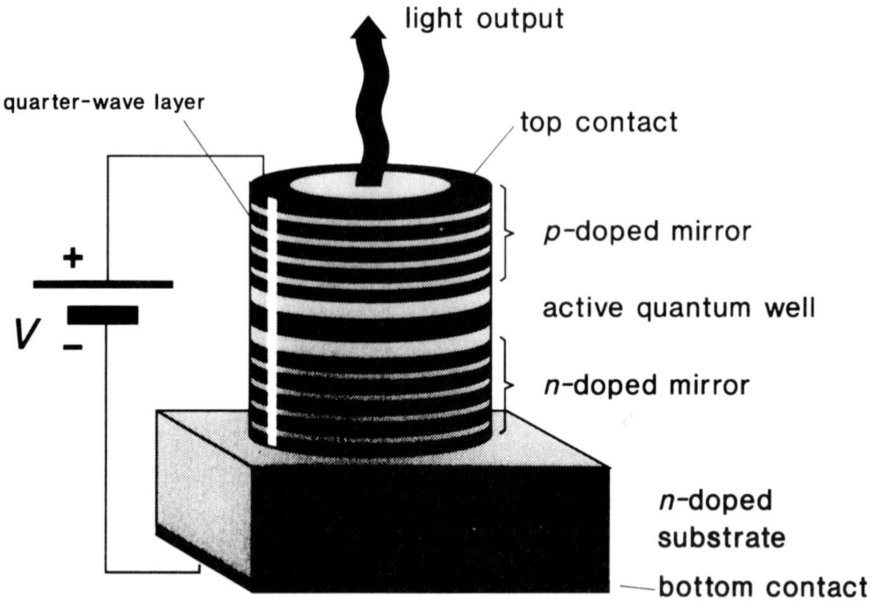

Figure 1.9 *Quantum-well vertical-cavity surface-emitting laser (VCSEL).*

multilayer mirrors or diffraction gratings. When they are placed externally to the laser cavity (hybrid configuration), linewidths of a few kHz can be attained. However, a monolithic approach that incorporates the wavelength-sensitive elements in the laser chip would be, in general, preferable. Unfortunately, the best results with distributed feedback (DFB) lasers, using etched gratings, only reach down to a width of a few MHz.

With the above technologies and taking into account the broad laser gain profile, tunable lasers are being developed [61]. In the monolithic approach, the tunability of the wavelength-sensitive element is achieved through the change in refractive index brought about by a tuning current.

1.5.2. Blue lasers

So far, successful realization of diode lasers has been restricted mostly to III–V semiconductors emitting in the infrared–yellow regions of the electromagnetic spectrum. The fabrication of diode lasers of shorter wavelength is technologically very important. For example, in display applications the whole visible spectrum could be covered. Also, the recording density in compact discs and other optical recording media or the speed and resolution in laser printing could be greatly improved. Finally, blue lasers would make possible optical communications through seawater and ice.

However, the use of wide-gap II–VI semiconductors with energy gap in the green–blue region is hindered by the difficulty of obtaining high-quality *n* and *p* doping for the same material, and therefore of obtaining good *p*–*n* junctions. Recently, new approaches such as growth by molecular beam epitaxy and use of quantum wells have allowed the fabrication of blue lasers of II–VI materials like Zn(S,Se)/(Zn,Cd)Se quantum wells [62].

1.5.3. Quantum well lasers

Quantum well lasers are diode lasers in which the active layer is a semiconductor quantum well. Quantum wells will be described in greater detail in Chapter 9 in connection with their electrooptic properties. A quantum well is formed when a thin layer (50–100 Å) of a semiconductor is surrounded by a semiconductor with a larger bandgap. In that sense, quantum well lasers are similar to double heterostructure lasers, but with a much thinner active layer. Like them they have a very small threshold current, which is proportional to the active volume. Because the well is so thin, quantum effects are very important and the movement of electrons in the direction perpendicular to the well plane is quantized. This manifests in a staircase-like density of states and a bandgap larger than the bulk material. Since the light cannot be confined in layers smaller than about 1000 Å a *separate confinement heterostructure* (SCH) is usually used. Here the well is separated from the outer layers by layers with an intermediate bandgap (and higher refractive index than the outer layers). They serve both to confine the light around the quantum well (in a slab waveguide) and to funnel injected carriers into the well [63]. An improved design is the *graded index separate confinement heterojunction* (GRINSCH) laser, in which the bandgap (refractive index) of the intermediate layers increases (decreases) in the outward directions [64]. The main advantages of quantum well lasers are a lower threshold current [38] due to the staircase density of states (Figure 1.7), an emission wavelength that can be tuned by varying the well thickness, a narrower and higher gain, which makes it easier to reach single-mode oscillation [65], a reduced chirp [66], and high modulation frequencies (close to 30 GHz) [63]. Moreover, in SCH quantum well lasers, degradation is smaller because the optical damage affects only the well layer and not the much larger optical cavity.

1.5.4. Surface-emitting lasers (SEL): microlasers

SELs are lasers emitting in a direction perpendicular to the layer planes [67–69], at variance with the common diode lasers. In some SELs the laser cavity is horizontal, and vertical emission is obtained by bending the output beam upwards with appropriate mirrors. *Vertical-cavity surface-emitting lasers* (VCSELs) (Figure 1.9) are made by etching a quantum well laser in a very narrow mesa structure. Because the optical cavity is very short the end mirrors must have a reflectivity close to 99%. They are made with the same epitaxial

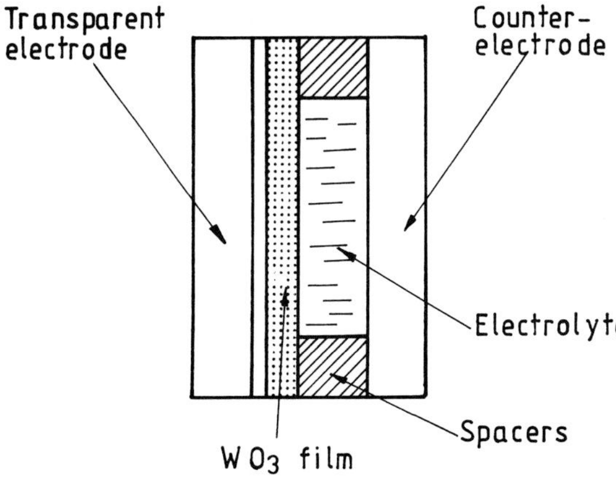

Figure 1.10 *Schematic diagram of an electrochromic display using* WO_3.

techniques employed to grow the quantum well, stacking multiple layers with different refractive indices on both sides of the quantum well. These Bragg mirrors reach the desired reflectivities for the emission wavelength [70,71].

Compared with edge-emitting lasers, VCSELs show some important advantages. They emit beams of circular section with no astigmatism. Their shorter cavity means larger mode spacing (>100 nm) and reduced mode hopping. They can be monolithically integrated in large 2-D arrays, even with microlenses. Packing densities higher than 2×10^6 devices/cm^2 have been achieved. Matrix-addressable surface emitting lasers (MASELAs) with VCSELs (already produced for the infrared wavelengths) could be used in displays, printing lasers, scanners, and so on.

However, the wall-plug efficiency is at most of 10% [72], compared with 30–50% of edge-emitting lasers, due to absorption of the beam in the highly doped top contact layer. Some solutions to this drawback are under way [73].

1.6. Electrochromism

Electrochromic materials [74–76] show a large change in absorption spectra (colour), associated to the injection or removal of electrons (and ions). Generally, electrochromic devices are made in the form of multilayer structures, essentially working like thin-film batteries. They typically consist of a transparent electrode covered with the electrochromic film, an electrolyte and a second electrode. The layout is schematically shown in Figure 1.10. The applied field injects the charge in the electrochromic film and induces the change in absorption.

Electrochromic materials find application in a variety of devices, including *displays* and electrically controlled optical windows and mirrors (*smart windows*). Many inorganic and organic materials have been described, although tungsten trioxide (WO_3) keeps a leading position for practical devices [77]. As an illustrative example, the solid-state reaction accounting for the change in colour in this material can be written as

$$WO_3 + xM^+ + xe^- \rightarrow M_xWO_3 \tag{1.3}$$

This reaction involves a change in lattice structure, accompanied by the partial conversion from W^{6+} to W^{5+} as a consequence of electron injection. Therefore, M_xWO_3 is a mixed-valence compound, having a very strong absorption at 900 nm caused by an electron transfer transition from W^{5+} to W^{6+}.

Other materials presenting the so-called cathodic coloration, associated with electron injection are MoO_3, V_2O_5, Nb_2O_5 and TiO_2. On the other hand, *anodic coloration* has been reported for IrO_2 [78], Rh_2O_3 [78], CoO_x [78] and NiO_x [79]. An increasing number of organic materials have been shown to be electrochromic. One may mention vialogens [80,81], metallophthalocyanines [82], and electroactive polymers (e.g. polyaniline [83]). For vialogens the change in colour results from the oxidation of molecules and their precipitation at the electrodes. Reduction of the precipitates traces back the process. On the other hand, the application of an electric field to a metalphthalocyanine in the presence of an electrolyte gives rise to a range of colours depending on the oxidation/reduction state. The same phenomenon is shown by some polymers, such as polyaniline.

Very recently, interest has arisen on biological polymers, such as bacteriorhodopsin [84]. This is the protein of the purple membrane in *Halobacterium halobium* and other bacteria. It shows a very high photochemical nonlinearity, presenting a variety of applications for dynamic holography and optical signal processing [85]. Bacteriorhodopsin is also electrochromic since electric fields induce strong changes in the absorption spectrum. The physical mechanisms are related to those responsible of the photochemical activity and appear to involve proton transfer among different molecular sites [86].

References

1. A. Yariv, *Optical Electronics*, Holt-Saunders Japan, Tokyo (1985).
2. D. Marcuse, *Principles of Quantum Electronics*, Academic Press, London (1980).
3. A. Yariv, *Quantum Electronics*, J. Wiley and Sons, Singapore (1989).
4. J. Wilson and J. F. B. Hawkes, *Optoelectronics. An Introduction*, Prentice-Hall, Englewood Cliffs (1983).
5. J. T. Verdeyen, *Laser Electronics*, Prentice-Hall, Englewood Cliffs (1989).
6. B. E. A. Saleh and M. C. Teich, *Fundamentals of Photonics*, J. Wiley, New York (1991).
7. J. Kerr, *Wied. Ann.* **18**, 213 (1875).
8. W. C. Röntgen, *Wied. Ann.* **18**, 213 (1883).
9. A. Kundt, *Wied. Ann.* **18**, 228 (1883).
10. F. Pockels, *Abh. Gottinger Ges. d. Wiss.* **39**, 169 (1884).
11. R. O. B. Carpenter, *J. Opt. Soc. Am.* **40**, 225 (1950).
12. R. C. Miller, D. A. Kleinman and A. Savage, *Phys. Rev. Lett.* **11**, 146 (1963).
13. P. V. Lenzo, E. G. Spencer and K. Nassau, *J. Opt. Soc. Am.* **56**, 633 (1966).
14. E. Bernal, G. D. Chen and T. C. Lee, *Phys. Lett.* **21**, 259 (1966).
15. A. R. Johnston and J. M. Weingart, *J. Opt. Soc. Am.* **55**, 828 (1965).
16. A. R. Johnston, *Appl. Phys. Lett.* **7**, 195 (1965).
17. G. H. Haertling and C. E. Land, *J. Am. Ceram. Soc.* **54**, 1 (1971).
18. F. Reinitzer, *Monastsh. Chem.* **9**, 421 (1888); English translation in *Liq. Crystals* **5**, 7 (1989).
19. P. G. de Gennes, *The Physics of Liquid Crystals*, Clarendon Press, Oxford (1975).
20. R. B. Meyer, L. Liebert, L. Strzelecki and P. Keller, *J. Phys. Lett.* **36**, L-69 (1975).
21. P. A. Franken, A. E. Hill, C. W. Peters and G. Weinreich, *Phys. Rev. Lett.* **7**, 118 (1961).
22. N. Bloembergen, *Nonlinear Optics*, Benjamin, New York (1965).
23. Y. R. Shen, *The Principles of Nonlinear Optics*, J. Wiley, New York (1984).
24. P. N. Prasad and D. J. Williams, *Introduction to Nonlinear Optical Effects in Molecules and Polymers*, J. Wiley, New York (1991).
25. G. Destriau, *J. Chem. Phys.* **33**, 587 (1936).
26. O. V. Losev, *Telegraf. Telefon.* **18**, 61 (1923); O. V. Losev, *Phil. Mag.* **6**, 1024 (1928).
27. R. J. Haynes, *Phys. Rev.* **98**, 1866 (1955).
28. C. J. Neue and J. I. Pankove, in *Display Devices* (ed. J. I. Pankove), Springer-Verlag, Berlin (1980).
29. E. O. Brimm, J. C. Brantley, J. H. Lorentz and M. H. Jellink, *J. Am. Chem. Soc.* **73**, 5427 (1951).
30. J. H. Compton and W. D. Schulman, *Color Centers in Solids*, Pergamon Press (1963).
31. S. K. Deb, *Appl. Opt. Suppl.* **3**, 193 (1969).
32. T. H. Maiman, *Nature* **187**, 493 (1960).
33. N. G. Basov, O. N. Krokhin and Yu. M. Popov, *Soviet Phys. JEPT* **40**, 1320 (1961).
34. R. N. Hall, G. E. Fenner, J. D. Kingsley, T. J. Soltys and R. O. Carlson, *Phys. Rev. Lett.* **9**, 366 (1962).
35. M. I. Nathan, W. P. Dumke, G. Burns, F. H. Dills and G. Lasher, *Appl. Phys. Lett.* **1**, 62 (1962).
36. D. Kales, *Laser Focus World*, January, **70**, (1993).
37. T. Tamir (ed.), *Guided-Wave Optoelectronics*, 2nd edn, Springer-Verlag, New York (1990).

38. J. P. van der Ziel, R. Dingle, R. C. Miller, W. Wiegmann and W. A. Nordland, Jr., *Appl. Phys. Lett.* **42**, 463 (1975).
39. W. T. Tsang, *Appl. Phys. Lett.* **39**, 786 (1981).
40. A. A. Bergh and P. J. Dean, *Light-Emitting Diodes*, Clarendon Press, Oxford (1976).
41. S. M. Sze, *Semiconductor Devices*, John Wiley, New York (1985).
42. R. Friend, D. Bradley and A. Holmes, *Phys. Wld* Nov., 42 (1992).
43. C. W. Tang and S. A. Van Slyke, *Appl. Phys. Lett.* **51**, 913 (1987).
44. J. H. Burroughes, D. D. C. Bradley, A. R. Brown, R. N. Marks, R. H. Frienal, P. L. Burn and A. B. Holmes, *Nature* **347**, 539 (1990).
45. D. Braun and A. Heeger, *Appl. Phys. Lett.* **58**, 1982 (1991).
46. Reported in *Photon. Spectra* March, 44 (1992).
47. L. T. Canham, *Appl. Phys. Lett.* **57**, 1046 (1990).
48. A. G. Cullis and L. T. Canham, *Nature* **353**, 335 (1991).
49. N. Koshida and H. Koyama, *Appl. Phys. Lett.* **60**, 347 (1992).
50. F. Namavar, H. P. Maruska and N. M. Kalkhoran, *Appl. Phys. Lett.* **60**, 2514 (1992).
51. H. K. Henisch, *Electroluminescence*, Pergamon, New York (1962).
52. R. Mach, *Solid State Sci.* **57**, 186 (1985).
53. H. Ohnishi, *Annu. Rev. Mater. Sci.* **19**, 83 (1989).
54. R. Törnqvist, in *Growth and Optical Properties of Wide-gap II–VI Low-dimensional Semiconductors* (eds T. C. McGill, C. M. Sotomayor-Torres and W. Gebhardt), Plenum Press, New York (1989).
55. J. I. Pankove (ed.), *Display Devices*, Springer-Verlag, Berlin (1980), 14.
56. A. Vecht, N. J. Werring, R. Ellis and P. J. F. Smith, *Proc. IEEE* **61**, 902–907 (1973).
57. B. Ray, in *Electronic Materials* (eds L. S. Miller and J. B. Mullin), Plenum, New York, (1991), 219.
58. G. H. B. Thomson, *Physics of Semiconductor Lasers*, J. Wiley, New York (1981).
59. H. C. Casey Jr and M. B. Panish, *Heterostructure Lasers*, Academic Press (1978).
60. G. P. Agrawal, in *Progress in Optics*, vol. 26 (ed. E. Wolf), Elsevier (1988).
61. S. Murata and I. Mito, *Optical and Quantum Electronics* **22**, 1 (1990).
62. R. Gunshor, A. Nurmikko and M. Kobayashi, *Phys. Wld* March, 46 (1992).
63. C. Weisbuch and B. Vinter, *Quantum Semiconductor Structures*, Academic Press, New York (1991), p. 165.
64. Ref. [1], p. 592.
65. H. Iwamura, T. Saku, T. Ishibashi, K. Otsuka and Y. Horikoshi, *Electron. Lett.* **19**, 180 (1983).
66. N. K. Dutta, N. A. Olsson and W. T. Tsang, *Appl. Phys. Lett.* **45**, 836 (1984).
67. H. Soda *et al.*, *Jpn. J. Appl. Phys.* **18**, 59 (1979).
68. J. Jewell, *Phys. Wld*, **3**, 28 (1990).
69. J. L. Jewell, A. Scherrer, S. C. McCall, Y. H. Lee, S. J. Walker, J. P. Harbison and L. T. Florez, *Optics News* **15**, 10 (1989).
70. J. L. Jewell, J. P. Harbison and A. Scherer, *Sci. Am.* Nov., 56 (1991).
71. J. L. Jewell and G. R. Olbright, *Laser Focus Wld* May, 217 (1992).
72. L. A. Coldren *et al.*, *Opt. Quantum Electron.* **2**, S105 (1992).
73. E. D. Jungbluth, *Laser Focus Wld* June, 15 (1992).
74. B. F. Faughan, R. S. Crandall and M. A. Lampert, *Appl. Phys. Lett.* **27**, 275 (1975).
75. H. N. Hersch, W. E. Kramer and J. H. McGee, *Appl. Phys. Lett.* **27**, 646 (1975).
76. K. Bange and T. Gambke, *Adv. Mater.* **2**, 10 (1990).
77. S. K. Deb, *Phil. Mag.* **29**, 801 (1973).
78. W. C. Dautremont-Smith, *Displays* **1**, 3, 67 (1982).
79. S. Yamada, T. Yoshioka, M. Miyashita, K. Urabe and M. Kitao, *J. Appl. Phys.* **63**, 2116 (1988).
80. I. F. Chang, B. L. Gilbert and T. J. Sun, *J. Electrochem. Soc.* **122**, 955 (1975).
81. C. J. Schoot, J. J. Ponjee, H. T. van Dam, R. A. van Doorn and P. T. Bolwijn, *Appl. Phys. Lett.* **23**, 64 (1973).
82. C. S. Frampton, J. M. O'Connor, J. Peterson and J. Silver, *Displays, Technol. Applic.* **9**, 174 (1988).

83. S. Kuwabatas, K. Mitsui and H. Yoneyama, *J. Electroanal. Chem., Interfacial Electrochem.* **281**, 97 (1990).
84. D. Oesterhelt and W. Stoeckenius, *Nature New Biol.* **233**, 149 (1971).
85. V. B. Taranenko, in *Nonlinear Optical Materials* (eds H. Kuhn and J. Robillard), CRC Press (1992).
86. K. Tsuji and B. Hess, *Ferroelectrics* **86**, 147 (1988).

Chapter Two

Light Propagation in Anisotropic Media

2.1. Introduction

Light propagation in anisotropic media (most of them single crystals) is characterized by a material response to the incident light which is dependent on the direction of the wave electric field. Natural anisotropic crystals such as *Iceland spar* ($CaCO_3$) and *quartz* (SiO_2) have been used in optics laboratories for a long time. On the other hand, modern optical devices based on anisotropic materials are prepared from crystals grown in the laboratory under strict conditions (high purity, homogeneity and crystal perfection are required). Typical examples are $NH_4H_2PO_4$ (ADP), commonly employed in second harmonic generation, and $LiNbO_3$, from which many optical and electrooptical waveguide devices are currently being produced. Optical anisotropy can also be induced in an otherwise isotropic medium by external agents such as stress, magnetic or electric fields, and this gives rise respectively to the subjects of *piezo-* and *acoustooptics*, *magnetooptics* and *electrooptics*. If the medium is anisotropic before the external agent is applied, it will somehow modify the medium's anisotropy.

It is then convenient to dedicate a chapter of this book to introducing the basic concepts of light propagation in anisotropic media, as well as the current notation and conventions. After a brief discussion of the classical microscopic picture of anisotropy, the dielectric tensor will be introduced and the modes of light propagation and their polarization properties will be studied using the macroscopic Maxwell equations. The two last sections of the chapter are devoted to the discussion of *optical activity*, a peculiar anisotropy arising from the spatial dispersion.

2.2. Birefringence and dichroism: a microscopic picture

Within a macroscopic frame of work, the optical response of an isotropic medium is characterized by the complex refractive index

$$\tilde{n}(\omega) = n(\omega) + i\kappa(\omega) \tag{2.1}$$

where the real part $n = c/v_f$ (v_f phase velocity) determines the propagation properties, and the imaginary part κ measures the attenuation of the wave intensity I through

$$I(z) = I(0)\,\mathrm{e}^{-\alpha z} = I(0)\,\mathrm{e}^{-(2\omega\kappa/c)z} \tag{2.2}$$

α being designated as absorption coefficient. Functions $n(\omega)$ and $\kappa(\omega)$ are commonly called *dispersion* and *absorption curves* of the material medium.

Let us now take a microscopic point of view and think of bounded electrons in a crystal as classical Lorentz oscillators. Then, the absorption $\kappa(\omega)$ and dispersion $n(\omega)$ curves are obtained in terms of the oscillator dynamics, and anisotropy arises from the fact that the spring constant depends on the oscillator direction. Figure 2.1 shows two such oscillators located in a crystal plane XY (Z oscillator is omitted for the sake of clarity). The electric field **E** of the light wave is assumed to lie in that plane, the Poynting vector **S** being perpendicular to it. If the spring constant K_x is assumed to be greater than K_y, the respective resonance frequencies given by $\omega_i = \sqrt{K_i/m}$ $(i = x, y)$ will accomplish $\omega_x > \omega_y$. This gives rise to two different absorption/dispersion curves for the two polarizations shown in Figure 2.2, and the absorption (resonance) peak lies at a higher frequency for X-polarized than for Y-polarized light.

For an incident wave with frequency ω_b well below both resonance frequencies, ω_x and ω_y, the material is transparent for either X- or Y-polarization (see Figure 2.2). However, since the refractive index n_x is smaller than n_y for this particular frequency, the X-polarized component of the wave will travel faster than the Y-polarized one. The material is then called *birefringent*, because, in general, the X- and Y-polarized components are separated inside the material. On the other hand, for an incident wave frequency ω_d falling within the absorption band with a peak at ω_y, but outside that with a peak at ω_x, the material will be transparent for the X-polarized component and will absorb the Y-polarized one. The material is then called *dichroic*.

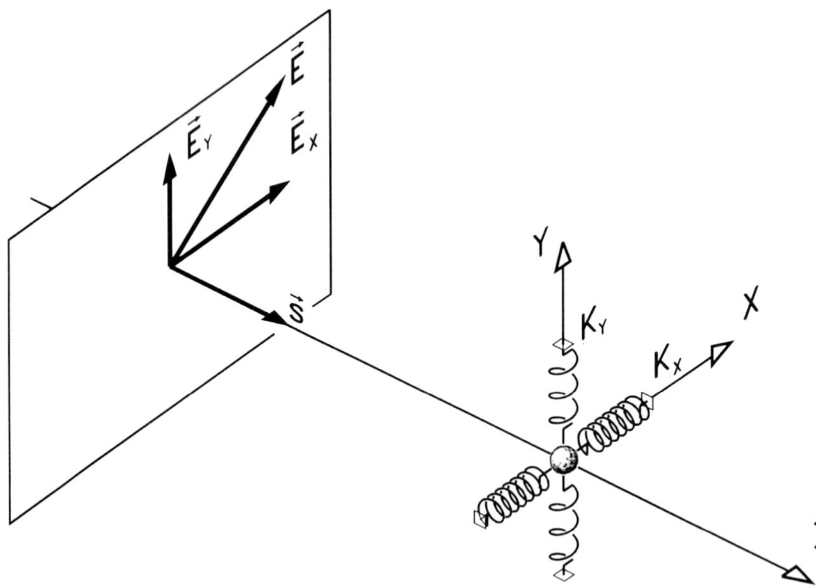

Figure 2.1 *Two oscillators lying in plane XY with different spring constants* ($K_x > K_y$) *exhibit different response to the electric field* **E** *of an incident wave.*

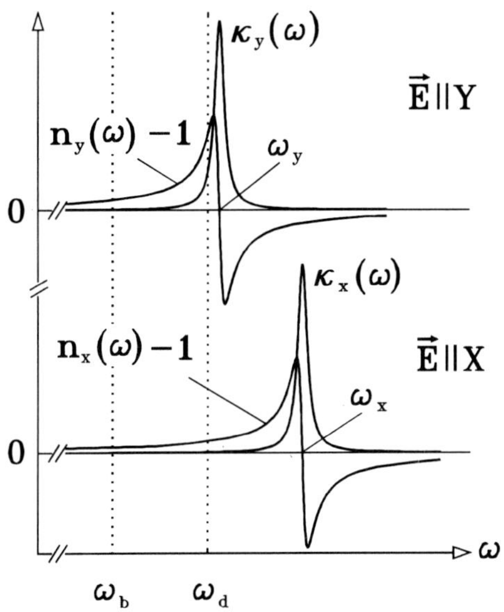

Figure 2.2 *Absorption–dispersion curves corresponding to two linear oscillators along* X *and* Y *axes with resonance frequencies* $\omega_x > \omega_y$.

Most current polarization devices are based on either birefringence or dichroism. In the following sections of this chapter we will limit ourselves to the study of birefringence, although in other chapters dichroism induced by an applied electric field will also be considered.

2.3. Macroscopic Maxwell equation

The propagation of an electromagnetic field in a dense medium, not containing free charges or currents, is described by the *macroscopic* Maxwell equations [1]

$$\nabla \cdot \mathbf{D} = 0, \qquad \nabla \cdot \mathbf{B} = 0 \tag{2.3a, b}$$

$$\nabla \times \mathbf{E} = -\frac{\partial \mathbf{B}}{\partial t}, \qquad \nabla \times \mathbf{H} = \frac{\partial \mathbf{D}}{\partial t} \tag{2.3c, d}$$

E and **H** respectively being the *electric* and *magnetic field intensity* vectors, whereas **D** and **B** are the associated *electric displacement* and *magnetic induction* vectors. All these vectors are termed *macroscopic* in the sense that they are average values of the microscopic fields (in a vacuum) over regions of the material which are large compared with interatomic distances, but small in comparison with wavelength (or distance one should travel to observe a substantial change in field value). Since only two out of the above equations are independent, additional relations among the field vectors are required to define the propagation problem univocally. These relations between **D** and **E**, and **B** and **H**, are known as *constitutive relations*. They are characteristic of the material and most often of an empirical nature. Theoretical models providing appropriate description of a number of materials have been developed. General schemes are briefly discussed in Chapters 3 and 4, whereas more specific phenomenological models are mentioned in Chapters 5 and 6.

2.4. The dielectric tensor

The constitutive relations are inferred from the general expressions

$$\mathbf{D}(\mathbf{E}) = \epsilon_0 \mathbf{E} + \mathbf{P}(\mathbf{E}) \tag{2.4a}$$

$$\mathbf{B}(\mathbf{H}) = \mu_0 \mathbf{H} + \mathbf{M}(\mathbf{H}) \tag{2.4b}$$

where ϵ_0 and μ_0 are the vacuum *electric* and *magnetic permittivities* respectively. **P** stands for the induced electrical *polarization* and **M** for the *magnetization*. Equations (2.4) are valid under the so-called dipolar approximation [2], that neglects the contribution of higher order multipoles.

Taking into account the finite response time of the material and the causality principle, the dependence $\mathbf{P}(\mathbf{E})$ is written as

$$P_i(t) = \epsilon_0 \int_0^{\infty} \chi_{ij}(\tau) E_j(t - \tau)\, \mathrm{d}t \tag{2.5}$$

where $\chi_{ij}(\tau)$ is a real response tensor and τ the delay time between the excitation **E** and the induced polarization **P**. The nonlocal spatial character of the response has been ignored, i.e. it is assumed that **P** and **E** refer to the same point. The breaking of this assumption leads to the interesting phenomenon of optical activity (see Sections 2.14 and 2.15).

By Fourier transform of the convolution product (2.5) one arrives at

$$P_i(\omega) = \epsilon_0 \chi_{ij}(\omega) E_j(\omega) \tag{2.6}$$

χ_{ij} being the *dielectric susceptibility* tensor defined through

$$\chi_{ij}(\omega) = \int_{-\infty}^{+\infty} \chi_{ij}(\tau)\, \mathrm{e}^{i\omega\tau}\, \mathrm{d}\tau \tag{2.7}$$

From (2.6) and the Fourier transform of (2.4a), it is established immediately that

$$D_i(\omega) = \epsilon_0 \tilde{\epsilon}_{ij}(\omega) E_j(\omega) \tag{2.8}$$

where

$$\tilde{\epsilon}_{ij}(\omega) = 1 + \chi_{ij}(\omega) \tag{2.9}$$

is the *complex dielectric tensor*. As with the complex refractive index defined in equation (2.1), the *tilde* notation has been used here to emphasize the difference with the real magnitude ϵ. With all other complex magnitudes appearing throughout the book no confusion is expected to arise, and so the *tilde* will be suppressed. In the above expressions $E_i(\omega)$, $P_i(\omega)$ and $D_i(\omega)$ refer to the complex amplitudes associated to a propagating field of frequency ω. Therefore, (2.8) extends the simple relation $D_i(t) = \epsilon_{ij} E_j(t)$, valid for instantaneous nondispersive response, to any monochromatic component of the propagating electromagnetic field. The ω dependence of the χ_{ij} and $\tilde{\epsilon}_{ij}$ accounts for the dispersive optical behaviour of the material. Taking into account the comments made in Section 2.2, the tensorial character of equations (2.6) and (2.8) can be understood in terms of different bond polarizabilities along different directions.

Regarding the magnetic constitutive relation (2.4b), one should note that at optical frequencies, much higher than those corresponding to any magnetization process, $\mathbf{M}(\mathbf{B}) \approx 0$, and any material behaves essentially as a vacuum.

Table 2.1 *Values of clamped and unclamped dielectric constants for $BaTiO_3$ and $LiNbO_3$.*

Material	ϵ^S	ϵ^T	Reference
$BaTi_3O$	$\epsilon_{11} = 2400$	$\epsilon_{11} = 3700$	[5]
	$\epsilon_{11} = 60$	$\epsilon_{33} = 135$	
$LiNbO_3$	$\epsilon_{11} = 44.3$	$\epsilon_{11} = 84.6$	[6]
	$\epsilon_{33} = 27.6$	$\epsilon_{33} = 29.1$	

2.4.1. Clamped and unclamped static dielectric constant

The polarization of a material in response to a static or low-frequency electric field should depend on the thermodynamical constraints during the process. One may keep the material at constant stress and temperature (strain free process) or at constant strain and temperature (stress free process). The former experiments lead to the so-called *unclamped* dielectric constant ϵ_T, which is generally measured. The latter type of experiments yield the *clamped* value ϵ_S. The difference between ϵ^T and ϵ^S (or χ^T and χ^S) can be understood as follows. One first applies the field $\mathbf{E}$ at constant stress to induce an unclamped polarization $\mathbf{P}^T$. Then, the material is mechanically stressed (stress tensor X_{ij}) to recover the original dimensions. For the first process

$$P_i^T = \chi_{ij}^T E_j \tag{2.10a}$$

The second process yields an additional polarization *only if* the medium lacks an inversion centre, i.e. is a *piezoelectric* crystal [3,4]. Then, the resulting final polarization $\mathbf{P}^S$ is

$$P_i^S = \chi_{ij}^S E_j = \chi_{ij}^T E_j - d_{ijk}^P X_{jk} \tag{2.10b}$$

where d_{ijk}^P is the piezoelectric tensor [3,4]

In order to establish the relation between χ_{ij}^T and χ_{ij}^S, one needs the additional equation

$$\mathscr{D}_{ij} = 0 = -s_{ijkl}^E X_{kl} + d_{ijm} E_m \tag{2.11}$$

expressing X_{kl} as a inear function of E_m. $\mathscr{D}_{ij}$ is the (null) strain tensor and s_{ijkl}^E the elastic compliance coefficients at constant elastic field.

As an example, the values of ϵ_{ij}^S and ϵ_{ij}^T for $BaTiO_3$ and $LiNbO_3$ are given in Table 2.1. In fact, even for centrosymmetrical materials (e.g. glass) there should be a small *electrostrictive* contribution to the polarization that would make ϵ_{ij}^T different from ϵ_{ij}^S. In the optical frequency range, ionic motion is frozen and $\epsilon_{ij}^T = \epsilon_{ij}^S$. However, unless explicitly stated, dielectric constants will refer to unclamped values.

2.5. General properties of the dielectric tensor

The dielectric tensor presents a number of useful properties of very general nature, that are here enumerated without proof [7].

(1) *Reality condition*:

$$\tilde{\epsilon}_{ij}^*(\omega) = \tilde{\epsilon}_{ij}(-\omega) \tag{2.12}$$

(2) *Onsager relation*:

$$\tilde{\epsilon}_{ij}(\omega) = \tilde{\epsilon}_{ji}(\omega) \tag{2.13}$$

It is valid whenever the system is invariant *vs* time inversion. This demands that the finite size of the wavevector **k** be ignored and that magnetic fields are not applied.

(3) For a *lossless medium*:

$$\tilde{\epsilon}_{ij}(\omega) = \tilde{\epsilon}^*_{ji}(\omega) \tag{2.14}$$

Consequently, in the absence of optical activity and magnetic fields (condition 2), the dielectric tensor is *real* and *symmetrical*. This is the situation normally considered in this chapter, except for the sections dealing with optical activity. The tensor ϵ_{ij} is then described in general by six independent real parameters. By choosing principal axes X, Y, Z, ϵ_{ij} can be written

$$\epsilon_{ij} = \epsilon_i \delta_{ij} = \begin{bmatrix} \epsilon_x & 0 & 0 \\ 0 & \epsilon_y & 0 \\ 0 & 0 & \epsilon_z \end{bmatrix} \tag{2.15}$$

the diagonal elements being the *principal dielectric constants*. Within this frame, relation (2.8) between **E** and **D** takes the form

$$D_x = \epsilon_0 \epsilon_x E_x, \qquad D_y = \epsilon_0 \epsilon_y E_y, \qquad D_z = \epsilon_0 \epsilon_z E_z \tag{2.16}$$

(4) *Crystal symmetries*: point group symmetry of the crystal lattice imposes additional relations among the ϵ_{ij} components, reducing the number of independent parameters. Group theory considerations allow to write the structure of the dielectric tensor for any crystal [8,9]. Relevant features of the ϵ_{ij} tensor for the various crystal classes are described in Section 2.10.

(5) *Kramers–Krönig relations*: the real (dispersive) and imaginary (dissipative) parts of the principal values of $\tilde{\epsilon}_{ij}$ are connected through the Kramers–Krönig relations [2,10],

$$\epsilon_i'(\omega) = 1 + \frac{2}{\pi}\mathbb{P}\int_0^\infty \frac{x\epsilon_i''(x)}{x^2 - \omega^2}\,\mathrm{d}x \qquad (i = 1, 2, 3) \tag{2.17a}$$

$$\epsilon_i''(\omega) = \frac{2\omega}{\pi}\mathbb{P}\int_0^\infty \frac{\epsilon_i'(x) - 1}{x - \omega}\,\mathrm{d}x \qquad (i = 1, 2, 3) \tag{2.17b}$$

They are a general outcome of the linear response theory together with the causality principle. A similar relation exists between the real n and imaginary κ parts of the refractive index

$$n(\omega) = 1 + \frac{2}{\pi}\mathbb{P}\int_0^\infty \frac{x\kappa(x)}{x^2 - \omega^2}\,\mathrm{d}x \tag{2.18}$$

although the proof proceeds along different lines from those used for the dielectric constant. Taking into account the relation between κ and the absorption coefficient α, inferred from equation (2.2), one may express formula (2.18) in the more useful form

$$n(\omega) = 1 + \frac{c}{\pi}\mathbb{P}\int_0^\infty \frac{\alpha(x)}{x^2 - \omega^2}\,\mathrm{d}x \tag{2.19}$$

Equations (2.17) and (2.19) are very useful since they permit us to obtain the real part of $\tilde{\epsilon}$ or $\tilde{n}$ from the knowledge of the imaginary part throughout the whole frequency range. Although this latter information is not generally available, a finite frequency range containing the dominant absorption bands is sufficient in many practical cases.

2.6. Fresnel and index ellipsoids

In lossless anisotropic media, the same expression as in isotropic media holds for the electric energy density, except that the tensorial character of the dielectric constant has to be taken into account [11]. By using principal axes and formulae (2.16), it can be written that

$$\begin{aligned} W_E &= \frac{1}{2}\mathbf{E}\cdot\mathbf{D} \\ &= \frac{\epsilon_0}{2}(\epsilon_x E_x^2 + \epsilon_y E_y^2 + \epsilon_z E_z^2) \\ &= \frac{1}{2\epsilon_0}\left(\frac{D_x^2}{\epsilon_x} + \frac{D_y^2}{\epsilon_y} + \frac{D_z^2}{\epsilon_z}\right) \end{aligned} \tag{2.20}$$

By making in (2.20) the substitutions $\epsilon_0 E_x^2/2W_E = x^2, \ldots$, and $D_x^2/2\epsilon_0 W_E = x^2, \ldots$, the following ellipsoids are obtained:

$$\epsilon_x x^2 + \epsilon_y y^2 + \epsilon_z z^2 = 1 \tag{2.21a}$$

$$\frac{x^2}{\epsilon_x} + \frac{y^2}{\epsilon_y} + \frac{z^2}{\epsilon_z} = 1 \tag{2.21b}$$

Therefore, the formula for the electric energy density defines the ellipsoids (2.21a) and (2.21b) respectively associated to the tensors ϵ_{ij} and $(\epsilon^{-1})_{ij}$ (inverse tensor, also called *impermeability tensor*). They are named *Fresnel ellipsoid* and *index ellipsoid* or *optical indicatrix* respectively. The semiaxes of the last one are the square root of the principal dielectric constants $\sqrt{\epsilon_x}$, $\sqrt{\epsilon_y}$ and $\sqrt{\epsilon_z}$; as will be seen below, they are equal respectively to the refractive index along each principal axis (see Figure 2.3). For this reason the inverse tensor $(\epsilon^{-1})_{ij}$ is also called *index tensor*, and it is often represented as $(1/n^2)_{ij}$. It is symmetric as ϵ_{ij} when time inversion invariance is assumed (i.e. no magnetic fields are present). The usefulness of these geometric representations will be made clear hereafter (Section 2.8).

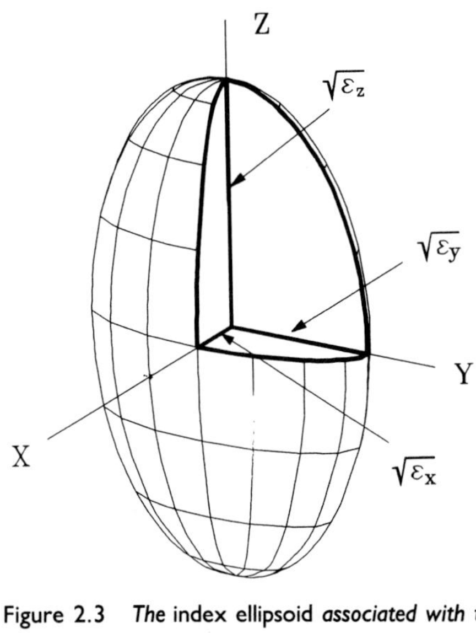

Figure 2.3 *The index ellipsoid associated with the inverse dielectric tensor* $(\epsilon^{-1})_{ij}$ *(index tensor).*

The intrinsic dispersive character of the dielectric permittivity makes all the tensor elements frequency dependent. Then, as a consequence of the diagonalization process, the principal dielectric constants (semiaxes of the index ellipsoid) and the directions of the dielectric axes (ellipsoid orientation relative to the crystal structure), are all frequency dependent. The last phenomenon is known as *dispersion of the axes*. It can only occur, however, in monoclinic and triclinic crystal systems, where the lattice symmetry does not previously determine three preferential orthogonal directions.

2.7. Propagation of a monochromatic plane wave in an anisotropic medium

The time and space evolution of a monochromatic field with angular frequency ω, $\mathbf{E} = \mathbf{E}_\omega(\mathbf{r})\,\mathrm{e}^{-i\omega t}$, is described by the wave equation [11]

$$\nabla \times (\nabla \times \mathbf{E}_\omega) - \omega^2 \mu_0 \epsilon_0 \epsilon(\omega) \mathbf{E}_\omega = 0 \tag{2.22}$$

obtained by applying the operator $\nabla\times$ on (2.3c) and taking (2.3d) and (2.8) into account. A linear superposition principle holds for the monochromatic components of an arbitrary field. For the particular case of an isotropic medium, (2.22) immediately leads to the Helmholtz equation

$$\Delta \mathbf{E}_\omega + \omega^2 \mu_0 \epsilon_0 \epsilon(\omega) \mathbf{E}_\omega = 0 \tag{2.23}$$

2.7.1. Geometrical considerations

Here we will consider the situation corresponding to a plane monochromatic wave in an anisotropic material. It will be seen below that the wave equation (2.22) takes a simpler form (equation (2.30)), but let us first analyse the peculiar relative orientations of the wavevectors in this case. The electric and magnetic fields in complex representation are

$$\mathbf{E}(\mathbf{r}, t) = \mathbf{E}\,\mathrm{e}^{i(\mathbf{k}\cdot\mathbf{r}-\omega t)}, \qquad \mathbf{H}(\mathbf{r}, t) = \mathbf{H}\,\mathrm{e}^{i(\mathbf{k}\cdot\mathbf{r}-\omega t)} \tag{2.24}$$

After substituting (2.24) into Maxwell equations (2.3c) and (2.3d), they become

$$\mu_0 \omega \mathbf{H} = \mathbf{k} \times \mathbf{E}, \qquad \omega \mathbf{D} = -\mathbf{k} \times \mathbf{H} \tag{2.25a, b}$$

The orthogonality relations (2.25) together with the Poynting vector expression characterizing the energy flow

$$\mathbf{S} = \mathbf{E} \times \mathbf{H} \tag{2.26}$$

determine the relative directions of all involved vectors. In the case of isotropic media, they form a unique trirectangular trihedron (**E**, **H**, **S**), because **E** is parallel to **D**, and **S** is parallel to **k** (see Figure 2.4a). In anisotropic media, however, there are no parallel vectors in general and they form two different trirectangular trihedra (see Figure 2.4b) (**H**, **D**, **k**) and (**H**, **E**, **S**). Also, since **H** is perpendicular to all other vectors, these are coplanar vectors. The fact that **S** and **k** are no longer parallel implies that the wavefront (**k**) and the energy (**S**) progress in different directions inside the crystal. These directions form the same angle as that formed by **D** and **E**. Also, only the vectors **D** and **H**, but not **E**, vibrate in the phase surfaces (perpendicular to **k**). We will return to this important point in Sections 2.8 and 2.9.

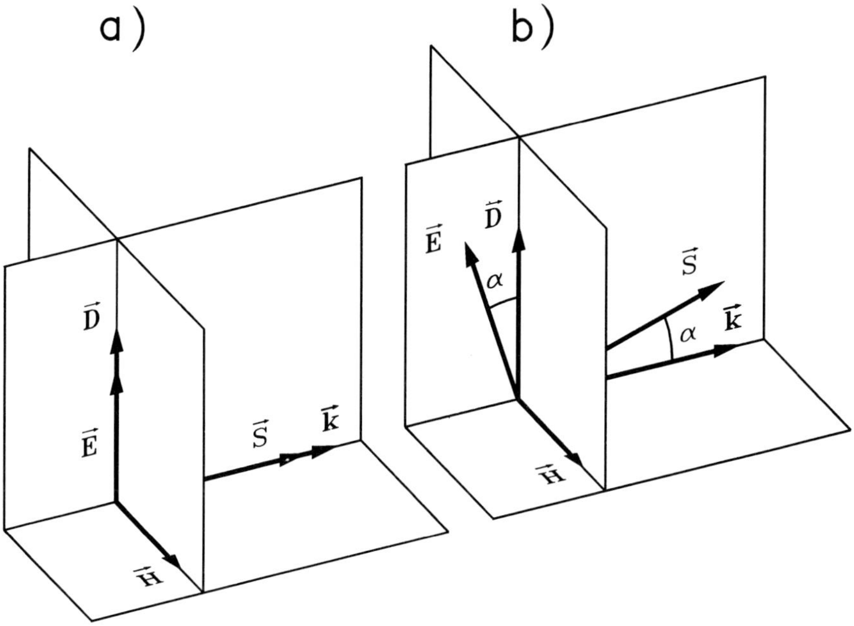

Figure 2.4 *Relative orientation of field vectors, wavevector and Poynting's vector for : (a) an isotropic medium; (b) an anisotropic medium.*

2.7.2. Wave equation: eigenmodes

Now, to obtain the wave equation for monochromatic plane waves, one substitutes (2.25a) into (2.25b) as follows:

$$\mathbf{D} = -\frac{1}{\mu_0\omega^2}\mathbf{k} \times (\mathbf{k} \times \mathbf{E}) = \frac{1}{\mu_0\omega^2}[k^2\mathbf{E} - (\mathbf{k} \cdot \mathbf{E})\mathbf{k}] \tag{2.27}$$

and uses $D_i = \epsilon_0 \epsilon_{ij} E_j$ to arrive at

$$\epsilon_{ij} E_j = \frac{1}{\epsilon_0\mu_0\omega^2}[k^2 E_i - (k_j E_j)k_i] \tag{2.28}$$

For convenience, in addition to $\epsilon_0\mu_0 = 1/c^2$, we will make the following substitutions:

$$\frac{\omega}{c} = k_0, \qquad \frac{\mathbf{k}}{k} = \mathbf{u}, \qquad \mathbf{k} = k_0 n\mathbf{u} = k_0\mathbf{n} \tag{2.29}$$

Then, equation (2.28) can be written in a more compact way:

$$[\epsilon_{ij} + n^2(u_i u_j - \delta_{ij})]E_j = 0 \tag{2.30}$$

which can also be directly obtained by substituting (2.24) into the wave equation (2.22). In the principal coordinate system X, Y and Z, (2.30) takes the form

$$\begin{bmatrix} \epsilon_x + n^2(u_x^2 - 1) & n^2 u_x u_y & n^2 u_x u_z \\ n^2 u_x u_y & \epsilon_y + n^2(u_y^2 - 1) & n^2 u_y u_z \\ n^2 u_x u_z & n^2 u_y u_z & \epsilon_z + n^2(u_z^2 - 1) \end{bmatrix} \begin{bmatrix} E_x \\ E_y \\ E_z \end{bmatrix} = 0 \tag{2.31}$$

For nontrivial solutions to exist, the determinant of the matrix in this equation must vanish, and leads to a relation between n and ϵ_x, ϵ_y, ϵ_z, for a given propagation direction $\mathbf{u}$:

$$\begin{vmatrix} \epsilon_x + n^2(u_x^2 - 1) & n^2 u_x u_y & n^2 u_x u_z \\ n^2 u_x u_y & \epsilon_y + n^2(u_y^2 - 1) & n^2 u_y u_z \\ n^2 u_x u_z & n^2 u_y u_z & \epsilon_z + n^2(u_z^2 - 1) \end{vmatrix} = 0 \qquad (2.32)$$

Notice that (2.31) is analogous to an equation of eigenvalues (n^2) and eigenvectors ($\mathbf{E}$) which determines the *normal modes* for light propagation inside anisotropic crystals. After developing the determinant (2.32), the characteristic equation giving the eigenvalues n^2, writes

$$\begin{aligned} &n^4(\epsilon_x u_x^2 + \epsilon_y u_y^2 + \epsilon_z u_z^2) \\ &\quad - n^2[u_x^2\epsilon_x(\epsilon_y + \epsilon_z) + u_y^2\epsilon_y(\epsilon_x + \epsilon_z) + u_z^2\epsilon_z(\epsilon_x + \epsilon_z)] + \epsilon_x\epsilon_y\epsilon_z = 0 \end{aligned} \qquad (2.33)$$

which is known as *Fresnel equation*, and can be considered as the fundamental equation of crystal optics. Note that the sixth degree terms in the refractive index n vanish upon developing the determinant, and only a second degree equation in n^2 is left. Therefore, there are only two eigenvalues for n^2 and two eigenvectors for $\mathbf{E}$. The physical origin of this result is the orthogonality relation $\mathbf{D} \perp \mathbf{n}$. It makes $\mathbf{D}$ oscillate in the phase plane, permitting only two independent vectors. Equation (2.33) then predicts the occurrence of two light waves travelling at different phase velocities (v_{f1}, v_{f2}) with different polarizations ($\mathbf{E}_1, \mathbf{E}_2$). (The indetermination in the sign of the n and v_f values associated to a given eigenvalue n^2, is removed by the direction $\mathbf{u}$ imposed by the incident wave).

Other forms for the Fresnel equation

After some algebraic manipulations the Fresnel equation can be written in a more compact form used in other texts:

$$\frac{u_x^2}{n^2 - \epsilon_x} + \frac{u_y^2}{n^2 - \epsilon_y} + \frac{u_z^2}{n^2 - \epsilon_z} = \frac{1}{n^2} \qquad (2.34)$$

It can also be expressed in terms of the phase velocity by substituting n^2 for c^2/v_f^2:

$$\frac{u_x^2}{v_f^2 - v_x^2} + \frac{u_y^2}{v_f^2 - v_y^2} + \frac{u_z^2}{v_f^2 - v_z^2} = 0 \qquad (2.35)$$

where the *principal propagation velocities*

$$v_x^2 = \frac{c^2}{\epsilon_x}, \qquad v_y^2 = \frac{c^2}{\epsilon_y}, \qquad v_z^2 = \frac{c^2}{\epsilon_z} \qquad (2.36)$$

have been introduced. Equation (2.35) is also a quadratic equation in v_f^2, so that there are two different values of v_f for each $\mathbf{u}$ direction. Note that v_x, v_y, v_z are not components of a vector and are defined only with reference to the principal axes.

2.7.3. Propagation along principal axes

For light propagating along one of the ellipsoid axes, e.g. X, the two refractive index values are the roots of the principal values of the dielectric tensor corresponding to the other axes, e.g.

$$n_1 = \sqrt{\epsilon_y}, \qquad n_2 = \sqrt{\epsilon_z} \tag{2.37}$$

For isotropic media, $\epsilon_x = \epsilon_y = \epsilon_z = \epsilon$, and equation (2.33) reduces to the scalar relation $n^2 = \epsilon$. This case is particularly simple and permits us to include the effect of absorption by means of the scalar complex refractive index (2.1), that obeys the equation

$$\tilde{n} = (n + i\kappa)^2 = \tilde{\epsilon} = \epsilon' + i\epsilon'' \tag{2.38}$$

This generalizes (2.37) and leads to simple relations between the index and dielectric tensor components,

$$n^2 - \kappa^2 = \epsilon', \qquad 2n\kappa = \epsilon'' \tag{2.39}$$

2.7.4. Propagation along arbitrary directions

To obtain the eigenvalues explicitly for any arbitrary propagation direction, let us write equation (2.33) in the abbreviated form

$$An^4 - Bn^2 + C = 0$$

where the coefficients A, B and C are easily inferred from (2.33). Then, the two *eigenvalues* associated to a given direction are

$$n_{1,2}^2 = \frac{B}{2A} \pm \frac{\sqrt{B^2 - 4AC}}{2A} \tag{2.40}$$

This expression also gives the material dispersion $n = n(\omega)$ via the dielectric constants $\epsilon_i(\omega)(i = x, y, z)$. From (2.31), the two corresponding *eigenvectors* can now be

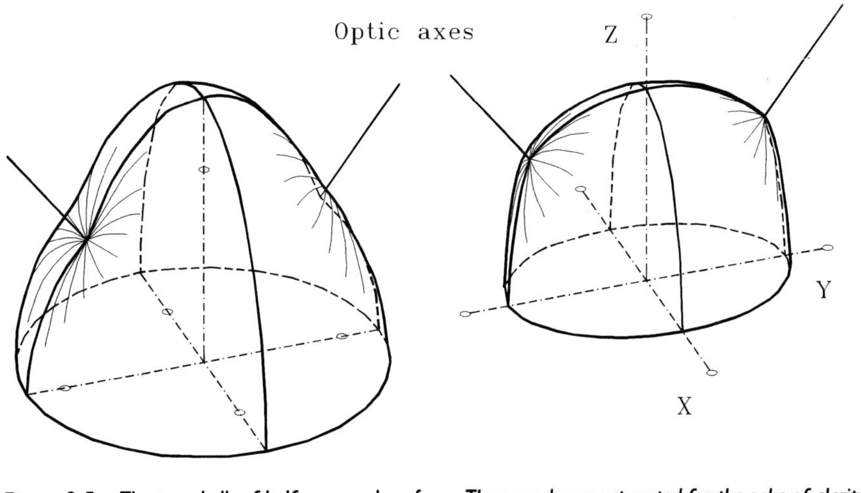

Figure 2.5 *The two shells of half a* normal surface. *They are drawn separated for the sake of clarity. The four points common to both shells determine the* optic axes of wave normals.

obtained:

$$\mathbf{E}_{1,2} = \text{const.} \begin{bmatrix} \dfrac{u_x}{n_{1,2}^2 - \epsilon_x} \\ \dfrac{u_y}{n_{1,2}^2 - \epsilon_y} \\ \dfrac{u_z}{n_{1,2}^2 - \epsilon_z} \end{bmatrix} \tag{2.41}$$

These two vectors determine the oscillation or polarization directions associated with each propagating mode. Since they are real vectors, planar or *linear polarization* is found for both modes and, in general, they are not orthogonal to each other. However, it will be shown in Section 2.8 that the corresponding displacement vectors, $\mathbf{D}_1$ and $\mathbf{D}_2$, are mutually orthogonal.

The Fresnel equation can be represented by a three-dimensional surface in $\mathbf{n}$ space (momentum space, since $\mathbf{n} = \mathbf{k}/k_0$). This surface is known as the *normal surface* and depends on the particular material, i.e. on the principal values ϵ_x, ϵ_y, ϵ_z. The surface consists of two shells which have only four points in common in the general case. In Figure 2.5 the inner shell has been drawn separated from the outer shell for the sake of clarity. For a certain propagating direction (u_x, u_y, u_z), the points at which the corresponding straight line intersects the two shells give the refractive index n_1 and n_2 associated with that direction. The two directions determined by the origin and the points common to both shells give only one value for the refractive index, $n_1 = n_2$. Along them, only one wave propagates and the two modes are degenerated just as in isotropic media. These directions are known as *optic axes*.*

2.8. Analysis of the polarization

The polarization of the two plane waves propagating in an anisotropic medium is determined by the eigenvectors $\mathbf{E}_1$ and $\mathbf{E}_2$ of equation (2.41). For a transparent medium, both vectors are real the two waves are linearly polarized. However, in order to examine in more detail the polarization properties, it is convenient to work with the displacement vector $\mathbf{D}$ instead of $\mathbf{E}$, since $\mathbf{D}$ (and not $\mathbf{E}$) is perpendicular to $\mathbf{k} = (\omega/c)n\mathbf{u}$.

2.8.1. Algebraic method

Let us take again equation (2.27)

$$\mathbf{D} = n^2\epsilon_0[\mathbf{E} - (\mathbf{u} \cdot \mathbf{E})\mathbf{u}] \tag{2.42}$$

where use has been made of relation $k^2/(\mu_0\omega^2) = n^2\epsilon_0$. In an axis system (X', Y', Z') such that $\mathbf{Z}'$ is parallel to $\mathbf{k}$, the $D_{z'}$ component vanishes and the problem is simplified to two dimensions ($D_{x'}$ and $D_{y'}$) (see Figure 2.6). Then, with $\mathbf{u} = \mathbf{u}_{z'} = (0, 0, 1)$, Equation (2.42) can be written in compact form,

$$D_{i'} = n^2\epsilon_0 E_{i'} \qquad (i' = x', y') \tag{2.43a}$$

$$\mathbf{D} = n^2\epsilon_0\mathbf{E}_\perp \tag{2.43b}$$

* More precisely they should be called the *optic axes of wave normals* in order to distinguish them from the *optic axes of rays* defined as the two directions along which only one energy velocity occurs.

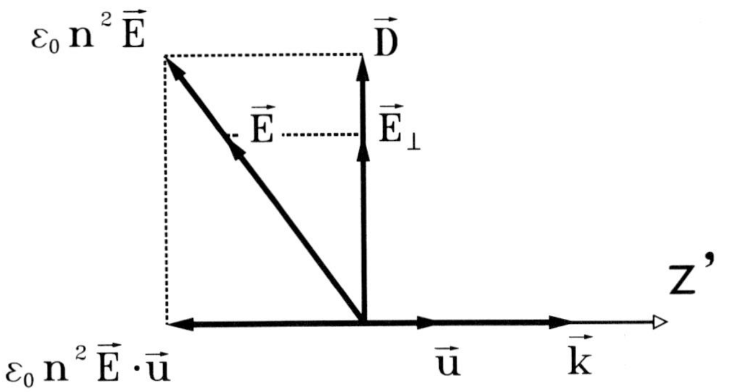

Figure 2.6 *Relative orientation of* **k**, **D** *and* **E**, *when choosing the* Z′ *axis along the* **k** *direction* ($D_{z'} = 0$).

where $\mathbf{E}_\perp$ is the $\mathbf{E}$ projection on $X'Y'$ plane. Now one can obtain the corresponding Fresnel equation for vector $\mathbf{D}$ in the $X'Y'$ plane,

$$\left[(\epsilon^{-1})_{i'j'} - \frac{1}{n^2}\delta_{i'j'}\right] D_{j'} = 0 \tag{2.44}$$

after substituting $(1/\epsilon_0)(\epsilon^{-1})_{i'j'}D_{j'}$ for $E_{i'}$ in (2.43a). By introducing the *transverse impermeability tensor* as

$$\eta_{i'j'} = \begin{bmatrix} (\epsilon^{-1})_{x'x'} & (\epsilon^{-1})_{x'y'} \\ (\epsilon^{-1})_{x'y'} & (\epsilon^{-1})_{y'y'} \end{bmatrix} \tag{2.45}$$

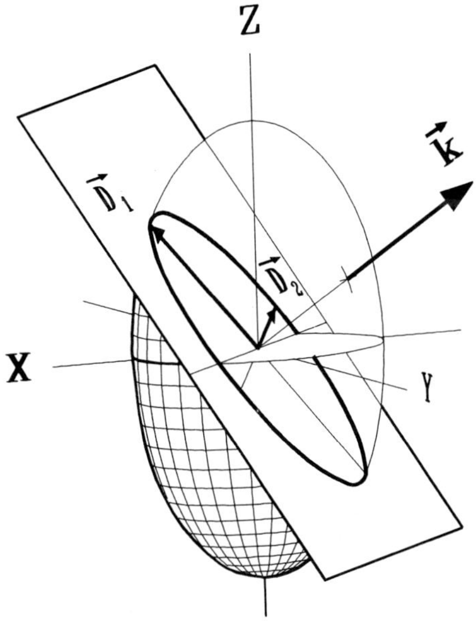

Figure 2.7 *In this figure, the ellipse resulting from the intersection of the index ellipsoid and the plane perpendicular to the wavevector* **k** *has been drawn. The principal axes of the ellipse determine the oscillation directions of vectors* $\mathbf{D}_1$ *and* $\mathbf{D}_2$, *and the lengths of its semiaxes determine the values* n_1 *and* n_2 *of the corresponding refractive indices.*

Equation (2.44) can be written

$$\begin{bmatrix} \eta_{x'x'} - \dfrac{1}{n^2} & \eta_{x'y'} \\ \eta_{x'y'} & \eta_{y'y'} - \dfrac{1}{n^2} \end{bmatrix} \begin{bmatrix} D_{x'} \\ D_{y'} \end{bmatrix} = 0 \tag{2.46}$$

Again, for nontrivial solutions to exist the determinant of the matrix must vanish:

$$\begin{vmatrix} \eta_{x'x'} - \dfrac{1}{n^2} & \eta_{x'y'} \\ \eta_{x'y'} & \eta_{y'y'} - \dfrac{1}{n^2} \end{vmatrix} = 0 \tag{2.47}$$

From this equation, two eigenvalues $1/n_1^2$ and $1/n_2^2$, corresponding to the two modes propagating along Z', are obtained. Equation (2.46) gives the two associated eigenvectors $\mathbf{D}_1$ and $\mathbf{D}_2$. On the other hand, since $\eta_{i'j'}$ is a symmetrical tensor, according to the general eigenvalue theory, the two eigenvectors must be orthogonal:

$$\mathbf{D}_1 \cdot \mathbf{D}_2 = 0 \tag{2.48}$$

In isotropic media, for any direction of propagation, there exists a unique (degenerated) eigenvalue n associated with the two eigenvectors. Hence, the two linear polarizations propagate at the same speed $v_f = c/n$ and mix to produce a general elliptic polarization.

2.8.2. Graphical method

The result just described permits us to use the index ellipsoid to graphically find the two indices of refraction and the two corresponding directions of **D** associated to an arbitrary direction **u** in a crystal. Since the inverse tensor $(\epsilon^{-1})_{ij}$ is associated with the index ellipsoid,

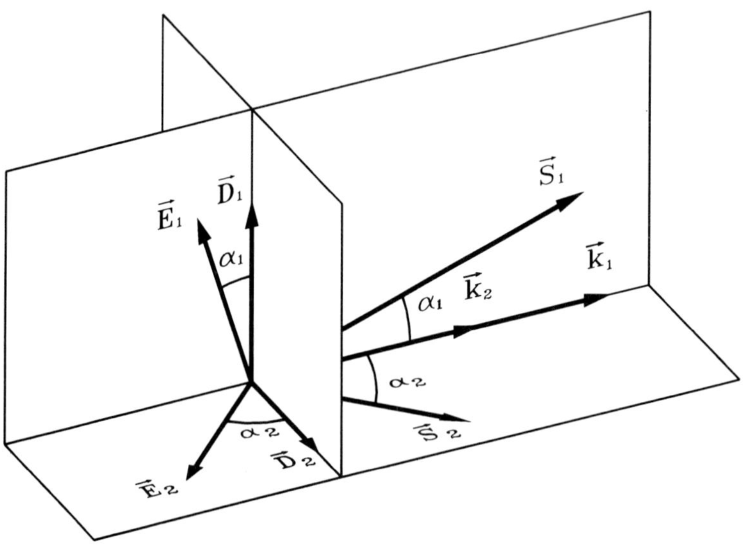

Figure 2.8 *Relative orientation of vectors **E**, **D**, **S** and **k**, for the two modes of propagation along the direction **u** = **k**/k. Whilst vectors **D**$_1$, **D**$_2$ are seen to be orthogonal, vectors **E**$_1$, **E**$_2$ are not.*

the transverse (bidimensional) tensor $\eta_{i'j'}$ in equation (2.45) will be associated with the ellipse resulting from the intersection of the index ellipsoid and the plane perpendicular to $\mathbf{k}||Z'$ (see Figure 2.7). Once this tensor has been diagonalized by equation (2.47), the two semiaxes of the ellipse are equal in length to n_1 and n_2 (as obtained from the eigenvalues $1/n_1^2$ and $1/n_2^2$), and the directions of the vectors $\mathbf{D}_1$, $\mathbf{D}_2$, are parallel to the ellipse axes. Figure 2.7 illustrates this graphical procedure.

Figure 2.8 illustrates the following additional orthogonality relations (they can also be proved algebraically):

$$\mathbf{D}_1 \cdot \mathbf{E}_2 = 0; \qquad \mathbf{D}_2 \cdot \mathbf{E}_1 = 0; \qquad \mathbf{u} \cdot (\mathbf{E}_1 \times \mathbf{H}_2) = 0 \tag{2.49}$$

2.9. Phase velocity, energy velocity and group velocity

A given normal surface at a given frequency ω contains information about the phase energy and group velocities at that particular ω. The *phase velocity* is obtained straightforwardly from the refractive index n at the considered direction $\mathbf{u}$:

$$\mathbf{v}_f = \frac{\omega}{k}\mathbf{u} = \frac{c}{n}\mathbf{u} \tag{2.50}$$

The *energy transport velocity*, also called the *ray velocity*, is defined by the Poynting vector as follows:

$$\mathbf{v}_E = \frac{\mathbf{S}}{W} = \frac{S}{W}\mathbf{s} \tag{2.51}$$

where W is the electromagnetic energy density. From Figure 2.4 it is easily inferred, and it can be confirmed by a detailed calculation, that

$$v_E = \frac{v_f}{\cos(\mathbf{u}, \mathbf{s})} \tag{2.52}$$

On the other hand, it can be proved that the *group velocity*, defined as $\mathbf{v}_g = \partial\omega/\partial\mathbf{k}$, is the same as the energy velocity [12]

$$\mathbf{v}_g = \frac{\partial\omega}{\partial\mathbf{k}} = \mathbf{v}_E \tag{2.53}$$

The definition of the group velocity implies that it is a vector perpendicular to the normal surface. Hence, the direction of energy propagation, also called *ray* and represented by **S** in Figure 2.4, is perpendicular to the normal surface and vectors **E** and **H** lie in the plane tangent to this surface, whereas vector **D** lies in the plane tangent to the phase surface.

In practice, the direction of propagation actually observed in a crystal is that of energy, i.e. **S**. By using the trihedron formed of vectors **E**, **H**, **s**, a development parallel to that carried out with trihedron **D**, **H**, **k**, can be made, giving rise to a *ray equation*, a *ray surface* and so on. In fact, there exists a *duality law* between the two groups of equations emphasizing the equivalence of both procedures. The ray surface is often used to generalize Huygens' principle to anisotropic refraction. Note, however, that Huygens' construction operates with wavelets coming from point sources instead of plane waves, and the propagation laws, in general, are not the same. The use of the normal surface method is then preferred as actually done in Sections 2.11 to 2.13.

2.10. The optical classification of crystals

The optical properties of a crystal depend, first of all, on the symmetry of its dielectric tensor ϵ_{ij} which determines the actual shape of the normal surface. By choosing as coordinate system the principal axes of ϵ_{ij}, its symmetry properties are characterized by the principal dielectric constants $\epsilon_x = n_x^2$, $\epsilon_y = n_y^2$, $\epsilon_z = n_z^2$. These properties can be illustrated by means of the index ellipsoid of equation (2.21b) (see Figure 2.3). In this respect, any crystal falls into one of the following three groups: *isotropic*, *uniaxial* and *biaxial* crystals.

Crystals belonging to the *cubic* system have three crystallographically equivalent directions which are mutually orthogonal. Therefore, the principal dielectric axes coincide with those directions, and the three principal values are equal to a unique scalar constant ($\epsilon_x = \epsilon_y = \epsilon_z = \epsilon = n^2$). The index ellipsoid and the normal surface degenerate to a sphere (Figure 2.9(a)) and the optical behaviour of a cubic crystal cannot be distinguished from that of an *isotropic* medium.

Crystals belonging to the *trigonal*, *tetragonal* and *hexagonal* systems have two or more crystallographically equivalent directions lying in a plane which is respectively perpendicular to the threefold, fourfold or sixfold axis. Then, one of the principal axes coincides with this high order symmetry axis (normally chosen as Z-axis) and the other two must be perpendicular to it and equivalent to each other (can be freely chosen within the XY-plane). This makes the principal values ϵ_x and ϵ_y equal, so it can be written that

$$\epsilon_x = \epsilon_y = \epsilon_\perp = n_o^2 \qquad \epsilon_z \epsilon_\| = n_e^2 \tag{2.54}$$

where transversal $\epsilon_\perp$ and longitudinal $\epsilon_\|$ dielectric constants, and ordinary n_o and extraordinary n_e refractive indices have been introduced for convenience. These denominations are justified below. It will also be seen that the high order Z-axis is the only optic axis left in these crystals. For this reason they are called *uniaxial crystals*. Figure 2.9(b) shows the index ellipsoid and the sections by the coordinate planes of the normal surface for this case.

Crystals belonging to the *orthorhombic*, *monoclinic* and *triclinic* systems have no crystallographically equivalent directions so that $\epsilon_x \neq \epsilon_y \neq \epsilon_z$. In the orthorhombic system the directions of the principal axes are determined by the three orthogonal axes of twofold symmetry. In the monoclinic system a principal axis must coincide with the unique axis of twofold symmetry whilst the orientation of the other two axes, within the plane perpendicular to the first one, depend on the frequency (*dispersion of the axes*). In the triclinic system the directions of the three axes are not linked to any crystallographical axis and are frequency dependent. These crystals always have two optic axes and are called *biaxial crystals*. Figure 2.9(c) shows the index ellipsoid and the sections by the coordinate planes of the normal surface for these crystals. The values of the principal refractive indices for several relevant crystals and wavelengths are given in Table 2.2.

2.11. Uniaxial crystals

Most optical devices based on anisotropic crystals are currently made from uniaxial crystals. Good examples are high quality polarizers, half- and quarter-wave plates and compensators made of quartz and calcite, or second harmonic generators using the phase matching method made of ADP, $LiNbO_3$ and others. According to (2.54) the index ellipsoid (2.21b)

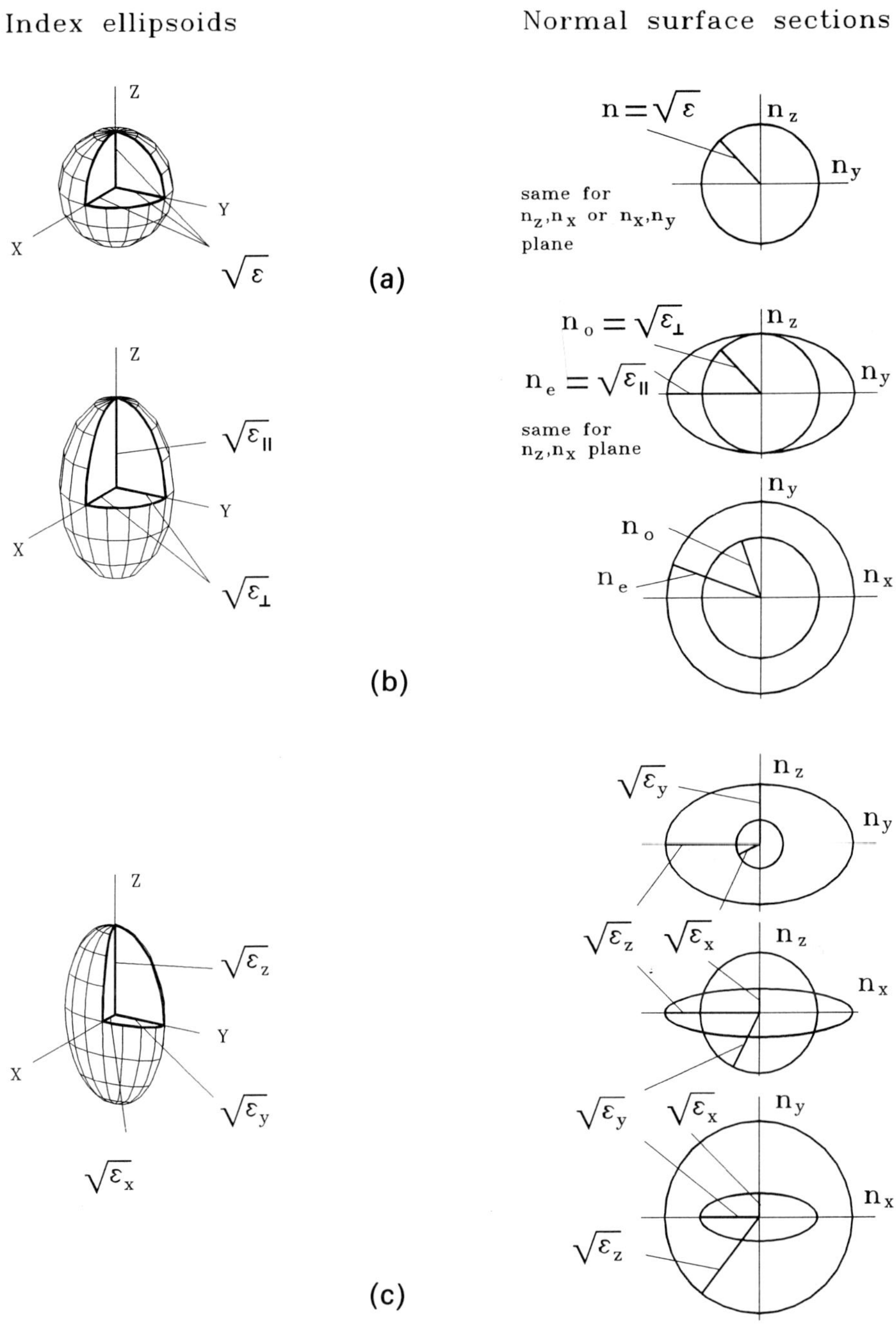

Figure 2.9 *Index ellipsoids and sections of the normal surface by the three coordinate planes for: (a) isotropic crystals; (b) uniaxial positive crystals; (c) biaxial crystals.*

Table 2.2 *Refractive indices of some relevant crystals.*

Crystal	λ (nm)	Refractive index			Reference
Isotropic		n_o			
CaF_2 (fluorite)	500	1.436			[13]
NaCl	500	1.552			[13]
C (diamond)	589	2.417			[13]
$Bi_{12}GeO_{20}$ (BGO)	633	2.530			[14]
$Bi_{12}SiO_{20}$ (BSO)	633	2.530			[14]
$Bi_{12}TiO_{20}$ (BTO)	633	2.580			[14]
CdTe	900	2.899			[13]
InP	1550	3.169			[13]
GaAs	918	3.569			[13]
Uniaxial		n_o		n_e	
Positive					
H_2O (ice)	589	1.309		1.313	[15]
MgF_2	500	1.380		1.392	[13]
Urea	633	1.48		1.59	[16]
SiO_2 (quartz)	589	1.544		1.553	[13]
BeO (beryllia)	589	1.718		1.734	[13]
ZrO_2 (zircon)		1.923		1.968	[17]
ZnS hexagonal	625	2.354		2.358	[13]
TiO_2 (rutile)	589	2.616		2.903	[13]
Negative					
KH_2PO_4 (KDP)	1064	1.49		1.46	[18]
	532	1.510		1.469	[18]
$(NH_4)H_2PO_4$ (ADP)	1064	1.51		1.47	[18]
	532	1.53		1.48	[18]
$NaNO_3$	589	1.585		1.337	[17]
β-BaB_2O_4 (BBO)	1064	1.6545		1.5423	[19,20]
	532	1.6743		1.5549	[19,20]
$CaCO_3$ (calcite)	589	1.658		1.486	[13]
Al_2O_3 (sapphire)	633	1.766		1.758	[13]
$LiIO_3$	1064	1.86		1.72	[18]
	532	1.90		1.75	[18]
$LiTaO_3$	633	2.176		2.180	[13]
$LiNbO_3$	1064	2.23		2.16	[13]
	532	2.33		2.23	[13]
$BaTiO_3$	600	2.430		2.372	[13]
Ag_3AsS_3 (proustite)	1064	2.82		2.58	[18]
Biaxial		n_x	n_y	n_z	
$NaNO_2$	1014	1.34	1.405	1.631	[17]
$CaSO_4 . 2H_2O$ (gypsum)		1.520	1.523	1.530	[17]
LiB_3O_5 (LBO)	1064	1.5656	1.5906	1.6055	[19,20]
	532	1.5785	1.6065	1.6212	[19,20]
KTP	1064	1.7367	1.7395	1.8205	[21]
$YAlO_3$	633	1.923	1.938	1.947	[17]
$KNbO_3$	633	2.279	2.329	2.167	[16]
SbSI	633	2.87	3.63	4.55	[17]

in uniaxial crystals degenerates to a spheroid given by

$$\frac{x^2+y^2}{\epsilon_\perp}+\frac{z^2}{\epsilon_\parallel}=1 \tag{2.55}$$

On the other hand, the normal surface will also simplify because it must be invariant under rotation around Z axis. After some algebraic manipulations, equation (2.33) can be written

$$(n^2-\epsilon_\perp)[\epsilon_\parallel n_z^2+\epsilon_\perp(n_x^2+n_y^2)-\epsilon_\perp\epsilon_\parallel]=0 \tag{2.56}$$

Thus, the fourth-order equation has factorized into two second-order equations:

$$n^2=\epsilon_\perp \tag{2.57}$$

$$\frac{n_x^2+n_y^2}{\epsilon_\parallel}+\frac{n_z^2}{\epsilon_\perp}=1 \tag{2.58}$$

(Note the resemblance between this spheroid and the index ellipsoid (2.55), although their axes are exchanged.) For practical use it is convenient to write the last equation in terms of the angle θ formed by the optic axis Z and the wavevector $\mathbf{k}=k_0\mathbf{n}$ (see Figure 2.10). By calling $n_e(\theta)$ the refractive index along this direction, equation (2.58) can be written

$$\frac{1}{n_e^2(\theta)}=\frac{\cos^2\theta}{n_o^2}+\frac{\sin^2\theta}{n_e^2} \tag{2.59}$$

The normal surface has now separated into two simpler surfaces (see Figure 2.11): sphere (2.57) and spheroid (2.58). The surfaces are tangential to each other at the common points where they intersect the symmetry axis (Z), i.e. $n_e(0^\circ)=\epsilon_\perp=n_o$. Therefore, the symmetry axis is the unique optic axis left and these crystals are named *uniaxial.* A uniaxial crystal is called *positive* if $n_o > n_e$, and *negative* if $n_o < n_e$.

The wave associated with equation (2.57) is called the *ordinary wave* because its refractive index $n_o=\sqrt{\epsilon_\perp}$, or phase velocity v_o, does not depend on the direction, just as in "ordinary" isotropic media. On the other hand, the wave associated with equations (2.58) and (2.59) is called the *extraordinary wave*, because its refractive index $n_e(\theta)$ is direction

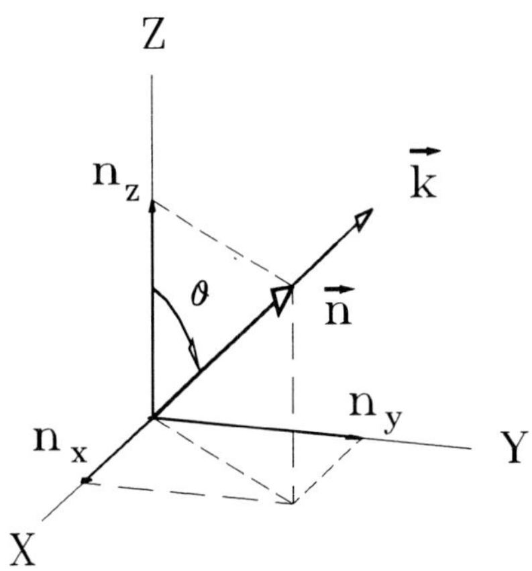

Figure 2.10 ***Orientation of wavevector k with respect to the optic axis Z.***

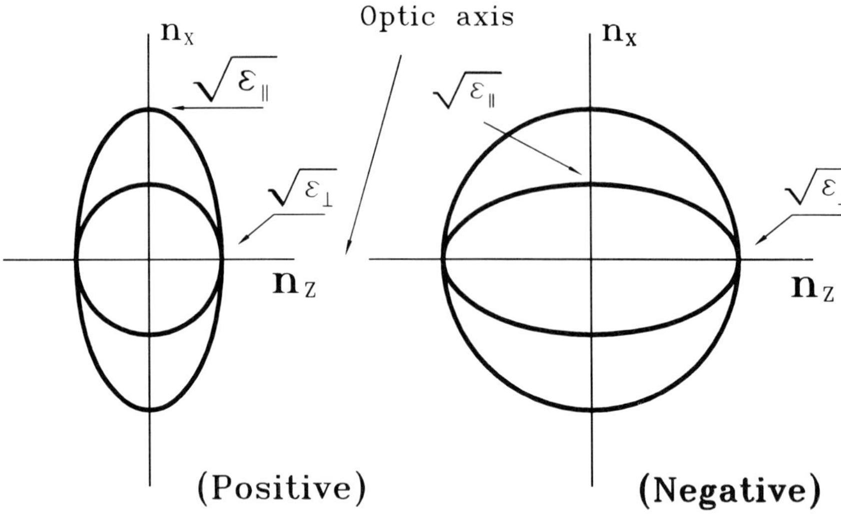

Figure 2.11 *Section of the normal surface by a plane containing the optic axis for a uniaxial positive crystal and a uniaxial negative crystal. The normal surface is generated by rotating this section about the optic axis (axis of symmetry).*

dependent. It goes from $n_e(0°) = \sqrt{\epsilon_\perp} = n_o$ for **k** parallel to the optic axis, to $n_e(90°) = \sqrt{\epsilon_{||}} = n_e$ for **k** perpendicular to the axis. In the ordinary wave, $\mathbf{S}_o$ is parallel to $\mathbf{k}_o$ due to the spherical symmetry. In the extraordinary wave they are no longer parallel, although they lie in the same section (see Figure 2.4). The angle $(\mathbf{k}_e, \mathbf{S}_e)$ formed by these vectors is easily obtained taking into account that **S** is perpendicular to the normal surface (see Section 2.6). Thus, from Figure 2.12 one gets

$$\tan\theta' = -\tan(\pi - \theta') = -\frac{\partial n_z}{\partial n_x} = \frac{\epsilon_\perp}{\epsilon_{||}}\frac{n_x}{n_z} = \frac{\epsilon_\perp}{\epsilon_{||}}\tan\theta \tag{2.60}$$

where equation (2.58) has been used to calculate $\partial n_z / \partial n_x$.

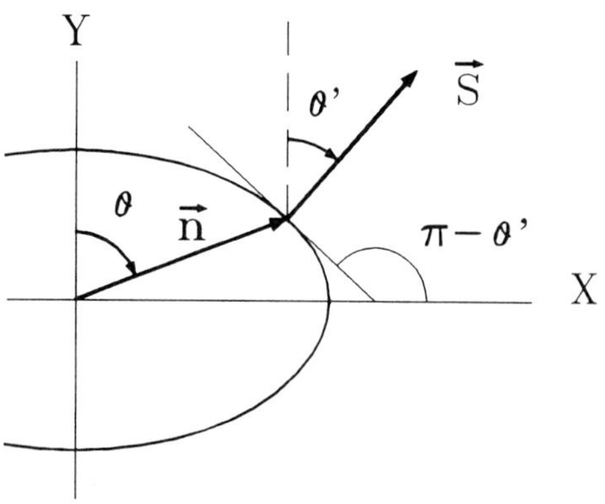

Figure 2.12 *Calculating the angle formed by vectors* **S** *and* **k** *from the extraordinary shell of the normal surface.*

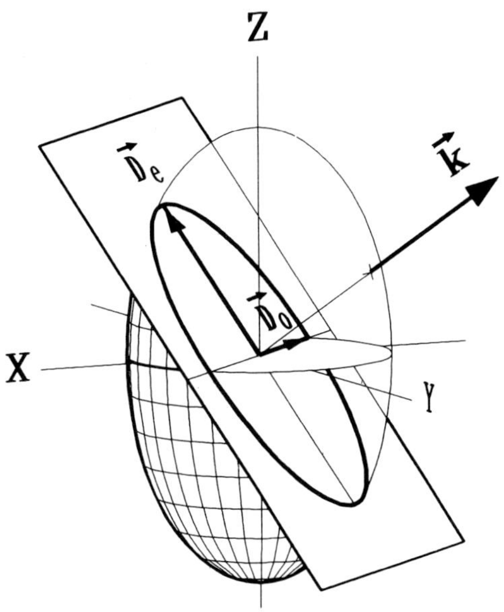

Figure 2.13 *Index ellipsoid and section through the origin by the wave plane for a uniaxial crystal (Z is a symmetry axis).*

The analysis of the polarization can easily be done on the index ellipsoid, using the method described in Figure 2.7. Figure 2.13 shows the index ellipsoid of an uniaxial crystal together with a plane wave propagating along the **u** direction. The plane defined by the optic axis Z and the wave vector is called the *principal plane*. Because the optic axis is a symmetry axis, the situation obtained by rotation around Z will be equivalent.

The section of the ellipsoid through the origin by the wave plane is an ellipse. The length of the ellipse semiaxis lying in the XY plane does not change on varying the angle θ (direction of **k**). Therefore, that length corresponds to the ordinary refractive index n_o, and the $\mathbf{D}_o$ vector defining the ordinary polarization is perpendicular to the principal plane. The length of the other semiaxis (lying in the principal plane) does change with θ and corresponds to the extraordinary refractive index $n_e(\theta)$. The extraordinary displacement vector $\mathbf{D}_e$, then, lies in the principal plane. Hence, the following relations hold:

$$\frac{\mathbf{D}_o}{D_o} = \mathbf{u} \times \mathbf{u}_z, \qquad \frac{\mathbf{D}_e}{D_e} = \frac{\mathbf{D}_o}{D_o} \times \mathbf{u} \tag{2.61}$$

where $\mathbf{u}_z$ is a unit vector along Z.

The polarization of the electric field vectors, $\mathbf{E}_o$ and $\mathbf{E}_e$, is given by expression (2.41). By choosing the principal plane as YZ plane, the extraordinary vector can be written:

$$\mathbf{E}_e = \text{const.} \begin{bmatrix} \dfrac{u_x}{n^2(\theta) - \epsilon_\perp} \\ \dfrac{u_y}{n^2(\theta) - \epsilon_\perp} \\ \dfrac{u_z}{n^2(\theta) - \epsilon_\parallel} \end{bmatrix} = \text{const.} \begin{bmatrix} 0 \\ \dfrac{\sin\theta}{n^2(\theta) - n_o^2} \\ \dfrac{\cos\theta}{n^2(\theta) - n_e^2} \end{bmatrix} \tag{2.62}$$

which explicitly shows the dependence on θ, as well as the fact that $\mathbf{E}_e$ lies in the principal plane (YZ). On the other hand, $\mathbf{E}_o$ must be parallel to $\mathbf{D}_o$ and so perpendicular to $\mathbf{E}_e$. Consequently, in uniaxial crystals the electric field vectors of the two propagating modes are perpendicular to each other (this is not true for biaxial crystals), although $\mathbf{E}_e$ is not perpendicular to the wavevector, i.e.

$$\mathbf{E}_o \cdot \mathbf{E}_e = 0, \qquad \mathbf{E}_e \cdot \mathbf{u} \neq 0 \tag{2.63}$$

2.12. Biaxial crystals

To study the normal surface in the case of biaxial crystals, the general form of the Fresnel's equation has to be considered. Let us assume that

$$\epsilon_x < \epsilon_y < \epsilon_z \tag{2.64}$$

The actual shape of the normal surface (shown in Figure 2.5) can be inferred by separately studying the sections by the three coordinate planes. The section by plane XY ($n_z = 0$) breaks equation (2.33) into two factors, giving

$$n_z = 0 \rightarrow \begin{cases} n^2 = \epsilon_z \\ \dfrac{n_x^2}{\epsilon_y} + \dfrac{n_y^2}{\epsilon_x} = 1 \end{cases} \tag{2.65}$$

Similarly, the sections by planes ZX and YZ are

$$n_y = 0 \rightarrow \begin{cases} n^2 = \epsilon_y \\ \dfrac{n_x^2}{\epsilon_z} + \dfrac{n_z^2}{\epsilon_x} = 1 \end{cases} \tag{2.66}$$

$$n_x = 0 \rightarrow \begin{cases} n^2 = \epsilon_x \\ \dfrac{n_y^2}{\epsilon_z} + \dfrac{n_z^2}{\epsilon_y} = 1 \end{cases} \tag{2.67}$$

These equations show that each section of the normal surface is made up of two simple curves: a circumference and an ellipse (see Figure 2.9(c)). The modes associated with the circumference are polarized perpendicular to the considered coordinate plane, whereas the modes associated with the ellipse are polarized in the plane. In plane XY the ellipse is inner to the circumference, in plane YZ it is outer, and in plane ZX the two curves intersect each other at four points. These points define the two optic axes. The three sections have been plotted altogether in Figure 2.14 to show an octant of the normal surface (compare also with Figure 2.5).

In order to determine the directions of the optic axes given by angle β in Figure 2.14, the *singularity* character of the four points defining them can be used. The singularity points of a surface are obtained as the solution of the system formed by the surface equation, Fresnel's equation $f(n_x, n_y, n_z) = 0$, and the first derivatives of f. We have also the further

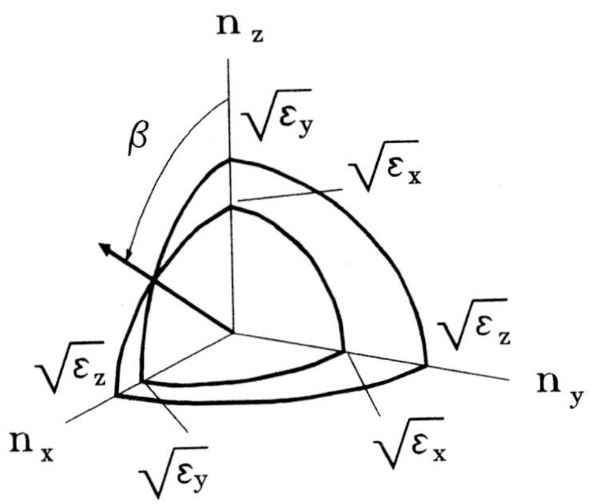

Figure 2.14 *Octant of the normal surface showing the three sections by the coordinate planes.*

simplification $n_y = 0$, because we know the four points are in plane ZX. After some algebraic manipulations the solution of that system is given by

$$n_x = \pm\sqrt{\frac{\epsilon_z(\epsilon_y - \epsilon_x)}{\epsilon_z - \epsilon_x}}, \qquad n_z = \pm\sqrt{\frac{\epsilon_x(\epsilon_z - \epsilon_y)}{\epsilon_z - \epsilon_x}} \tag{2.68}$$

These four points define the directions of the two optic axes by the corresponding angle β as follows:

$$\frac{n_x}{n_z} = \pm\tan\beta = \pm\sqrt{\frac{\epsilon_z(\epsilon_y - \epsilon_x)}{\epsilon_x(\epsilon_z - \epsilon_y)}} \tag{2.69}$$

The same result is obtained by using the geometrical construction of the index ellipsoid.

2.13. Refraction at a boundary of an anisotropic crystal

Refraction at a boundary of an anisotropic crystal is expected to be more complex and richer than in isotropic materials. The more striking difference is the appearance of the so-called *phenomenon of double refraction*. This appears as a natural consequence of the occurrence of two waves travelling at different velocities inside the crystal. More subtle and striking is the phenomenon known as *conical refraction* observed in biaxial crystals for propagation along the optic axes. We will restrict the discussion to the directions of the refracted waves, without considering the amplitudes of the refracted and transmitted fields.

Let $\mathbf{k}_{\text{inc}} = (\omega/c)\mathbf{n}_{inc}$ be a plane wave incident on the surface of an anisotropic medium (see Figure 2.15). In general, the transmitted wave will consist of a mixture of the two modes of propagation $\mathbf{k}_1 = (\omega/c)\mathbf{n}_1$ and $\mathbf{k}_2 = (\omega/c)\mathbf{n}_2$ (ordinary and extraordinary waves in the case of uniaxial crystals) that must end at certain points of the normal surface. As for isotropic materials, in anisotropic media the boundary condition implies that: (i) all the wavevectors lie in the incidence plane (as shown in Figure 2.15); (ii) their tangential components along the boundary are the same, namely

$$n_{\text{inc}}\sin\theta_{\text{inc}} = n_1\sin\theta_1 = n_2\sin\theta_2 \tag{2.70}$$

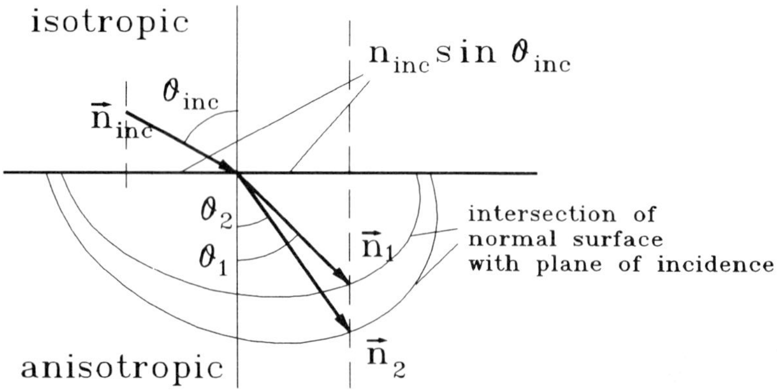

Figure 2.15 *Section by the plane of incidence of the normal surface of a general anisotropic medium, showing the boundary with an isotropic medium and the wavevectors involved in refraction. The graphical procedure to obtain the refracted wavevector is illustrated.*

From this equation the directions θ_1, θ_2 for the refracted wavevectors can be obtained. Although this equation looks like the simple Snell's law for isotropic media, note that n_1, n_2 are not constant but direction dependent, which makes the calculation less simple. Figure 2.15 illustrates graphically the procedure in a general case. Figure 2.16 illustrates the case of a uniaxial crystal when the optic axis lies in the incidence plane and when it is perpendicular to it; for the ordinary wave n_o is constant so that a true Snell's law applies. In all cases, the Poynting's vector (direction of energy propagation) has also been drawn. It is determined by the perpendicular to the normal surface, according to the discussion of Section 2.9, and, in general, the corresponding angles θ_1', θ_2' are different from those of the wavevectors. Even for normal incidence, where $\mathbf{k}_1 || \mathbf{k}_2$, the rays may follow off-normal directions and double refraction appears (see Figure 2.16).

For the privileged directions along the optic axes only one wavevector (a unique phase velocity) appears regardless of the state of polarization. These directions correspond to the singular points of the normal surface, where the perpendicular to the surface giving the direction of energy flow is not defined. The direction in this case can be found by drawing vectors perpendicular to the normal surface at points in an infinitesimal neighbourhood of the singular point. This defines an infinite number of directions, which form a cone, corresponding to the direction of energy flow associated to the optic axes. This phenomenon, called *conical refraction**, can be experimentally observed by launching a light beam at normal incidence along an optic axis. Since the energy flow is actually visualized inside the crystal, the cone of directions can be observed and measured.

2.14. Optically active materials

In a material with chiral symmetry, i.e. lacking any mirror plane (like quartz), the polarization vector $\mathbf{E}$ of a linearly polarized wave experiences a rotation during light propagation. The rotation θ is proportional to the penetration depth in the material and so

* More precisely it should be called *internal* conical refraction to distinguish it from the *external* conical refraction that appears when considering the directions of the optic ray axes.

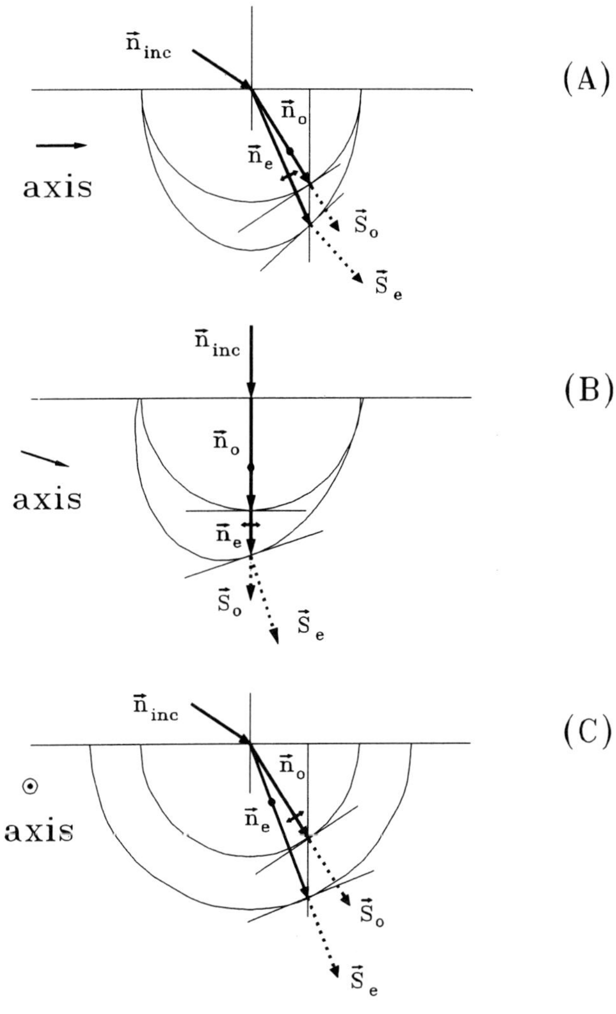

Figure 2.16 *Section by the plane of incidence of the normal surface of a uniaxial (positive) medium showing the boundary with an isotropic medium and the wavevectors involved in refraction. In (a) (oblique incidence) and (b) (normal incidence), the optic axis lies in the plane of incidence. In (c) the optic axis is perpendicular to the incidence plane.*

a *specific rotatory power* can be defined:

$$\rho = \frac{\theta}{d} \tag{2.71}$$

which is dependent on wavelength (dispersive behaviour). According to the sense of the rotation with regard to the propagation vector, one distinguishes *dextrogyre* (**E** rotates clockwise for an observer receiving the light beam) and *levogyre* materials, with ρ positive and negative respectively. For a number of substances and wavelength ranges the dependence $\rho(\lambda)$ can be approximated by a Cauchy relation:

$$\rho = A + \frac{B}{\lambda^2} \tag{2.72}$$

Values of ρ for a few materials are listed in Table 2.3. Much larger values than those listed are observed in the helicoidal molecular arrangements obtained in cholesteric liquid crystals (see Chapter 6). A useful and didactic collection of reprints dealing with all aspects of optical activity, from fundamentals to applications, can be found in reference [23].

The phenomenon of optical activity is a consequence of the so-called *spatial dispersion*, i.e. the dependence of the dielectric tensor $\tilde{\epsilon}_{ij}$ on the light wave vector $\mathbf{k}$. It is expected that the dependence is sufficiently smooth and then one can use a linear expansion:

$$(\epsilon^{-1})_{ij}(\mathbf{k},\omega) = (\epsilon^{-1})_{ij}(0,\omega) + G_{ijl}k_l \tag{2.73}$$

where G_{ijl} is a tensor responsible for the optical activity.

The physical reason behind (2.73) is that the polarization $\mathbf{P}$ and the electric field $\mathbf{E}$ depend, not only on the local value of $\mathbf{D}$ but on its space variation (nonlocal response), so that

$$E_i = \frac{1}{\epsilon_0}(\epsilon^{-1})_{ij}D_j + \frac{1}{\epsilon_0}\gamma_{ijl}\frac{\partial D_j}{\partial x_l} \tag{2.74}$$

with coefficients $(\epsilon^{-1})_{ij}$ and γ_{ijl} which are real. For excitation with a plane wave, $\mathbf{D}\,e^{i(\mathbf{k}\cdot\mathbf{r}-\omega t)}$, it is found that

$$E_i = \frac{1}{\epsilon_0}(\epsilon^{-1})_{ij}D_j + i\frac{1}{\epsilon_0}k_l\gamma_{ijl}D_j \tag{2.75}$$

in accordance with (2.73). Moreover, it immediately follows that

$$G_{ijl} = i\gamma_{ijl} \tag{2.76}$$

i.e. the tensor G_{ijl} is purely imaginary. It can be also shown that G_{ijl} is hemisymmetrical in the i, j indices:

$$G_{ijl}(\omega) = -G_{jil}(\omega) \tag{2.77}$$

implying that the tensor $(\epsilon^{-1})_{ij}$ is not symmetrical but obeys the condition (2.12).

The optical activity correction $\Gamma_{ij} = G_{ijl}k_l$ to the index tensor $(\epsilon^{-1})_{ij}$, can be written [24]

$$\Gamma_{ij} = i\gamma_{ijl}k_l = \tfrac{1}{2}\gamma_{ijl}\,\delta_{ijs}\,\delta_{sij}\,k_l = iG_{sl}\delta_{ijs}k_l/k??? \tag{2.78}$$

Table 2.3 *Specific rotatory power of some substances.*

Substance	ρ (deg./mm)	λ (nm)	Reference
Quartz (SiO_2)	+49	400	[12]
	+17	650	[12]
BGO ($Bi_{12}GeO_{20}$)	±45	488	[14]
BSO ($Bi_{12}SiO_{20}$)	±44	488	[14]
BTO ($Bi_{12}TiO_{20}$)	±14	488	[14]
Cinnabar (HgS)	+32.5	—	[22]
$NaBrO_3$	+2.8	—	[22]
$NaClO_3$	+3.1	—	[22]
$AgGaS_2$	950	485	[12]
	430	505	[12]

δ_{ijs} being the Kronecker symbol and

$$G_{sl} = \tfrac{1}{2}\gamma_{ijl}\,\delta_{sij} \tag{2.79}$$

the so-called *gyrotropic* tensor.

The explicit relation between Γ_{ij} and G_{ij} is

$$\Gamma_{ij} = \begin{bmatrix} 0 & iG_{33} & -iG_{23} \\ -iG_{33} & 0 & iG_{13} \\ iG_{23} & -iG_{13} & 0 \end{bmatrix} \tag{2.80}$$

with the Z-axis along $\mathbf{k}$.

The number of nonzero and linearly independent components of G_{ij} (and so of Γ_{ij}) is determined by the local symmetry of the medium. If this is isotropic, the structure of the tensor is

$$\Gamma_{ij} = \begin{bmatrix} 0 & iG_{33} & 0 \\ iG_{33} & 0 & 0 \\ 0 & 0 & 0 \end{bmatrix} \tag{2.81}$$

If, additionally, the medium possesses inversion symmetry, then $G_{33} = 0$ and $\Gamma_{ij} \equiv 0$.

2.15. Plane monochromatic waves in optically active media

Let us consider a monochromatic plane wave with wavevector $\mathbf{k} = k\mathbf{u}$ propagating in an isotropic optically active medium. The bidimensional eigenvalue equation (2.44) is

$$\left[(\epsilon^{-1})_{ij} \quad \frac{1}{n^2}\delta_{ij}\right] D_j = 0 \tag{2.82}$$

For an isotropic medium and $\mathbf{u}$ along Z,

$$(\epsilon^{-1})_{ij} = \begin{bmatrix} \dfrac{1}{\epsilon_0} & iG_{33} & 0 \\ -iG_{33} & \dfrac{1}{\epsilon_0} & 0 \\ 0 & 0 & \dfrac{1}{\epsilon_0} \end{bmatrix} \tag{2.83}$$

and so (2.82) can be written

$$\begin{bmatrix} \dfrac{1}{\epsilon_0} - \dfrac{1}{n^2} & iG_{33} \\ -iG_{33} & \dfrac{1}{\epsilon_0} - \dfrac{1}{n^2} \end{bmatrix} \begin{bmatrix} D_1 \\ D_2 \end{bmatrix} = 0 \tag{2.84}$$

The compatibility equation is

$$\begin{vmatrix} \dfrac{1}{\epsilon_0} - \dfrac{1}{n^2} & iG_{33} \\ -iG_{33} & \dfrac{1}{\epsilon_0} - \dfrac{1}{n^2} \end{vmatrix} = 0 \tag{2.85}$$

whose solutions are

$$\begin{aligned}\frac{1}{n_1^2} &= \frac{1}{\epsilon_0} - G_{33}\\ \frac{1}{n_2^2} &= \frac{1}{\epsilon_0} + G_{33}\end{aligned} \tag{2.86}$$

Taking into account that $G_{33} \ll 1/\epsilon_0$, (2.86) becomes

$$\begin{aligned}n_1 &= n_0 + \tfrac{1}{2} n_0 \epsilon_0 G_{33}\\ n_2 &= n_0 - \tfrac{1}{2} n_0 \epsilon_0 G_{33}\end{aligned} \tag{2.87}$$

n_0 being the refractive index in the absence of optical activity.

Note that only the G_{33} component of the gyrotropic tensor, belonging to its symmetrical part, determines the index values. Therefore, one often refers to that symmetrical part of G_{ij}, as the gyrotropic tensor.

The corresponding eigenvectors are as follows. For $n = n_1$:

$$D_1 + iD_2 = 0 \tag{2.88}$$

representing a dextrogire rotating **D** on the XY plane. Similarly, for $n = n_2$:

$$D_1 - iD_2 = 0 \tag{2.89}$$

corresponding to a left-handed rotating **D**. In summary, the normal modes of polarization for a plane monochromatic wave, are two circularly polarized waves of opposite sign. The circular birefringence $\Delta n = n_1 - n_2$ is

$$\Delta n = n_1 - n_2 = n_0 \epsilon_0 G_{33} \tag{2.90}$$

If the crystal is uniaxial and light propagates along the optic axis, the same result applies. For off-axis propagation, the eigenmodes correspond to elliptically polarized waves. As long as one moves from the axis, the eccentricity increases and above a few degrees from it the eigenmodes are, very approximately, linearly polarized waves.

References

1. L. D. Landau and E. M. Lifshitz, *Electrodynamics of Continuous Media*, Mir, Moscow (1969), Pergamon Press, London, and Addison Wesley, Reading, MA (1965).
2. J. D. Jackson, *Classical Electrodynamics*, John Wiley, New York (1975).
3. E. Fatuzzo and W. J. Merz, *Ferroelectricity*, North-Holland, Amsterdam (1967).
4. C. Kittel, *Introduction to Solid State Physics*, John Wiley, New York (1971).
5. M. B. Klein, in *Electrooptic and Photorefractive Materials* (ed. P. Günter), Springer-Verlag, Berlin (1987), p. 236.
6. *Properties of* $LiNbO_3$, EMIS Datareview Series N. 5, INSPEC, London (1989).
7. M. Cardona, *Modulation Spectroscopy*, Academic Press, New York (1969).
8. J. F. Nye, *Physical Properties of Crystals*, Clarendon Press, Oxford (1957).
9. S. Bhagavantam, *Crystal Symmetry and Physical Properties*, Academic Press (1966).
10. F. Stern, in *Solid State Physics* (eds F. Seitz and D. Turnbull), Vol. 15, Academic Press (1973).
11. A. Yariv, *Quantum Electronics*, John Wiley, New York (1989).
12. A. Yariv and P. Yeh, *Optical Waves in Crystals*, John Wiley, New York (1984).
13. E. D. Palick (ed.), *Handbook of Optical Constants of Solids*, Vol. 1 (1985) and Vol. 2 (1991), Academic Press, London.

14. L. Arizmendi, J. M. Cabrera and F. Agulló-López, *Int. J. Optoelectron.* **7**(2), 149 (1992).
15. R. C. Weast (ed.), *Handbook of Chemistry and Physics*, The Chemical Rubber Co., Cleveland (1972).
16. P. Günter, *Ferroelectrics* **75**, 5 (1987).
17. Landolt-Börstein, *Group III: Crystal and Solid State Physics* (New Series, ed. O. Madelung), *Vol. 28: Ferroelectrics and Related Substances*, Springer-Verlag, Berlin (1990).
18. D. N. Nikogosyan, *Sov. J. Quantum Electron.* **7**(1), 1 (1977).
19. C. Chen, *Laser Focus Wld* Nov., 129 (1989).
20. C. Chen, Y. X. Fan, R. C. Eckardt and R. L. Byer, *SPIE Laser Nonlin. Opt. Mater.* **681**, 12 (1986).
21. C. L. Tang, W. R. Bosenberg, T. Ukachi, R. J. Lane and L. K. Cheng, *Laser Focus Wld* Sep., 87 (1990).
22. R. D. Guenter, *Modern Optics*, John Wiley, New York (1990).
23. A. Lakhtakia (ed.), *Selected Papers on Natural Optical Activity*, SPIE Milestone, MSI5, Washington (1990).

Chapter Three

Electrooptics: Concepts, Phenomena and Techniques

3.1. Introduction

The optical properties of a material can be modified through the action of an external agent such as a magnetic field, a mechanical stress or an electric field. This latter effect is the objective of electrooptics, whose fundamentals have been authoritatively reviewed in some classical books [1–6] and review papers [7]. As described in Chapter 2, optical properties are characterized by the refractive index tensor $(1/n^2)_{ij} = (\epsilon^{-1})_{ij}$, whose principal axes and eigenvalues determine the propagation of a light wave in the material. For absorbing media the principal index values are complex and the two components are influenced by the applied electric field. In fact, the electric field-induced change Δn in the real refractive index, *electrorefraction*, and $\Delta\kappa$ in the imaginary part, *electroabsorption*, are not independent but are related by the Kramers–Krönig relations (2.17). Therefore, they are simultaneously present in all electrooptic experiments. Electrorefraction provides a physical method to modify in a controllable way the phase, the direction and the polarization of a light beam. On the other hand, electroabsorption permits the direct control of the light intensity passing through the material. This latter effect is only appreciable near optical resonances, where attenuation is high and consequently it has been much less used for devices. However, useful miniature high-speed absorptive light modulators have been made [8,9].

In accordance with the general introduction given in Chapter 1, the electrooptic response considered in this book is the local effect associated with the distortion of the electronic clouds (assuming atoms fixed at their equilibrium positions) by the applied electric field. In fact, electrons are the only charge carriers that can be excited at light frequencies. In a quantum approach to the problem, the field modifies the electron wavefunctions and energy levels of the system and consequently the dielectric response.

Atomic relaxation and molecular reorientation modify that purely electronic picture and determine indirect additional contributions to the electrooptic effect. This is so because the field-induced atomic or molecular rearrangement causes a change in the electronic charge distribution and so in the optical behaviour of the material. In some ferroelectric oxides, such as $BaTiO_3$ or $LiNbO_3$, this contribution is even dominant over the direct (electronic) one [10,11].

More complex phenomena appear in ceramics and liquid crystals. For ceramics one has to take into account the effect of the field on the crystalline grain structure. In liquid crystals, the electric field causes a reorientation of the long organic molecules and leads to a variety of structural arrangements and optical responses. In particular, it can induce dynamic instabilities that modify the light scattering properties, and consequently the *extinction coefficient*.

One way to differentiate among the various contributing mechanisms relies on their different response times (see Section 3.6). For example, molecular reorientation processes at room temperature often require times τ of the order of 10^{-10} s and so their contribution is quenched when the applied field is modulated above the corresponding eigenfrequency $1/\tau$. Similarly, the piezoelectric contribution in accentric dielectric materials can be inhibited for modulation frequencies higher than those associated to the mechanical resonances of the sample.

3.1.1. Electrogyration

If one takes into account space-dispersion effects due to the finite size of the light wavevector **k**, as in optically active crystals, the gyration tensor has to be added in the expression for the dielectric tensor (Chapter 2), i.e.,

$$(\epsilon^{-1})_{ij} = (\epsilon_0^{-1})_{ij} + G_{ijl} k_l \tag{3.1}$$

ϵ_0^{-1} stands for the index tensor corresponding to $k = 0$. In this case, the applied field can also modify G_{ijl} (i.e. the optical rotatory power) or even induce it, causing an additional effect known as *electrogyration*. Experiments are difficult to analyse due to the interfering contribution of the field-induced birefringence, and reliable data are not abundant. Reviews on the subject are those by Vlokh [12] and Fousek [13] on dielectric crystals. Specific rotations for many linear electrogyration crystals are around $\Delta\rho = 0.01\,^\circ\mathrm{C/mm}$, for an electric field $E = 1\,\mathrm{kV/cm}$. Higher electrogyration effects are, indeed, observed in liquid crystals, which are the subject of Chapter 6. No further discussion of the effect will be offered in this book.

3.2. Electrooptic coefficients

Let us consider a transparent (nonabsorbing) medium, characterized by a real index tensor $(1/n^2)_{ij}$. The application of an electric field **E** induces a change in the tensor components. Since it is expected that the changes will be relatively small (see below for some numerical estimates), one may expand $\Delta(1/n^2)$ in powers of **E** and keep only the lowest order terms. In fact, the most prominent effects arise from the linear and quadratic terms so that up to the second order one can write

$$\Delta(1/n^2)_{ij} = r_{ijk} E_k + s_{ijkl} E_k E_l \tag{3.2}$$

The linear term determines the so-called *Pockels effect*, whereas the quadratic term is responsible for the *Kerr effect*. The corresponding linear and quadratic electrooptic coefficients r_{ijk} and s_{ijkl} constitute third- and fourth-rank tensors, respectively, whose components depend on the light frequency ω due to material dispersion. Moreover, in most experiments the applied electric field **E** is modulated at a certain frequency Ω. Because of the finite response time of the active electrooptic mechanisms, a dependence of r_{ijk} and s_{ijkl}

Table 3.1 *Units and conversion factors for the electrooptic coefficients.*

Coefficient	MKS unit	esu unit	Conversion factor
r	m/V	cm/statV	$r(\text{esu}) = 3 \times 10^4\, r(\text{MKS})$
s	m^2/V^2	$\text{cm}^2/\text{statV}^2$	$s(\text{esu}) = 9 \times 10^8\, s(\text{MKS})$

on Ω is also expected (see later sections). Therefore, $r_{ijk}(\omega, \Omega)$ and $s_{ijkl}(\omega, \Omega)$ are, in general, functions of the two frequencies ω and Ω.

One should remark that expansion (3.2) is useful for operation at optical frequencies well below the relevant electronic transitions (e.g. the main absorption edge in semiconductors or insulators) as is the case in most applications. However, when the light frequency is very close to those of electronic transitions, more complex dependencies of n on the electric field have to be considered, which are markedly dependent on the specific physical mechanism. In fact, the study of these effects either in the transmission, or even better, in the reflectance spectra, is a very useful tool to investigate the optical properties of solids, particularly around the critical points (*Van Hove singularities*) [14,15].

Units

Table 3.1 lists the units and conversion factors for r and s. Although the MKS system will be used throughout this book, some data will be given in esu following common practice.

A remark on notation

Apparently, there is universal agreement to designate the Pockels coefficients by r_{ijk}. However, for Kerr coefficients other notations are often employed in the literature, e.g. R_{ijk} [7], ρ_{ijk} [16] and p_{ijk} [17].

3.2.1. Contracted matrix representation

Due to the symmetry condition $(1/n^2)_{ij} = (1/n^2)_{ji}$, it is usual to adopt a second-rank 6×3 matrix r_{mk} to represent the r_{ijk} tensor. The row index m designates all distinct ij pairs (irrespective of the ordering), according to the convention

$$\begin{gathered} 11 = XX = 1, \qquad 22 = YY = 2, \qquad 33 = ZZ = 3, \\ YZ = ZY = 23 = 4, \qquad XZ = ZX = 13 = 5, \qquad XY = YX = 12 = 6 \end{gathered} \tag{3.3}$$

The column index k (k = 1, 2, 3) has the same meaning as in r_{ijk}.

Similarly, for the Kerr tensor s_{ijkl} the ordering within the pairs ij and kl is irrelevant. Therefore, it is customary to use a 6×6 matrix s_{ij}, with the two indices i and j defined according to (3.3), and applying the rule

$$\left.\begin{aligned} s_{ab} &= s_{ijkk} && \text{if } (ij) = (ji) = a, \quad (kk) = b \\ s_{ab} &= s_{ijkl} + s_{ijlk}, && \text{if } (ij) = (ji) = a, \quad (kl) = (lk) = b \quad (k \neq l) \end{aligned}\right\} \tag{3.4}$$

3.3. Equivalent formulations

The electrooptic description given by expansion (3.2) is, indeed, the most commonly used, since the r_{ij} and s_{ij} coefficients are accessible to direct measurement (see Section 3.13).

However, there exist other equivalent ways to formulate the electrooptic effect involving the dependence of the refractive index tensor on the electric displacement vector $\mathbf{D} = \epsilon_0 \epsilon \mathbf{E}$ or the polarization $\mathbf{P}$. As an example, for materials undergoing phase transitions to polar phases, it is more convenient to use the **P** *formulation*

$$\Delta(1/n^2)_{ij} = f_{ijk} P_k + g_{ijkl} P_k P_l \tag{3.5}$$

since it provides a better insight into the physics of the electrooptic response. Tensors f_{ijk} and g_{ijkl} are designated *polarization-optics coefficients*. As for r_{ijk} and s_{ijkl}, contracted matrix representations are normally adopted. Taking into account that

$$P_l = \epsilon_0 \chi_{lk} E_k = \epsilon_0 (\epsilon_{lk} - \delta_{lk}) E_k \tag{3.6}$$

the following relations between the electrooptic and polarization-optic coefficients can easily be established:

$$r_{ijk} = \epsilon_0 f_{ijl} (\epsilon_{lk} - \delta_{lk}) \tag{3.7a}$$

$$s_{ijkl} = \epsilon_0^2 g_{ijmn} (\epsilon_{mk} - \delta_{mk})(\epsilon_{nl} - \delta_{nl}) \tag{3.7b}$$

One further advantage of the **P**-formulation is that coefficients f and g are less dependent on the specific material than r and s.

3.4. Electroabsorption

The field-induced change in the refractive index n (electrorefraction) is essentially linked to an associated change in the absorption coefficient α (electroabsorption). As it has been shown in Chapter 2, the refractive index n and the absorption coefficient α are related by the Kramers–Krönig dispersion relation

$$n(\omega, 0) = 1 + \frac{c}{\pi} \int_0^\infty \mathrm{d}\omega' \, \frac{\alpha(\omega', 0)}{\omega'^2 - \omega^2} \tag{3.8}$$

where the absence of any applied field ($\mathbf{E} = 0$) has been explicitly indicated for the n and α functions.

By applying an electric field $\mathbf{E}$, this equation can be written

$$n(\omega, E) = 1 + \frac{c}{\pi} \int_0^\infty \mathrm{d}\omega' \, \frac{\alpha(w', 0)}{\omega'^2 - \omega^2} + \frac{c}{\pi} \int_0^\infty \mathrm{d}\omega' \, \frac{\Delta\alpha(\omega', E)}{\omega'^2 - \omega^2} \tag{3.9}$$

$\Delta\alpha(\omega', E)$ being the change in absorption coefficient caused by the field E (a scalar formulation has been assumed). Subtracting (3.8) from (3.9), one obtains

$$\Delta n(\omega, E) = \frac{c}{\pi} \int_0^\infty \mathrm{d}\omega' \, \frac{\Delta\alpha(\omega', E)}{\omega'^2 - \omega^2} \tag{3.10}$$

relating electrorefraction Δn to electroabsorption $\Delta\alpha$. Relation (3.10) is completely general and applies to any type of material and physical mechanism. For zinc-blende semiconductors it has been used [18,19] to determine Δn from the $\Delta\alpha$ calculated with a well-established (Franz–Keldysh) model (see Section 5.10.3). In principle, the information on the change in absorption for all wavelengths should be included in the integral (3.10). Fortunately, one of the electronic transitions is often dominant and the calculation of the integral is markedly simplified. In particular, for photon energies in the transparency region

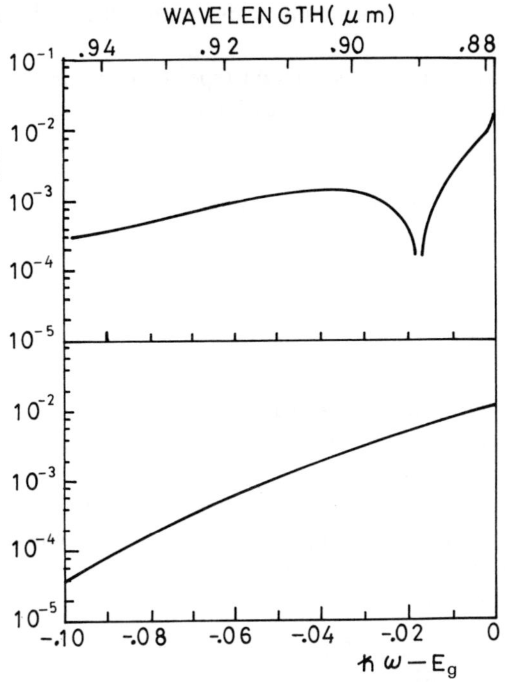

Figure 3.1 *Dependence of the field-induced* $|\Delta n|$ *and* $\Delta\kappa$ *on wavelength for GaAs* ($E = 10^5$ V/cm). *Adapted from reference [19].*

of semiconductors or insulators, only the fundamental absorption edge has generally to be considered [20]. As an example, the calculated dependences of $|\Delta n|$ and $\Delta\kappa$ on photon energy (near the band edge) for GaAs under a field $E = 10^5$ V/cm are shown in Figure 3.1. Anyhow, direct measurements of Δn are always preferable.

3.5. Pockels effect: electronic and ionic contributions

Let us consider a dielectric solid, operating in the transparency region. The Pockels effect is described by the first-order term in equation (3.2)

$$\Delta(1/n^2) = r_{ijk}E_k \tag{3.11}$$

However, this relationship is not univocally defined unless the experimental conditions are appropriately specified. The situation is similar to that described in Section 2.4.1, in relation to the dielectric tensor. Depending on whether the material is kept free (*unclamped*) or *clamped* during the experiment, one obtains different responses described by coefficients respectively designated by r^{T} or r^{S}.

The difference between r^{T} and r^{S} arises from the fact that in clamped experiments the crystal is not deformed (*constant strain conditions*). On the other hand, for unclamped experiments (*constant stress conditions*) one has to consider the deformation brought about by the electric field, as a consequence of the piezoelectric properties of Pockels materials. This deformation will, in turn, induce a change in refractive index via the piezooptic tensor. For small deformations this contribution simply adds to the direct electronic response.

To go deeper into the argument, one has to consider two types of ionic motion in a crystal lattice, respectively corresponding to optical and acoustical modes. For *optical modes*, positive and negative ions inside the unit cell move in opposite directions and lead to a net polarization that couples with the applied electric field. However, since these modes do not modify the size of the unit cell, they can be activated in either clamped or unclamped experiments. If q designates the normal coordinate for a given mode, the change in refractive index caused by such ionic relaxation can be written

$$\Delta\left(\frac{1}{n^2}\right)_{ij} = \frac{\partial(1/n^2)_{ij}}{\partial q}\frac{\partial q}{\partial \mathrm{E}_k}\mathrm{E}_k \tag{3.12}$$

where the factor $\partial(1/n^2)_{ij}/\partial q$ is directly related to the change in polarizability associated with the distortion q, which is responsible for the *Raman effect* (see Section 3.5.2).

On the other hand, *acoustical modes* change the dimensions of the unit cell and so cannot be activated in clamped experiments. In the unclamped geometry, because of the lack of an inversion centre in Pockels crystals, they generate a net polarization that contributes to the electrooptic response. If $\mathscr{D}_{ij}$ is the local elastic strain tensor, the acoustical mode contribution to the electrooptic effect is written

$$\Delta\left(\frac{1}{n^2}\right)_{ij} = \frac{\partial(1/n^2)_{ij}}{\partial \mathscr{D}_{lm}}\frac{\partial \mathscr{D}_{lm}}{\partial \mathrm{E}_k}\mathrm{E}_k \tag{3.13}$$

The factor $\partial\mathscr{D}_{lm}/\partial\mathrm{E}_k = d^{\mathscr{D}}_{lmk}$ is the third-rank *piezoelectric strain tensor* and $\partial(1/n^2)_{ij}/\partial\mathscr{D}_{lm} = \pi_{ijlm}$ the fourth-rank *strain-optic tensor*. Expression (3.12) shows that the electrooptic effect associated with the acoustical mode distortions results from both piezoelectric and piezooptic phenomena. One should note here that this contribution will be relevant at the mechanical resonance frequencies determined by material and sample geometry. Above those resonances, lying typically around 10^6 Hz, the effect is quenched.

In the absence of any acoustical or optical distortion, one has the purely electronic or direct response described by the *electronic electrooptic coefficient* r^{e}_{ijk}

$$r^{\mathrm{e}}_{ijk} = \left\{\frac{\partial(1/n^2)_{ij}}{\partial\mathrm{E}_k}\right\}_{\substack{\mathscr{D}=0\\ q=0}} \tag{3.14}$$

In accordance with the above discussion the electrooptic coefficients r^{S}_{ijk} and r^{T}_{ijk} can be expressed as follows:

$$r^{\mathrm{S}}_{ijk} = r^{\mathrm{e}}_{ijk} + \frac{\partial(1/n^2)_{ij}}{\partial q_m}\frac{\partial q_m}{\delta\mathrm{E}_k} \tag{3.15a}$$

$$\begin{aligned} r^{\mathrm{T}}_{ijk} &= r^{\mathrm{e}}_{ijk} + \frac{\partial(1/n^2)_{ij}}{\partial q_m}\frac{\partial q_m}{\partial\mathrm{E}_k} + \frac{\partial(1/n^2)_{ij}}{\partial\mathscr{D}_{lm}}\frac{\partial\mathscr{D}_{lm}}{\partial\mathrm{E}_k} \\ &= r^{\mathrm{e}}_{ijk} + \frac{\partial(1/n^2)_{ij}}{\partial q_m}\frac{\partial q_m}{\partial\mathrm{E}_k} + \pi_{ijlm}d^{\mathrm{E}}_{lmk} \end{aligned} \tag{3.15b}$$

and so they are related by

$$r^{\mathrm{T}}_{ijk} = r^{\mathrm{S}}_{ijk} + \pi_{ijlm}d^{\mathrm{E}}_{lmk} \tag{3.16}$$

In practice, unclamped experiments are most commonly performed and, unless otherwise specified, we will always refer to r^{T}.

3.5.1. Response times

The response time τ for the various contributions to the electrooptic effect can be roughly estimated as the reciprocal of the corresponding eigenfrequencies. For the direct electronic mechanism, $\tau \approx 10^{-14}$ s, whereas for the optical mode contribution, $\tau \approx 10^{-12}$ s. For the piezoelectric resonance effect (acoustical mode contribution), the eigenfrequencies depend on sample geometry and mechanical parameters, but typically lie below or around 1 MHz, leading to $\tau \approx 10^{-6}$ s. However, in many experiments the response speed is limited by the time constant of the lumped electrical circuit used to apply the field. Considering that the modulating source has an internal resistance R_{S} and that C is the capacitance of the electroded sample, the response time of the electrical circuit is $\tau_{\mathrm{c}} = R_{\mathrm{S}}C$. In other words, the modulator bandwidth is limited by a highest frequency $\Omega_{\mathrm{c}} = 1/R_{\mathrm{S}}C$. For higher frequencies the voltage applied to the sample is reduced and the electrooptic response drops to zero. Values of Ω_{c} are in the range 10^6–10^8 Hz. Therefore, in the absence of piezoelectric resonances, electrical parameters set a limit to the maximum modulation that can be attained.

Since the capacitance C is directly proportional to the dielectric constant ϵ of the sample, higher modulation frequencies can be reached with lower ϵ materials, e.g. organic compounds. In other words, ϵ is a relevant parameter in modulator design.

3.5.2. Relationship between Raman and Pockels effects

The connection between Raman and Pockels effects pointed out above, will be now more explicitly formulated [21]. First, it can be shown (see Section 4.9 for details) that

$$\delta(1/n^2)_{ij} = -\epsilon_{ii}\epsilon_{jj}(\delta\epsilon)_{ij} = -\epsilon_{ii}\epsilon_{jj}(\delta\chi)_{ij} \tag{3.17}$$

Then, relations (3.14) and (3.15a) can be written

$$r^{\mathrm{e}}_{ijk} = -\epsilon_{ii}\epsilon_{jj}\left(\frac{\partial\chi_{ij}}{\partial E_k}\right)_{\mathscr{D}=0,q=0} \tag{3.18a}$$

$$r^{\mathrm{S}}_{ijk} = r^{\mathrm{e}}_{ijk} - \epsilon_{ii}\epsilon_{jj}\frac{d\chi_{ij}}{dq}\frac{dq}{dE_k} \tag{3.18b}$$

On the other hand, the first-order Raman scattering cross-section S, corresponding to an incident u_i and a scattered u_j wave polarization, is [22,23]

$$S^{\mathrm{TO}}_{ij} \approx |\delta\chi_{ij}|^2 \approx |\partial\chi_{ij}/\partial q|^2 \tag{3.19a}$$

for a transversal (TO) mode. However, for a longitudinal (LO) mode one has to take into account (as for the electrooptic response), the electric field generated by the mode distortion and then,

$$S^{\mathrm{LO}}_{ij} \approx \left|\frac{\partial\chi_{ij}}{\partial q} + \frac{\partial\chi_{ij}}{\partial E_k}\frac{\partial E_k}{\partial q}\right|^2 \tag{3.19b}$$

Therefore, the expressions for r^{e}, r^{S} and S_{ij} depend on the same tensor magnitudes, $\partial\chi_{ij}/\partial q$ (*Raman tensor*), $\partial\chi_{ij}/\partial E_k$ and $\partial q/\partial E_k$. In particular, equation (3.18b) shows that a given optical mode may contribute to r^{S}, if it is *Raman-active* ($\partial\chi_{ij}/\partial q \neq 0$) and *polar*

($\partial q/\partial E_k \neq 0$). This can only occur in noncentrosymmetrical crystals. A discussion of mode contributions to electrooptic coefficients is given in references [13,24].

3.6. Frequency dispersion behaviour

Due to the finite response time of the electrooptic mechanisms, the coefficients $r_{ij}(\omega, \Omega)$ depend on the modulation frequency Ω of the electric field as well as on the light frequency ω. Therefore, when a value is given for r_{ij}, both ω and Ω should be specified. The dependence on ω is associated with the optical dispersion of the material, which relates to the electronic response time. It is determined by the frequency and strength of the active electronic transitions. On the other hand, the dependence on Ω arises from the additional ionic or lattice contribution to the electrooptic effect (see Section 3.5.1). An illustrative diagram for the variation of r_{ij} with Ω is given in Figure 3.2, for both clamped and unclamped geometries. In clamped experiments, the piezoelectric term is not operative and so for steady fields or modulation frequencies up to 10^{11} Hz, somewhat below the optical mode band frequencies, no dependence on Ω should be observed. Above those modulation frequencies, the optical mode contribution is inhibited and the purely electronic response is obtained. This limit situation can, then, be more appropriately described in terms of nonlinear optical mixing of two optical fields. This illustrates the connection between electrooptics and nonlinear optics, to be discussed at length in Chapter 4. For frequencies above the acoustical mode resonances, unclamped experiments yield an identical Ω-dispersion curve to that obtained for a clamped geometry. However, for lower modulation

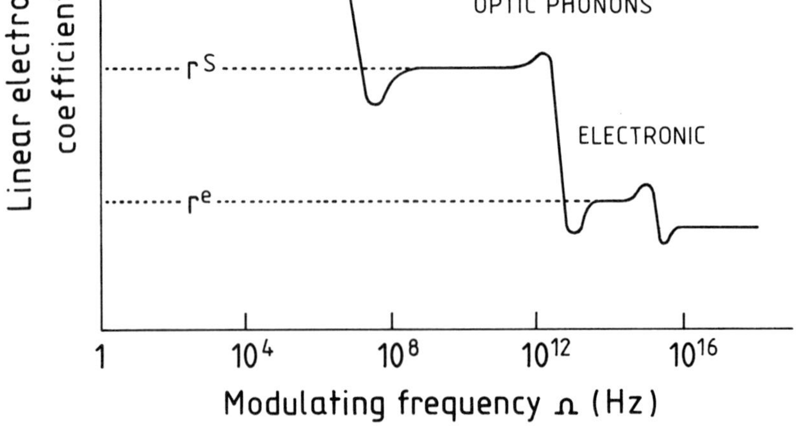

Figure 3.2 *Illustrative dependence of r^e, r^S and r^T on the modulation frequency Ω of the electric field [7].*

frequencies, one should take into account the piezoelectric term which gives an additional contribution to the unclamped values.

3.6.1. Miller's parameter

The variation of the electrooptic coefficients with light frequency can be markedly smoothened if one uses a related empirical parameter Δ_{ijk} proposed by Miller [25] and defined as

$$\begin{aligned}\Delta_{ijk}(\omega) &= -\frac{\epsilon_{ii}(\omega)\epsilon_{jj}(\omega)r_{ijk}(\omega)}{4\{\epsilon_{ii}(\omega)-1\}\{\epsilon_{jj}(\omega)-1\}\{\epsilon_{kk}(0)-1\}} \\ &= -\frac{\epsilon_{ii}(\omega)\epsilon_{jj}(\omega)r_{ijk}(\omega)}{4\chi_{ii}(\omega)\chi_{jj}(\omega)\chi_{kk}(0)}\end{aligned} \tag{3.20}$$

Moreover, Δ_{ijk} is less dependent on temperature and specific material than r_{ijk}. This behaviour suggests a correlation between large electrooptic coefficients and strong polarizabilities (or dielectric constants), providing a thumb rule in the search for efficient electrooptic materials. Further comments on the Miller's parameter are deferred to the next chapter on nonlinear optics where it finds a broader meaning.

3.7. Kerr effect

Although all media present the Kerr (quadratic) electrooptic effect, it is the dominant term in centrosymmetrical materials such as liquids or glasses, where the Pockels contribution is absent. Therefore, when speaking of Kerr materials one is normally referring to those lacking a linear electrooptic response. In Kerr devices, an organic liquid, nitrobenzene, has been very often used. However, very efficient transparent electrooptic ceramics and semiconductor nanostructures have inclined the balance towards solid-state materials.

In unclamped or constant stress experiments, one has to take into account (as for Pockels media) the effect on the refractive index caused by the electric field-induced strain. This contribution to the electrooptic response is additive with that corresponding to a constant volume situation (clamped experiments). Therefore, one should write, ignoring the Pockels term,

$$\Delta\left(\frac{1}{n^2}\right)_{ij} = \sum s^{\mathrm{T}}_{ijkl}E_kE_l = \sum s^{\mathrm{S}}_{ijkl}E_kE_l + \sum \pi_{ijmn}\mathscr{D}_{mn} \tag{3.21}$$

Alternatively, in terms of the polarization $\mathbf{P}$,

$$\Delta\left(\frac{1}{n^2}\right)_{ij} = \sum g^{\mathrm{T}}_{ijkl}P_kP_l = \sum g^{\mathrm{S}}_{ijkl}P_kP_l + \sum \pi_{ijmn}\mathscr{D}_{mn} \tag{3.22}$$

where the superscripts T and S stand for unclamped and clamped, respectively. The strain tensor $\mathscr{D}_{mn}$ in centrosymmetrical media is quadratically related to the electric field (or polarization) through a fourth-rank *electrostriction* tensor $\mathbf{Q}$:

$$\mathscr{D}_{mn} = \sum \mathscr{Q}^{\mathrm{E}}_{mnrs}E_rE_s \tag{3.23a}$$

or

$$\mathscr{D}_{mn} = \sum \mathscr{Q}^{\mathrm{P}}_{mnrs} P_r P_s \tag{3.23b}$$

so that

$$s^{\mathrm{T}}_{ijkl} = s^{\mathrm{S}}_{ijkl} + \sum \pi_{ijmn} \mathscr{Q}^{\mathrm{E}}_{mnkl} \tag{3.24}$$

For some materials the contribution of the strain-optic term to s^{T} is comparable to s^{S}. The advantages of the **P**-formulation will stand out when dealing with materials undergoing a ferroelectric–paraelectric transition.

The analysis of a Kerr experiment is particularly simple for isotropic materials, which is the only case to be discussed in Section 3.12.

3.7.1. Kerr effect in liquids

In pure liquids and solutions the Kerr effect has, generally, a dominant contribution due to the orientation of the active molecules by the applied field and so it presents a number of peculiar features. The mathematical formalism describing the effect is not going to be presented here and can be consulted in some classical papers, e.g. [26]. For nonpolar substances, the physical basis for the Kerr response is the field-induced orientation of molecules having a dielectric anisotropy. Often, molecules can be assimilated into a revolution ellipsoid, having a dielectric tensor with two different principal values $\epsilon_{\|}$ and $\epsilon_{\perp}$ (and referring to the long axis direction). The coupling energy to the field is, then, dependent on molecular orientation and so a certain orientational distribution is reached at thermodynamical equilibrium for an applied field **E** and temperature T. This anisotropic distribution gives rise to a birefringence Δn between light with linear polarization parallel and perpendicular to the field. For polar molecules an additional energy coupling term, $W = -\boldsymbol{\mu} \cdot \mathbf{E}$ (μ being the molecular dipole moment), contributes to determine the anisotropic distribution of molecular orientations.

The orientational mechanism yields $\Delta n \approx E^2$ (Kerr behaviour), although at very high fields saturation effects may appear. The response time is determined by rotational diffusion of the molecules and is markedly slower than that associated with electronic mechanisms. Although it depends on the viscosity of the liquid, typical relaxation frequencies are in the radio-frequency and microwave ranges.

3.8. Symmetry restrictions on electrooptic coefficients

The number of nonzero and linearly independent components of a tensor are determined by the point symmetry of the material. Detailed group theory methods to obtain these components are described in several general references [27,28], already mentioned in Chapter 2. Of particular interest is the case of materials lacking an inversion centre, i.e. belonging to a noncentrosymmetrical crystal class. It can then easily be shown that all polar odd-rank tensors, such as r_{ijk}, are identically zero, whereas no similar restriction applies to even-rank polar tensors as s_{ijkl}. This result implies that gases, ordinary liquids, and amorphous solids do not present any Pockels effect, and the dominant electrooptic response is Kerr-type. Appendix 3A shows the independent nonzero Pockels coefficients in contracted notation for the 21 noncentrosymmetrical point-group symmetries. The results for the axial group ∞mm, corresponding to a poled polymer film have also been included (there is no difference with groups $4mm$ and $6mm$). On the other hand, the structure of the Kerr matrices s_{ij} for all 32 crystal point groups and for an isotropic medium is given in Appendix 3A. A detailed review showing the implications of crystal symmetry on electrooptic device performance has been published by Eimerl [29].

Another symmetry relation, due to Kleinmann and discussed in Chapter 4, may hold that further reduces the number of independent nonzero components of the electrooptic tensors. It states that any permutation of the subindices *ijk* or *ijkl* does not alter the value of the coefficients r_{ijk} and s_{ijkl}, respectively. This symmetry is expected to apply only to the direct electronic contributions r^{e}_{ijk} and s^{e}_{ijkl}, when light frequency is far from resonances. As an example, for the Pockels tensor in an axial symmetry (∞mm) medium, only *two* independent nonzero elements remain: r^{e}_{13} and r^{e}_{33}. On the other hand, for an isotropic medium there is only *one* independent nonzero Kerr coefficient s^{e}_{11}. Full expressions for the r_{ij} matrices, taking Kleinmann symmetry into account, are given in Appendix 4C.

3.9. A microscopic model: the anharmonic oscillator

As mentioned in Section 3.1, there is a large variety of microscopic mechanisms leading to electrooptic and nonlinear phenomena. However, from a didactic point of view, the anharmonic oscillator model is particularly general and useful, since it illustrates some of the key physical ideas, such as the resonance enhancement and dispersive behaviour of the electrooptic response. Moreover, it provides a classical simple description for the electrooptic behaviour associated with the nonlinear distortion of an electron cloud when excited by an electromagnetic field, and yields a definite expression for the *electronic* or *direct* electrooptic coefficients. More advanced and realistic general theoretical approaches will be summarily presented in Chapter 4, whereas phenomenological models applying to specific types of materials will be described in Chapters 5 and 6.

Let us consider a medium consisting of a density N of noninteracting electronic anharmonic oscillators. In each oscillator the electron is simultaneously subjected to a constant field $E(0) = E_0$, and an oscillatory field at frequency ω (light wave)

$$E(\omega) = (E_{\omega}\,\mathrm{e}^{-i\omega t} + E^{*}_{\omega}\,\mathrm{e}^{+i\omega t})/2 \tag{3.25}$$

where $\tilde{E}_{w}$ stands for the complex amplitude. Neglecting local field effects, the force experienced by the electron is $qE = q\{E_0 + E(\omega)\}$, in addition to the oscillator binding force. Then, one uses the classical motion equation to obtain the electron dynamics and electrooptic response corresponding to either the Pockels or Kerr situations. Local field effects can be easily included in the model [30], but they do not essentially alter the physical picture.

3.9.1. Pockels effect

The anharmonic binding potential has to be noncentrosymmetrical to permit the occurrence of a Pockels effect. Assuming an electronic oscillation along X, the simplest odd potential is $V = -ax^3/3$, which implies a quadratic anharmonic binding force $f = max^2$. Then, the equation of motion for the electron can be written as

$$\frac{\mathrm{d}^2x}{\mathrm{d}t^2} + \Gamma\frac{\mathrm{d}x}{\mathrm{d}t} + \omega_0^2x + ax^2 = \frac{qE}{m} \tag{3.26}$$

where x measures the electron displacement, Γ is a friction or dissipation coefficient, ω_0 the eigenfrequency of the oscillator in the harmonic approximation, ax^2 the anharmonic term in the elastic force and E the total field $E = E(\omega) + E_0$.

The solution of (3.26) is conveniently obtained by using a perturbation expansion for x:

$$x = x^{(1)} + x^{(2)} + x^{(3)} + \dots \tag{3.27}$$

At first order (harmonic approximation) the ax^2 term is neglected and then one easily obtains

$$x^{(1)} = \left\{ \frac{qE_\omega \, e^{-i\omega t}}{2m(\omega_0^2 - \omega^2 - i\omega\Gamma)} + \text{c.c.} \right\} + \frac{qE_0}{m\omega_0^2} \tag{3.28}$$

where c.c. stands for the complex conjugate of the preceding expression. The solution is just the sum of those corresponding to independent excitation of the oscillator by $E(\omega)$ and $E(0)$, in accordance with the additivity principle valid for a linear model. When considering a material made up of a concentration N of noninteracting oscillators, the linear polarization is

$$P^{(1)} = \tfrac{1}{2}\{P_\omega^{(1)} \, e^{-i\omega t} + \text{c.c.}\} + P_0^{(1)} \tag{3.29}$$

with

$$P_\omega^{(1)} = \frac{Nq^2 E_\omega}{m(\omega_0^2 - \omega^2 - i\omega\Gamma)} \tag{3.30}$$

and

$$P_0^{(1)} = \frac{Nq^2 E_0}{m\omega_0^2} \tag{3.31}$$

Using the linear complex susceptibility at frequency ω,

$$\chi(\omega) = \frac{P_\omega^{(1)}}{\epsilon_0 E_\omega} = \frac{Nq^2}{\epsilon_0 m(\omega_0^2 - \omega^2 - i\omega\Gamma)} \tag{3.32}$$

and the real static susceptibility

$$\chi(0) = \frac{P_\omega^{(1)}}{\epsilon_0 E_0} = \frac{Nq^2}{\epsilon_0 m\omega_0^2} \tag{3.33}$$

one can express (3.29) as

$$P^{(1)} = \epsilon_0 \chi(\omega) E_\omega \, e^{-i\omega t} + \text{c.c.} + \epsilon_0 \chi(\omega) E_0 \tag{3.34}$$

As expected, the polarization, as well as the linear susceptibility at frequency ω, are not influenced by the simultaneous application of a constant electric field E_0.

The second-order solution is obtained by substituting (3.28) into the anharmonic term ax^2 in (3.26). This substitution gives rise to displacements x oscillating at frequencies 2ω, ω and 0. For the oscillation at frequency ω one arrives, after some algebra, at

$$x^{(2)}(\omega) = \frac{-aq^2 E_\omega E_0 \, e^{-i\omega t}}{2m^2\omega_0^2(\omega_0^2 - \omega^2 - i\omega\Gamma)^2} + \text{c.c.} \tag{3.35}$$

as the correction to the harmonic (linear) displacement oscillating at frequency ω.

The associated nonlinear polarization is

$$P^{(2)}(\omega) = \frac{-aNq^3 E_\omega E_0 \,\mathrm{e}^{-i\omega t}}{2m^2\omega_0^2(\omega_0^2 - \omega^2 - i\omega\Gamma)^2} + \text{c.c.}$$

$$= \tfrac{1}{2}\{P_\omega^{(2)}\,\mathrm{e}^{i\omega t} + \text{c.c.}\} \tag{3.36}$$

Therefore, the change in the complex dielectric constant due to the anharmonic term is

$$\Delta\tilde{\epsilon}(\omega) = \frac{P_\omega^{(2)}}{\epsilon_0 E_\omega} = \frac{-aNq^3 E_0}{\epsilon_0 m^2\omega_0^2(\omega_0^2 - \omega^2 - i\omega\Gamma)^2} \tag{3.37}$$

whose real part is

$$\Delta\epsilon(\omega) = \frac{-aNq^3 E_0\{(\omega_0^2 - \omega^2)^2 - \omega^2\Gamma^2\}}{\epsilon_0 m^2\omega_0^2\{(\omega_0^2 - \omega^2)^2 + \omega^2\Gamma^2\}^2} \tag{3.38}$$

Consequently, in terms of the refractive index, one may write

$$\Delta\left(\frac{1}{n^2}\right) = -\frac{\Delta\epsilon}{\epsilon^2} = \frac{aNq^3 E_0\{(\omega_0^2 - \omega^2)^2 - \omega^2\Gamma^2\}}{\epsilon_0\epsilon^2 m^2\omega_0^2\{(\omega^2 - \omega_0^2)^2 + \omega^2\Gamma^2\}^2} \tag{3.39}$$

From this expression, the electrooptic coefficient r^{e}, which is more properly written as $r^{\mathrm{e}}(\omega, \Omega = 0)$, is

$$r^{\mathrm{e}}(\omega, \Omega = 0) = \frac{aNq^3\{(\omega_0^2 - \omega^2)^2 - \omega^2\Gamma^2\}}{\epsilon_0\epsilon^2 m^2\omega_0^2\{(\omega_0^2 - \omega^2)^2 + \omega^2\Gamma^2\}^2} \tag{3.40}$$

clearly showing the resonant behaviour at $\omega_0 \approx \omega$, as for the linear susceptibility $\chi(\omega)$. Far enough from the resonance ω_0, where one can neglect the friction terms, ($\Gamma \approx 0$), equation (3.40) can be expressed as

$$r^{\mathrm{e}}(\omega) = \frac{aNq^3}{\epsilon_0\epsilon^2 m^2\omega_0^2(\omega_0^2 - \omega^2)^2} = \frac{C\chi^2(\omega)\chi(0)}{\{\epsilon(\omega)\}^2} \tag{3.41}$$

where

$$C = \frac{4am\epsilon_0^2}{\epsilon^2 N^2 q^3} \tag{3.42}$$

One may then write for the Miller parameter (a scalar in this model)

$$\Delta(\omega) = \frac{-\epsilon^2 r^{\mathrm{e}}(\omega)}{4\chi^2(\omega)\chi(0)} = -\frac{am\epsilon_0^2}{N^2 q^3} \tag{3.43}$$

providing some support for the empirical rule on the approximated frequency independence of the Miller parameter.

Modulated applied field

In many practical situations the applied electric field is modulated at a frequency $\Omega < 10^{10}$ Hz (i.e. $\Omega \ll \omega$). Then, the anharmonic term mixes ω and Ω and gives rise to polarizations oscillating at $\omega \pm \Omega$ (*nonlinear wave-mixing*). By repeating the mathematical procedure outlined above for the case $\Omega = 0$, one obtains for frequencies ω far from ω_0,

$$r^{e}(\omega \pm \Omega) = \frac{(aNq^3/2\epsilon_0\epsilon^2 m^2)}{(\omega_0^2 - \omega^2)(\omega_0^2 - \Omega^2)(\omega_0^2 - [\omega \mp \Omega]^2)} \tag{3.44}$$

presenting resonances at $\omega = \omega_0$, $\Omega = \omega_0$, and $\omega \mp \Omega = \omega_0$. Since $\Omega \ll \omega$, ω_0, in all electrooptic experiments, one can easily check that $r^{e}(\omega + \Omega) \simeq r^{e}(\omega - \Omega) \simeq r^{e}(\omega)$, i.e. no Ω-dispersion should be observed in the electronic electrooptic response. The dependence $r(\omega, \Omega)$ becomes more complex when ionic motions are taken into account, although it cannot be described by any practicable model. Anyhow, one should note that the schematic dependence illustrated in Figure 3.2 corresponds to cutting of the $r^{e}(\omega, \Omega)$ surface by a plane $\omega =$ constant.

3.9.2. Linear electroabsorption

The model also permits to obtain the electroabsorption $\Delta\alpha(E)$ induced by the electric field E_0. Recalling that

$$\alpha = \frac{\omega\epsilon''}{nc} \tag{3.45}$$

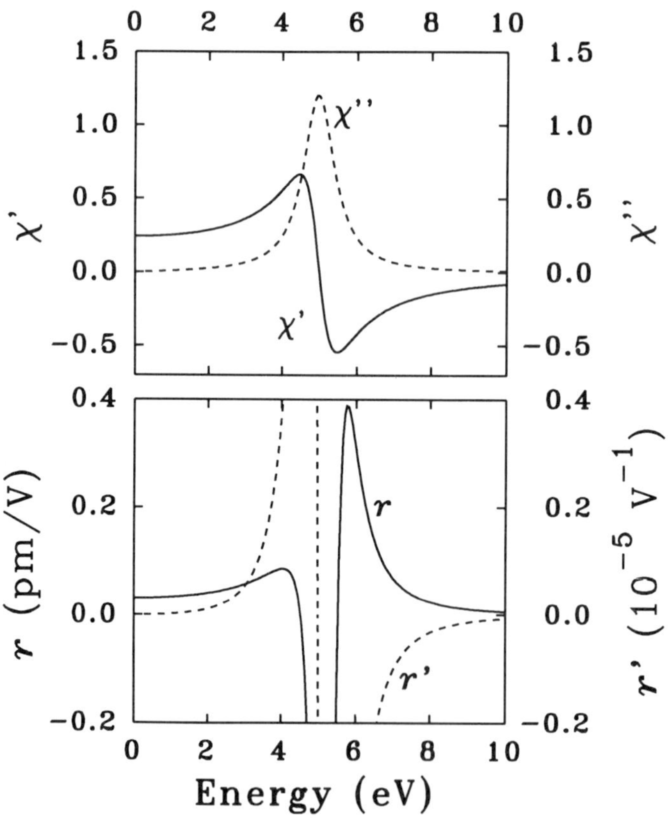

Figure 3.3 *Dependences* r(ω) (electrorefraction) *and* r′(ω) (electroabsorption) *for an anharmonic oscillator model.* $\chi'(\omega)$ *and* $\chi''(\omega)$ *are also shown for comparison.*

and calculating $\Delta\epsilon''(E_0)$ from the imaginary part of (3.37) one arrives at

$$\Delta\alpha = r' E_0 \tag{3.46}$$

r' being a linear electroabsorption coefficient given by

$$r' = \left(\frac{2aNq^3\Gamma}{nc\epsilon_0 m^2\omega_0^2}\right)\left(\frac{\omega^2(\omega_0^2-\omega^2)}{(\omega^2-\omega_0^2)^2+\omega^2\Gamma^2}\right) \tag{3.47}$$

also showing the typical resonant behaviour. One may check that the expressions obtained in the model for Δn and $\Delta\alpha$ are, indeed, consistent with the Kramers–Krönig relation (2.19).

Figure 3.3 illustrates the behaviour of $r(\omega)$ and $r'(\omega)$ for an anharmonic oscillator model with $N = 10^{22}\,\mathrm{cm}^{-3}$, $\omega_0 = 8\times 10^{15}$ Hz ($\hbar\omega_0 = 5$ eV), $\hbar\Gamma = 1$ eV and $a = 4\times 10^{35}/(\mathrm{m\,s}^2)$. One should note the strong enhancement of both r and r' at resonance.

3.9.3. Kerr response

The oscillator model can be now applied to account for the second-order (Kerr) effect. One should use the iterative process used in Section 3.9.1 and proceed to the next order by substituting the second-order solution into the ax^2 anharmonic force term in (3.26). However, it is preferable to start from an odd anharmonic force term bx^3, instead of ax^2, in the motion equation. No Pockels effect is then operative and the mathematical procedure is somewhat simplified. By following exactly the same steps as in Section 3.9.1 (the details are left to the reader) one finally obtains

$$s^{\mathrm{e}}(\omega, \Omega = 0, \Omega = 0) = \frac{3bNq^4\{(\omega_0^2-\omega^2)^2-\omega^2\Gamma^2\}}{\epsilon_0\epsilon^2 m^3\omega_0^4\{(\omega_0^2-\omega^2)^2+\omega^2\Gamma^2\}^2} \tag{3.48}$$

3.10. Simple estimate of the nonresonant electronic coefficients r and s

A very simple estimate of the electronic contribution to r and s can be now made by assuming a binding Coulomb potential for the electrons. The equilibrium distance R from the atomic nucleus will be taken as the Bohr radius. Then, for small oscillations x of the electron around R, the Coulomb force can be written

$$f(\mathrm{R}+x) = f_0 - \left(\frac{2f_0}{\mathrm{R}}\right)x + \left(\frac{6f_0}{\mathrm{R}^2}\right)x^2 - \left(\frac{24f_0}{\mathrm{R}^3}\right)x^3 + \ldots \tag{3.49}$$

with

$$f_0 = \frac{q^2}{4\pi\epsilon_0 \mathrm{R}^2} = q\mathrm{E_i} \tag{3.50}$$

being the equilibrium binding force. This can be considered as a typical *internal force* for electrons in matter, whereas $\mathrm{E_i}$ is an *internal electric field*. From (3.50) one immediately obtains the *harmonic* and the *anharmonic* a and b force constants appearing in the motion equation for the oscillator. They become

$$m\omega_0^2 = \frac{2f_0}{R} \tag{3.51a}$$

$$a = \frac{6f_0}{mR^2} \tag{3.51b}$$

$$b = \frac{-24f_0}{mR^3} \tag{3.51c}$$

3.10.1. Pockels coefficients

Substituting (3.51b) into formula (3.40) for r^e and making $\Gamma = 0$, as it corresponds for non-resonant conditions, one finally reaches

$$r = \left(\frac{\omega_P}{\omega_0}\right)^2 \frac{q}{\epsilon f_0} \sim E_i^{-1} \tag{3.52}$$

where $\omega_P = \{Ne^2/m\epsilon\epsilon_0\}^{1/2}$ is the *plasma frequency* for the electron gas $(\omega_P^2/\omega_0^2\epsilon \approx 1)$. In other words, the electrooptic coefficient r^e is estimated to be of the order of the reciprocal of the *internal electric field* experienced by electrons in matter, i.e. $r^e \approx 10^{-11}$ m/V. This is in good qualitative accordance with experiment.

The same result can be also obtained without recourse to equation (3.40), by simply calculating the Stark shift on the electron eigenfrequency ω_0 and using the nonresonant expression for the dielectric constant in an ideal Lorentz model.

3.10.2. Kerr coefficients

A similar estimate for s, based on equation (3.48), can also easily be given. Under nonresonant conditions ($\Gamma = 0$), one arrives at

$$s = \frac{\omega_P^2}{\omega_0^2}\frac{9q^2}{\epsilon f_0^2} \approx \frac{q^2}{f_0^2} = E_i^{-2} \tag{3.53}$$

i.e. the order of magnitude of the Kerr coefficient is the reciprocal of the square of internal fields in matter or $s \approx 10^{-22}$ m^2/V^2. Consequently, the expected index change in centrosymmetrical materials, for a typical field $E = 10^4$ V/cm is

$$\Delta n = \frac{sn^3E^2}{2} \approx n^3\left(\frac{E}{E_i}\right)^2 \approx 10^{-9} \tag{3.54}$$

i.e. markedly lower than that caused by the linear effect. However, for some Kerr materials, such as ceramics and semiconductor multiple quantum wells, s reaches values in the range 10^{-15}–10^{-16} m^2/V^2 (see Chapters 5, 6 and 9) and so index changes can be markedly enhanced.

One can extrapolate relations (3.52) and (3.53) and conclude, as a thumb rule, that the estimated value for the n-order electrooptic coefficients is approximately E_i^{-n}.

3.11. Optical wave propagation through a Pockels medium

Let us consider the propagation of a plane monochromatic wave through a Pockels medium under an applied field **E**. By using equation (3.11) with the appropriate r_{ij} coefficients, the field-dependent index tensor is first obtained. Then, the formalism of Chapter 2 allows one to determine the effective refractive index for any specified propagation and polarization directions. We will now successively illustrate the two stages of the problem, by using a crystal of KDP (KH_2PO_4) as a relevant example. The crystal belongs to the $\bar{4}2m$ class and is, therefore, optically uniaxial, the optic axis being the quaternary crystallographic axis.

3.11.1. Determination of the new principal indexes and directions

The mathematical complexity of the problem depends on the symmetry of the crystal. For unperturbed KDP, the index tensor can be written

$$\left(\frac{1}{n^2}\right)_{ij} = \begin{vmatrix} 1/n_o^2 & 0 & 0 \\ 0 & 1/n_o^2 & 0 \\ 0 & 0 & 1/n_e^2 \end{vmatrix} \tag{3.55}$$

when the reference system is taken so that the Z direction is along the quaternary axis and the X direction along the binary crystal axis. n_o and n_e are respectively, the ordinary and extraordinary indices of the crystal. On the other hand, the electrooptic tensor in the contracted notation has the structure (Appendix 3A)

$$r_{ij} = \begin{vmatrix} 0 & 0 & 0 \\ 0 & 0 & 0 \\ 0 & 0 & 0 \\ r_{41} & 0 & 0 \\ 0 & r_{41} & 0 \\ 0 & 0 & r_{63} \end{vmatrix} \tag{3.56}$$

with only two independent nonzero components.

By applying the electric field along Z ($E_x = E_y = 0$) the field-dependent index tensor is

$$\left(\frac{1}{n^2}\right)_{ij} = \begin{vmatrix} 1/n_o^2 & r_{63}E_z & r_{41}E_y \\ r_{63}E_z & 1/n_o^2 & r_{41}E_x \\ r_{41}E_x & r_{41}E_y & 1/n_e^2 \end{vmatrix} \tag{3.57}$$

The equations determining the eigenvectors or principal optical directions $\mathcal{V}_i$ are

$$\left(\frac{1}{n^2}\right)_{ij} \mathcal{V}_j - \nu_i \mathcal{V}_i = 0 \tag{3.58}$$

where ν_i are the corresponding *eigenvalues* or principal indexes. From the compatibility condition for this set of homogeneous equations

$$\det\left|\left(\frac{1}{n^2}\right)_{ij} - \nu_i \delta_{ij}\right| = 0 \tag{3.59}$$

one obtains the roots or eigenvalues

$$\left.\begin{aligned} \nu_1 &= \frac{1}{n_e^2} \\ \nu_2 &= \frac{1}{n_o^2} + r_{63}E \\ \nu_3 &= \frac{1}{n_o^2} - r_{63}E \end{aligned}\right\} \tag{3.60}$$

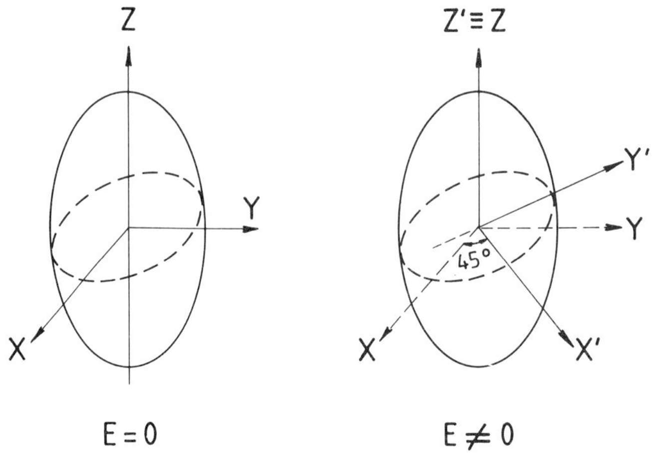

Figure 3.4 *Index ellipsoid for KDP, before and after application of an electric field along the optic axis. Note the rotation of axes from* XY *to* X′Y′.

Going back now to equations (3.58), the new principal directions are obtained:

$$\left.\begin{array}{lll} \text{For } \nu_1\text{: } X = 0, Y = 0 & (Z' \equiv Z \text{ axis}) \\ \text{For } \nu_2\text{: } X + Y = 0, Z = 0 & (Y' \text{axis}) \\ \text{For } \nu_3\text{: } X - Y = 0, Z = 0 & (X' \text{axis}) \end{array}\right\} \tag{3.61}$$

The relation between the new $(X'Y'Z')$ and old (XYZ) axes is schematically illustrated in Figure 3.4. Since the perturbing term $r_{63}E$ in (3.60) is generally very small in comparison with ν_i, one can write for the new index values along the modified principal axes

$$\left.\begin{array}{l} n_{x'} = \dfrac{n_o}{(1 + n_o^3 r_{63} E)^{1/2}} \approx n_o - \frac{1}{2} n_o^3 r_{63} E \\ n_{y'} = \dfrac{n_o}{(1 - n_o^3 r_{63} E)^{1/2}} \approx n_o + \frac{1}{2} n_o^3 r_{63} E \\ n_{z'} = n_z \approx n_e \end{array}\right\} \tag{3.62}$$

In summary, the field perturbed index tensor (or equivalently, the index ellipsoid) is now completely determined. Other directions for the applied field are handled in a similar way. On the other hand, the method can be easily extended to any other crystal structure.

Pockels figure of merit

Expressions (3.62) show that the field-induced changes in refractive index are governed by the proportionality coefficients $n_o^3 r_{63}$. This is a general result, independent of material symmetry, which has to do with the definition of r through $\Delta(1/n^2) = rE$, immediately leading to $\Delta n = -n^3 rE/2$ in a naive scalar formulation. Therefore, the electrooptic efficiency of a material is characterized by the *figure of merit* $n^3 r_{ij}$ (using the r_{ij} coefficient adequate to the prevailing experimental conditions).

3.11.2. Field-induced birefringence

As discussed in Chapter 2 the optical behaviour of an arbitrary plane wave is determined by that of the two polarization eigenmodes corresponding to its propagation direction.

They present linear polarization along the two principal axes of the elliptical section of the index ellipsoid by a plane normal to the wavevector. The distortion of the ellipsoid caused by the applied field induces a *rotation* θ of the ellipse and a *change* in the *magnitude* of the principal axes with regard to the zero-field situation. From the point of view of device operation the key parameter is the *birefringence* Δn associated with the selected propagation direction, i.e. the difference between the indices n_1 and n_2 for the two eigenmodes. It determines the so-called *retardation* δ or dephasage between the eigenmodes after traversing the crystal slab. δ is given by

$$\delta = \frac{2\pi \Delta n l}{\lambda} = \frac{\omega l \Delta n}{c} \tag{3.63}$$

l being the slab length and λ the light wavelength in vacuum. It can now be calculated by the method of Section 3.11.1.

In most experiments, the propagation direction is along one of the principal axes because it minimizes the effects of beam divergence and crystal misalignment. Propagation along the optic axis in uniaxial crystals has the further advantage of avoiding any interfering temperature dependence of the natural birefringence during experiments. Therefore, propagation along principal axes will be always assumed in the following analyses.

In order to be specific, let us again consider the case of KDP with the field applied along the optic axis (Z), as an illustrative example. Two alternative configurations for beam propagation will next be discussed: (a) along the field direction (*longitudinal configuration*), and (b) perpendicular to the field (*transverse configuration*).

(a) Longitudinal configuration

An incident linearly polarized wave can be decomposed into two components linearly polarized along the X', Y' axes of the perturbed index tensor, Figure 3.5(a). If the length of the crystal is l, the *retardation* between the two waves at the end face will be

$$\delta = \phi_{y'} - \phi_{x'} = \omega n_o^3 r_{63} \frac{V}{c} = 2\pi n_o^3 r_{63} \frac{V}{\lambda} \tag{3.64}$$

where $\phi_{x'}$ and $\phi_{y'}$ are, respectively, the phase changes of the two component waves, polarized along X' and Y', induced by the transit through the crystal. One sees that δ depends exclusively on the applied potential V and not on the crystal length l. The potential V_π necessary to induce a dephasage $\delta = \pi$, designated as *half-wave voltage*, is a very important parameter in the design of electrooptic modulators. It is immediately obtained from (3.64):

$$V_\pi(\text{longitudinal}) = \frac{\lambda}{2n_o^3 r_{63}} = \frac{\pi c}{\omega n_o^3 r_{63}} \tag{3.65}$$

and permits us to write the retardation in the simple form

$$\delta = \frac{\pi V}{V_\pi} \tag{3.66}$$

By using the r_{63} coefficient for KDP, V_π amounts to 8.5 kV at $\lambda = 0.633\,\mu$m (Table 7.1). The same order of magnitude applies to other Pockels crystals. However, it will be shown that much smaller and so useful values of V_π can be achieved by other electrooptic mechanisms (e.g. molecular reordering in liquid crystals).

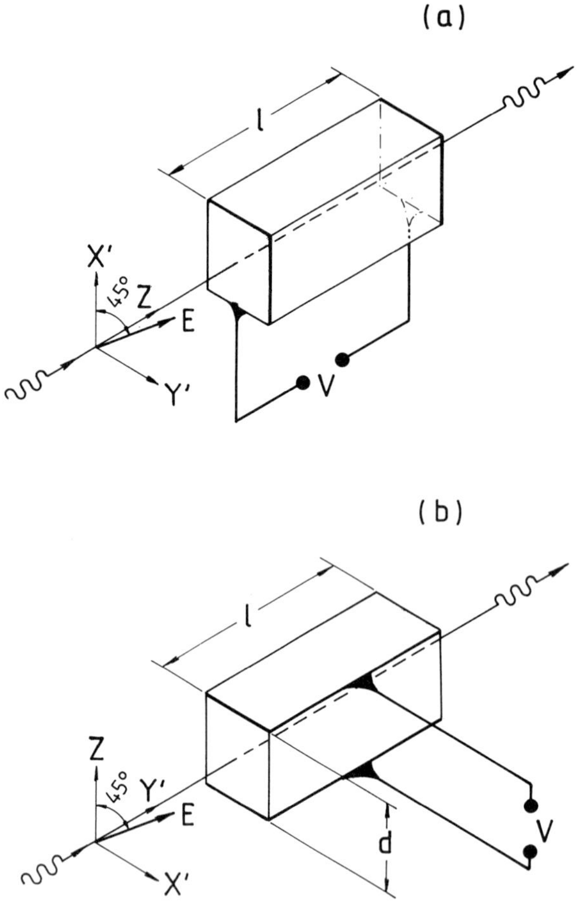

Figure 3.5 *Experimental electrooptic configurations: (a) longitudinal; (b) transverse.*

(b) Transverse configuration

It is a frequently used practical scheme, with the applied field (along Z), i.e. perpendicular to the light propagation vector (along Y'). As in the previous case, the incident linearly polarized beam is decomposed into two, respectively polarized along X' and Z', Figure 3.5b. At the end face of the crystal of length l, the dephasage δ between the two components is

$$\delta = \frac{\omega l(n_{z'} - n_{x'})}{c} = \frac{\omega l}{c}\left\{(n_e - n_o) + n_o^3 r_{63}\frac{V}{2d}\right\} \tag{3.67}$$

where d is the crystal dimension along Z. This expression includes the dephasage associated to the intrinsic crystal birefringence (first term on the right side) and that caused by the applied field (second term). At variance with the longitudinal case, the dephasage δ is proportional to l. The half-wave voltage V_π for a field-induced $\delta(v) = \pi$, is

$$V_\pi(\text{transverse}) = \frac{\lambda d}{n_o^3 r_{63} l} = \frac{2d}{l}\, V_\pi(\text{longitudinal}) \tag{3.68}$$

>and similarly to the longitudinal case, one can write

$$\delta = \frac{\pi V}{V_\pi} \tag{3.69}$$

Values of V_π can be now markedly lower than those obtained for the longitudinal configuration, if $d/l \ll 1$. This situation is easily reached in channelled waveguides where $l > 5\,\text{mm}$, and $d < 5\,\mu\text{m}$ (see chapter 8).

In addition to V_π, other relevant parameters of the material will emerge when discussing modulator design considerations in Chapter 7.

3.11.3. Effective electrooptic coefficients

In an uniaxial Pockels material, the birefringence caused by an electric field $\boldsymbol{E}$ on a propagating plane wave can be written

$$\Delta n = \frac{n_o^3 r_{\text{eff}} E}{2} \tag{3.70}$$

defining an *effective electrooptic coefficient* r_{eff} depends on both propagation and field directions. Table 3.2, taken from reference [29], summarizes the r_{eff} values for longitudinal and transversal light propagation in all uniaxial crystal classes, assuming the field applied along the optic axis. Moreover, for the longitudinal configuration, the field-induced rotation θ of the relevant index ellipse is also given. On the other hand, since the transverse

Table 3.2 ***Effective electrooptic coefficients for uniaxial crystals [29].***

	Longitudinal		Transversal	
Crystal class	r_{eff}	θ	r_{eff}	ϕ
4			$r_{13} - (n_O/n_e)^3 r_{33}$	arbitrary
$\bar{4}$	$2\dfrac{r_{41}^2 - r_{63}^2}{(r_{41}^2 + r_{63}^2)^{1/2}}$	$\tan 2\phi = (r_{63}/r_{13})$	$(r_{13}^2 + r_{63}^2)^{1/2}$	$\tan 2\phi = (r_{63}/r_{13})$
422				
$4mm$			$r_{13} - (n_O/n_e)^3 r_{33}$	arbitrary
$\bar{4}2m$	$2r_{63}$	45°	r_{63}	45°
3			$r_{13} - (n_O/n_e)^3 r_{33}$	arbitrary
32				
$3m$			$r_{13} - (n_O/n_e)^3 r_{33}$	arbitrary
6			$r_{13} - (n_O/n_e)^3 r_{33}$	arbitrary
$\bar{6}$				
622				
$6mm$			$r_{13} - (n_O/n_e)^3 r_{33}$	arbitrary
$\bar{6}m2$				
23, $\bar{4}3m$	$2r_{63}$	45°	r_{63}	45°

Electric field applied parallel to the optic axis. Blank spaces indicate zero field-induced birefringence ($r_{\text{eff}} = 0$). r_{eff} and ϕ values for the transversal configuration correspond to the largest birefringence.

configuration is not univocally defined, only the maximum r_{eff} value and the corresponding angle ϕ between the light wavevector and the X crystal axis, are listed.

One may draw some relevant conclusions from this table. For longitudinal configuration, birefringence is only observed for classes $\bar{4}$, $\bar{4}2m$, 23 and $\bar{4}3m$. For these classes the field has converted the crystal from uniaxial to biaxial, with the new axes X' and Y' *locked* at fixed positions, regardless of the applied field magnitude. In particular, for the previous example of KDP, the locked axes are at 45° from the unperturbed X and Y axes.

For transverse configuration, no field-induced birefringence is produced for the 422, 32, $\bar{6}$, 622 and $\bar{6}m2$ classes, which consequently do not show birefringence in any of the two configurations.

One should also note that for some configurations and crystal classes, the measurement of the birefringence Δn permits the unequivocal determination of one particular electrooptic coefficient. In some other cases, one can only determine a linear combination of electrooptic coefficients. As an example, for $BaTiO_3$ ($4mm$) in transversal configuration the measured birefringence determines r_c given by

$$r_c = r_{13} - \left(\frac{n_e}{n_o}\right)^3 r_{33} \tag{3.71}$$

Measurement of individual coefficients requires additional experiments.

For biaxial crystals with the field applied along Z (or X_3) and light propagation along principal axes X, Y or Z, the effective electrooptic coefficients are often designated [29] as r_a, r_b and r_c, respectively. The corresponding expressions are as follows:

$$r_a = r_{33} - \left(\frac{n_2}{n_3}\right)^3 r_{23} \tag{3.72a}$$

$$r_b = r_{33} - \left(\frac{n_1}{n_3}\right)^3 r_{13} \tag{3.72b}$$

$$r_c = r_{13} - \left(\frac{n_2}{n_1}\right)^3 r_{23} \tag{3.72c}$$

in accordance with the following expressions for the birefringence,

$$\left.\begin{aligned} \Delta n\,(X\,\text{axis}) &= \frac{n_3^3 r_a}{2} \\ \Delta n\,(Y\,\text{axis}) &= \frac{n_3^3 r_b}{2} \\ \Delta n\,(Z\,\text{axis}) &= \frac{n_1^3 r_c}{2} \end{aligned}\right\} \tag{3.73}$$

3.12. Light propagation in isotropic Kerr media

In an isotropic material the refractive index tensor reduces to a scalar, i.e. to a multiple of the unit tensor, $(1/n^2)_{ij} = n\delta_{ij}$, so that the index ellipsoid is a sphere. When the field is applied along Z, the symmetry becomes axial, with two principal values for n, n_e along Z and n_o for the directions in the XY plane. From the definition of s_{ijkl} one immediately obtains the relations

$$\frac{1}{n_e^2} - \frac{1}{n^2} = s_{zzzz}E^2 = s_{33}E^2 \tag{3.74a}$$

$$\frac{1}{n_o^2} - \frac{1}{n^2} = s_{xxzz}E^2 = s_{13}E^2 \tag{3.74b}$$

which can be approximately written

$$n_e - n \approx -\tfrac{1}{2}\, s_{33}n^3E^2 \tag{3.75a}$$

$$n_o - n \approx -\tfrac{1}{2}\, s_{13}n^3E^2 \tag{3.75b}$$

where it has been assumed that $n_e \approx n_o \approx n$.

The field-induced birefringence $n_e - n_o$ is then obtained by subtracting equations (3.75), and taking into account that $s_{33} - s_{13} = s_{66}$ (Appendix 3B),

$$\Delta n = n_e - n_o = -\tfrac{1}{2}\, s_{66}n^3E^2 \tag{3.76}$$

Introducing the parameter B,

$$B = \frac{1}{2}\, s_{66}\, \frac{n^3}{\lambda} \tag{3.77}$$

finally, one arrives at

$$\Delta n = -\lambda BE^2 \tag{3.78}$$

From equations (3.76) and (3.78), it is clear that the parameters $n^3 s_{66}$ or B are an adequate *figure of merit* to measure the index change induced in an isotropic Kerr material.

3.12.1. Propagation of a plane monochromatic wave

For a Kerr medium in the longitudinal configuration (i.e. **E** and **k** along Z), any wave polarization is maintained during propagation. Moreover, the field does not induce any birefringence effect.

On the other hand, for the transversal configuration, the situation is similar to that discussed in Section 3.11. For a linearly polarized wave, the two components with polarizations along Z and perpendicular, propagate at different speeds c/n_o and c/n_o, respectively and a dephasage δ appears at the exit face. This dephasage turns the initial linear polarization into an elliptical polarization. For a length l of the medium

$$\delta = 2\pi(n_o - n_e)\,\frac{1}{\lambda} \approx 2\pi BlE^2 \tag{3.79}$$

and the *half-wave voltage* V_π is

$$V_\pi = \frac{d}{(2Bl)^{1/2}} \tag{3.80}$$

d being the separation between electrodes $(E = V/d)$. As for Pockels media, the change in the polarization state of the beam can be converted into an intensity change and a modulator device can be constructed.

3.13. Measurement of electrooptic coefficients

This section will deal with the experimental techniques appropriate to bulk samples and films. Methods designed for waveguide configurations are deferred to Chapter 8. Pockels and Kerr coefficients can be determined with the same setup and so the two effects will be jointly considered. The basis of most methods is the measurement of the phase change $\Delta\phi = \omega l \Delta n / c$ introduced by the electric field, across a light path l in the material. To that end, these $\Delta\phi$ are converted into intensity changes (as in modulator design), either by polarimetric or interferometric devices. In some other methods, one exploits the dependence of the polarization of a reflected beam on the complex refractive index of the material. Some less conventional methods should be also mentioned. For example, near or at resonances, the field-induced change in the absorption coefficient is measured and then the change in refractive index is calculated through the Kramers–Krönig relations. Finally, the clamped and electronic electrooptic Pockels coefficients can be obtained from absolute measurements of Raman intensities. A summary of this technique, which also yields the nonlinear mixing coefficients, will be given in Section 3.13.3.

Relative accuracy of most methods is typically 10% or lower, although in careful experiments values of a few percent are claimed. One has to take into account that several factors may limit such accuracy and even lead to erroneous results. In particular, a key parameter that should be precisely controlled is temperature. In many cases, a thermal stability better than 0.1°C is required. The situation may, of course, be critical in the vicinity of phase transitions. Anyhow, measurements of the dependence of the electrooptic coefficients on temperature as well as on field modulation and light frequencies provide very useful complementary information to elucidate electrooptic mechanisms.

3.13.1. Single beam methods (Noninterferometric)

Polarimetric techniques (Transmission) [31–36]

For the measurement of the field-induced dephasage in single-beam type experiments, the instrumental arrangement is as illustrated in Figure 3.6. A *transverse* configuration is shown for convenience. A Soleil compensator has been included, although it is not used in all cases. A linear polarizer P orients the electric vector of the incident beam at 45° with regard to the applied field. After traversing the length l of the crystal, a dephasage δ has been introduced between the two component waves with polarizations parallel and perpendicular to the field. By placing an analyser A, crossed to P, the outgoing light intensity will be

$$I = I_0 \sin^2\left(\frac{\delta}{2}\right) \tag{3.81}$$

where $\delta = \delta(0) + \delta(V)$, with $\delta(0)$ referring to the intrinsic birefringence of the material, including additional effects that may arise from residual stresses, or from a given compensator setting. $\delta(V)$ is the electric field-induced contribution, which in general may include linear (Pockels) and quadratic (Kerr) terms. For a Pockels material, the Kerr contribution can be, generally, ignored and the analysis is markedly simplified. The dependence $I(V)$ (or equivalently $I(\delta)$) is schematically illustrated in Figure 3.7. It can be determined from measurements of the outgoing intensity I as a function of a variable applied potential V. By fitting the measured δ to a theoretical expression such as (3.81), one

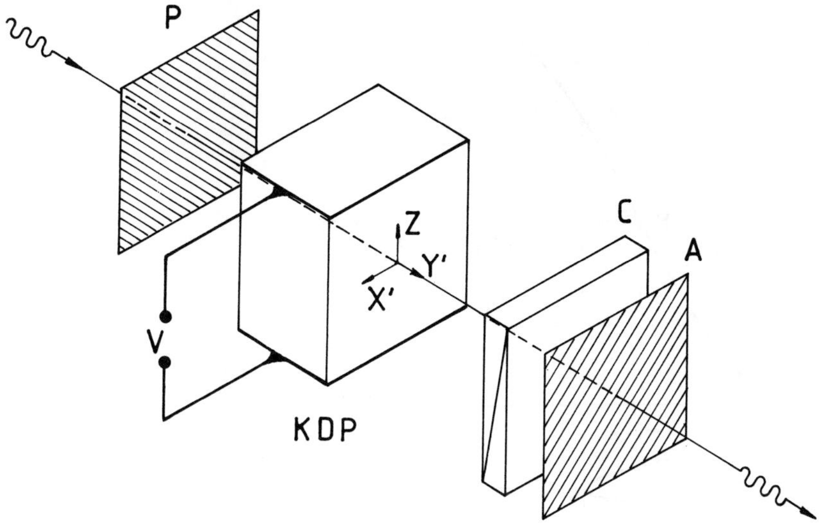

Figure 3.6 *Schematic diagram of a transversal modulator based on KDP. P: polarizer; A: analyser; C: compensator.*

may determine the appropriate r coefficient, including its sign. The same information can be obtained through direct measurements of δ by means of a Soleil compensator. $\delta(V)$ is just the compensator-induced dephasage that cancels the effect of the field. Alternatively, one may measure V_π by determining the two consecutive field values yielding minimum (null) and maximum intensity (see below).

In most practical cases one uses a limited range of field values in comparison to V_π/d, and operates in a small linear region of the $I(V)$ curve. It is, then, useful to introduce an initial dephasage $\delta(0) = \pi/2$, by using a d.c. bias or preferably with the Soleil compensator, in order to work at the inflexion point, $(I = I_0/2)$. That assures a maximum dI/dV slope, i.e. the highest sensitivity. Under these conditions, one may write

$$I = \frac{I_0}{2}\left(1 + \sin\frac{\pi V}{V_\pi}\right) \tag{3.82}$$

For small applied potentials $V \ll V_\pi$

$$\Delta I = \frac{\pi I_0 V}{2V_\pi} \tag{3.83}$$

i.e. from a linear $\Delta I \approx V$ plot, V_π and so the appropriate r coefficient can be immediately determined.

Modulation methods

Higher sensitivity can be obtained by applying an oscillatory (a.c.) potential $V = V_m \sin \Omega t$, $(V_m \ll V_\pi)$. This has the further advantage of reducing or even eliminating space-charge problems, if no d.c. bias is applied. Other deleterious effects of d.c. fields are sample damage (high fields) and heating. Moreover, by studying the Ω-dependence of the electrooptic coefficients across the piezoelectric resonances, one may determine the piezoelectric contribution to r^T.

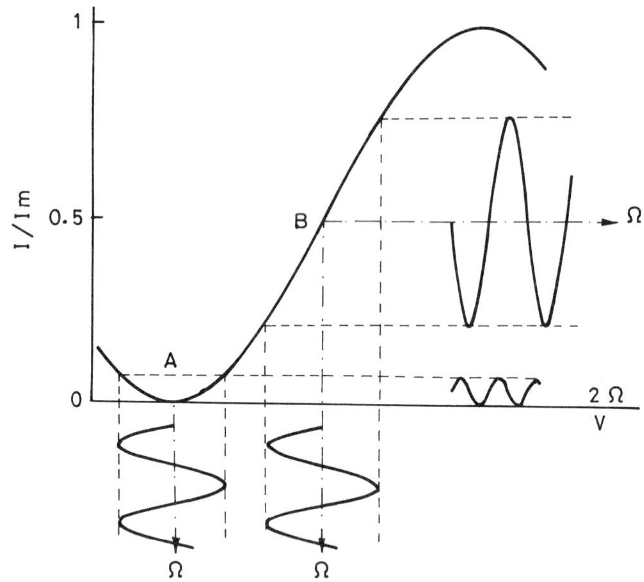

Figure 3.7 *I(V) curve for a modulator such as that shown in Figure 3.6.*

As noted above, it is useful to work at the inflexion point A of $I(V)$ for enhanced sensitivity. Then formula (3.81) leads to a modulated transmitted intensity $\mathrm{d}I/\mathrm{d}t$,

$$\frac{\mathrm{d}I}{\mathrm{d}t} = \frac{I_0}{2}\frac{\mathrm{d}\delta}{\mathrm{d}V} V_m \Omega \cos(\Omega t) \cos\{\delta(V)\} \tag{3.84}$$

In the general case, δ will include both the Pockels and Kerr contributions

$$\delta(V) = \delta_{\mathrm{P}} \sin \Omega t + \delta_{\mathrm{K}} \sin^2 \Omega t \tag{3.85}$$

with δ_{P} and δ_{K} depending on crystal structure and experimental configuration.

Let us consider a Pockels medium (e.g. KDP) and ignore the Kerr contribution. After substituting (3.85) into (3.84), one may expand $\mathrm{d}I/\mathrm{d}t$ into a Fourier series. Recalling the general formula

$$\cos(a \sin \Omega t) = J_0(a) + 2J_2(a) \cos 2\Omega t + 2J_4(a) \cos 4\Omega t + \ldots \tag{3.86}$$

J_m being the Bessel function of order m, one obtains a lowest-order oscillating term at frequency Ω,

$$\frac{\mathrm{d}I}{\mathrm{d}t} = I_m \cos \Omega t \tag{3.87}$$

with an amplitude

$$I_m = \frac{\pi}{2} I_0 \frac{V_m}{V_\pi} \Omega J_0 \frac{\pi V_m}{V_\pi} \tag{3.88}$$

where use has been made of the relation $\delta_{\mathrm{P}} = \pi V_m / V_\pi$. By measuring I_m, r is immediately obtained from (3.88).

The working point in the modulation experiments can be chosen in a different way so that $I(0)$ is either a minimum or a maximum of the $I(V)$ curve, i.e. $\delta(0) = p\pi$ (p integer). For a

Pockels material the lowest-order Fourier term in dI/dt oscillates at frequency 2Ω. As for the case discussed above, the amplitude of the oscillation yields the r coefficient. On the other hand, intensity measurements at this frequency 2Ω have been sometimes used for a precise identification of the minima and maxima of $I(V)$ and so of the half-wave voltage [32].

For a Kerr medium, a similar analysis can be carried out, although it becomes more cumbersome. The lowest-order term also oscillates at frequency Ω and its amplitude permits the determination of the coefficient s.

Ellipsometric techniques (reflection)

Ellipsometry [37] is a very well established technique to determine the optical parameters n and κ of a given material. An incident beam linearly polarized (generally at 45° with regard to the incident plane) is directed onto a slab of the investigated material and its polarization state analysed after reflection. This outgoing polarization state is determined by the dephasage and relative attenuation introduced during reflection between the two components of the incident beam which are linearly polarized parallel (p wave) and perpendicular (s wave) to the incident plane. From this information n and κ can be obtained. In principle, one might use the technique to measure the change in n induced by an applied electric field and consequently the electrooptic coefficient. However, the sensitivity of the method is very poor and has not been used in practice. Since one is exclusively interested in the field-induced changes and not in the absolute values, a rather simple and useful method has been recently proposed by Teng and Man [38], and applied to poled polymer films. The experimental set-up is essentially the same used in conventional ellipsometry and is schematically illustrated in Figure 3.8. The polymer film sample, with axial symmetry around the perpendicular direction and ordinary and extraordinary indices n_o and n_e, respectively, is placed between electrodes on a glass substrate. In the absence of the field, the dephasage δ_{sp} between the s and p waves is

$$\delta_{sp}(0) = \Delta\phi_s(0) - \Delta\phi_p(0) = \left(\frac{4\pi l}{\lambda_o}\right)\left\{n_o \cos\theta_r - \frac{n_o}{n_e}\,(n_e^2 - \sin^2\theta_i)^{1/2}\right\} \tag{3.89}$$

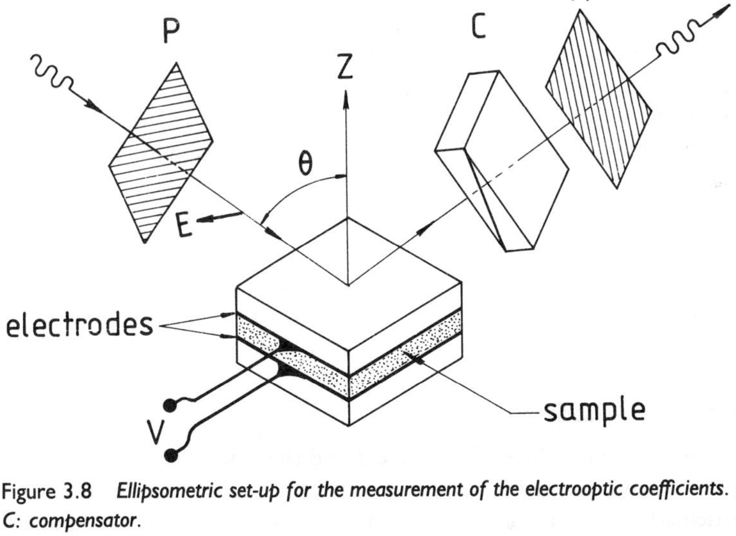

Figure 3.8 *Ellipsometric set-up for the measurement of the electrooptic coefficients. P: polarizer; A: analyser; C: compensator.*

θ_i being the incidence angle, θ_r the refraction angle inside the sample, and l the film thickness.

The additional dephasage introduced by a field E is obtained by derivation of expression (3.89) with regard to the ordinary and extraordinary indices and becomes

$$\begin{aligned}\Delta\delta_{sp}(E) &= \delta_{sp}(E) - \delta_{sp}(0)\\ &= \frac{4\pi l}{\lambda}\left\{\left[\left(1 - \frac{\sin^2\theta_i}{n_e^2}\right)^{1/2} - \frac{n_o}{(n_o^2 - \sin^2\theta_i)^{1/2}}\right]\Delta n_o\right.\\ &\quad\left. + \frac{n_o \sin^2\theta_i}{n_e^2(n_e^2 - \sin^2\theta_i)^{1/2}}\Delta n_e\right\}\end{aligned} \tag{3.90}$$

which is a rather cumbersome expression. Neglecting optical anisotropy ($n_e = n_o = n$), and assuming a modulated field $E = E_o \cos\Omega t$, it turns into

$$\begin{aligned}\Delta\delta_{sp} &= \frac{4\pi l \sin^2\theta_i}{\lambda(n^2 - \sin^2\theta_i)^{1/2}}\left(\frac{\Delta n_e - \Delta n_o}{n}\right)\\ &= \frac{4\pi l \sin^2\theta_i}{\lambda(n^2 - \sin^2\theta_i)^{1/2}}\frac{E_o\cos(\Omega t)n^2(r_{33} - r_{13})}{2}\end{aligned} \tag{3.91}$$

which permits the determination of $r_{33} - r_{13}$, from the dephasage $\Delta\delta_{sp}$. This is obtained, similarly to polarimetric experiments, from the reflected light intensity I_R, after passing through a crossed analyser. For incident polarization at 45° from the incidence plane, I_R is given by

$$I_R = I_{R,m}\sin^2\left(\frac{\delta_{sp}}{2}\right) = I_{R,m}\sin^2\left\{\frac{[\delta_{sp}(0) + \Delta\delta_{sp}(E)]}{2}\right\} \tag{3.92}$$

$I_{R,m}$ being the maximum intensity. The dependence of I_R on δ is as illustrated in Figure 3.7. Generally, the field is modulated at a frequency Ω and the operating conditions are set at the inflexion point $I_R(0) = I_{R,m}/2$. Detailed formulae can be found in the quoted references.

In order to determine each of the two nonzero electrooptic coefficients separately, one needs an additional relation between r_{33} and r_{13}. Generally, it is assumed that, $r_{33} = 3r_{13}$. In fact, this relation is expected to apply when the polymer is made up of linear molecules and it is poled with a field low enough to ensure that the orientational distribution corresponds to the linear region of the Langevin plot (see Section 4.15).

Further applications of the technique, including more rigorous and detailed analysis of the data have been recently published [39–41]. Moreover, Kerr measurements have been successfully performed [40].

Reflectivity measurements. Some workers have directly measured the electric field-induced change on the sample reflectivity [42], instead of on the polarization state of the reflected beam. More indirect and difficult to analyse are some reported measurements [43,44] of the reflectivity anisotropy caused by the applied field in cubic crystals, such as GaAs.

3.13.2. Interferometric methods

Other methods to measure the field-induced dephasage ϕ and the index change Δn involve double or multiple beam interferometric techniques. They ensure, in general, a better sensitivity and are particularly advantageous when the material is in the form of a thin film.

Mach–Zehnder interferometer

A Mach–Zehnder interferometric configuration has been often used [45–48] (Figure 3.9). The operation principles will be now briefly discussed. The light intensity at the detector is

$$I_D = \frac{I_o\{1 + \cos(\Phi - \phi)\}}{2} \tag{3.93}$$

where I_o is the laser intensity, which is equally divided between the two interferometer arms. Φ is an adjustable bias phase introduced on one of the light paths and ϕ the dephasage introduced by the sample. The analysis becomes particularly simple when one is dealing with a homogeneous uniaxial slab, such as a poled polymer film, with the optic axis normal to the film faces. Then

$$\phi = 2\pi(n_o - 1)\frac{l}{\lambda} \tag{3.94}$$

l being the film thickness, λ the wavelength (in vacuum) and n_o the ordinary refractive index.

Let us apply an electric field that modifies both the refractive index n_o and the thickness l. This latter effect is associated with the piezoelectric (*linear*) and electrostrictive (*quadratic*) contributions to the electrooptic response. In the general case, one should write for the field-induced change in the detected intensity

$$\Delta I_D = \frac{I_o}{2}\sin(\Phi - \phi)\left(\frac{\delta\phi}{\delta n}\Delta n + \frac{\delta\phi}{\delta l}\Delta l\right) \tag{3.95}$$

where

$$\frac{\delta\phi}{\delta n} = \frac{2\pi l}{\lambda} \tag{3.96a}$$

$$\frac{\delta\phi}{\delta l} = \frac{2\pi(n_o - 1)}{\lambda} \tag{3.96b}$$

In many experiments the field-induced size effects are ignored ($\Delta n/n \gg \Delta l/l$), although their relevance has been ascertained in thin organic films [49]. Moreover, it is convenient to work at $\Phi - \phi = p\pi/2$ (p odd integer). Then, recalling the Pockels and Kerr expressions for Δn, one obtains

$$\Delta I_D = \left(\frac{\pi I_o l n_o^3 r}{2\lambda}\right)E + \left(\frac{\pi I_o l n_o^3 s}{2\lambda}\right)E^2 \tag{3.97}$$

By measuring the dependence $\Delta I_D(E)$, one could, in principle, separate the linear and quadratic contributions and determine r and s.

In practice, the applied field is modulated at a frequency Ω. Then, the amplitude of the detected intensity that oscillates at the same frequency can be written

$$\Delta I_D(\Omega) = \frac{\pi I_o V_m n_o^3}{2\lambda} r \tag{3.98}$$

V_m being the amplitude of the applied modulated potential. From ΔI_D measured with a lock-in amplifier, r is easily derived.

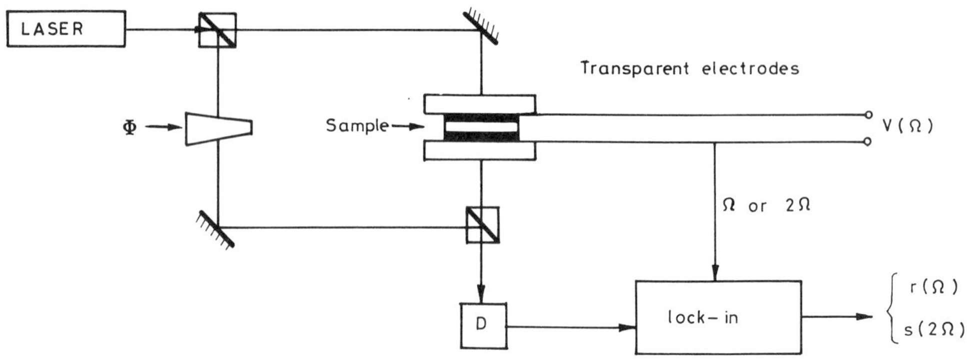

Figure 3.9 *Mach–Zehnder interferometer set-up for the measurement of the electrooptic coefficients. D: light detector. (Adapted from [49].)*

On the other hand, by tuning the lock-in amplifier to a frequency 2Ω, the amplitude of the detected signal is

$$\Delta I_D(2\Omega) = \frac{\pi I_o n_o^3 V_m^2}{4\lambda l} s \tag{3.99}$$

which allows s to be determined.

It should be stressed here that the independent determination of r and s is only possible if the field-induced dimensional changes can be ignored in equation (3.95).

Michelson interferometer. A rather similar technique using a Michelson interferometer has also been recently demonstrated [50].

Fabry–Perot interferometer

Multiple-beam techniques, such as those based on the Fabry–Perot interferometer (Figure 3.10), offer an interesting alternative [51,52] in the case of thin films. The sample is sandwiched between transparent electrodes acting as cavity mirrors. The application of an electric field modifies the refractive index of the sample and consequently the transmittance of the interferometer. The intensity of the transmitted light beam can be written

$$I_S = \frac{(1-R)^2}{(1-R)^2 + 4R\sin^2\phi} \tag{3.100a}$$

R being the reflectivity of each mirror. ϕ is the phase-shift on each transit of the interferometer thickness l, given by

$$\phi = \frac{2\pi n l}{\lambda} \tag{3.100b}$$

Application of a voltage $V_m \cos \Omega t$ to the electrodes will induce a modulation in the refractive index and thickness of the film. The associated variation in the transmitted intensity can be written

$$\Delta I_S = \frac{\partial I_S}{\partial \phi}\left\{\left(\frac{\partial \phi}{\partial n}\right)\Delta n + \left(\frac{\partial \phi}{\partial l}\right)\Delta l\right\} \tag{3.101a}$$

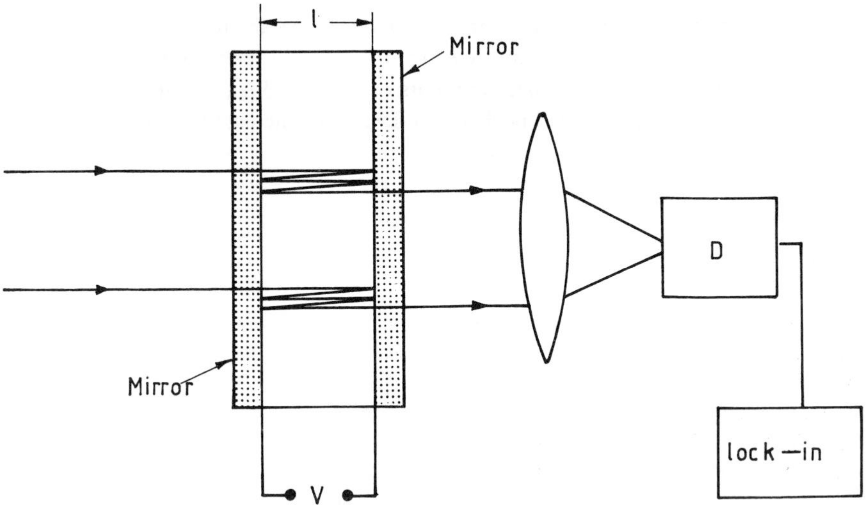

Figure 3.10 *Fabry–Perot interferometer arrangement for measuring the electrooptic coefficients.*

Since both n and R are not appreciably dependent on λ around the transmission peak of the interferometer, one may find $\partial I_S/\partial\phi$ from the wavelength response $I_S(\lambda)$ and write (3.101a) as

$$\Delta I_S = -\lambda \frac{\partial I_S}{\partial \lambda}\left(\frac{\Delta n}{n} + \frac{\Delta l}{l}\right) \tag{3.101b}$$

By measuring I_S with a lock-in amplifier, tuned either at Ω or 2Ω, one may determine r and s, as discussed for the Mach–Zehnder interferometer (assuming Δl can be neglected or is known).

3.13.3. Raman spectroscopy measurements

The connection between Raman spectroscopy and electrooptic response has already been established in Section 3.5.2. As briefly discussed next, the measurement of *absolute Raman scattering efficiencies* allows the determination of the clamped electrooptic coefficients as well as the electronic or nonlinear mixing contribution [21,24,53]. The method can be useful only when small and irregular crystals are available or the electrical resistivity is too low. Moreover, it is normally restricted to solids with a simple structure and a small number of atoms per unit cell.

The situation is particularly simple for cubic zinc-blende crystals (e.g. GaAs), where most experiments have been performed. They present a single Γ_{15} optical mode, split into one LO and two TO modes, with frequencies Ω_{LO} and Ω_{TO}, respectively. Furthermore, there is only one nonzero electrooptic coefficient r_{41}. By proceeding with the formalism of Section 3.5.2, one arrives at the following final relations:

$$\left(\frac{S_{LO}}{S_{TO}}\right)^{1/2} \approx \left(1 - \frac{\Omega_{LO}^2 - \Omega_{TO}^2}{\Omega_{TO}^2 \mathscr{F}}\right) \tag{3.102a}$$

$$r_{41}^S \sim (S_{TO})^{1/2}\left(1 + \frac{1}{\mathscr{F}}\right) \tag{3.102b}$$

$$r_{41}^S = r_{41}^e(1 + \mathscr{F}) \tag{3.102c}$$

where $\mathscr{F}$ is the so-called *Faust–Henry* or *electron-lattice* parameter [54]. It measures the relative strengths of the lattice and electronic contributions to the electrooptic coefficients. Therefore, from absolute measurements of S_{LO} and S_{TO}, $\mathscr{F}$ can be determined by means of equation (3.102a). Then, both r_{41}^S and r_{41}^e are calculated from (3.102b) and (3.102c).

References

1. A. F. Harvey, *Coherent Light*, Wiley-Interscience, London (1970).
2. I. P. Kaminow and E. H. Turner, in *Handbook of Lasers* (ed. R. J. Pressley), Chemical Rubber Publ. Co., Cleveland, (1971), pp. 447–459.
3. S. H. Wemple and M. di Domenico Jr, in *Applied Solid State Science* (ed. R. Wolfe), Vol. 3, Academic Press, New York (1972), pp. 263–383.
4. I. P. Kaminow, *An Introduction to Electrooptic Devices*, Academic Press, New York (1974).
5. T. S. Narasimhamurty, *Photoelastic and Electrooptic Properties of Crystals*, Plenum Press, New York (1981).
6. L. J. Pinson, *Electrooptics*, John Wiley, New York (1985).
7. P. Günter, *Ferroelectrics* **75**, 5 (1987).
8. Y. Noda, M. Suzuki, Y. Kushiro and S. Akiba, *J. Lightwave Technol.* **LT-4**, 1445 (1986).
9. A. Alping, X. S. Wu, T. R. Hausken and L. A. Coldren, *Appl. Phys. Lett.* **48**, 1243 (1986).
10. M. B. Klein, in *Electrooptic and Photorefractive Materials* (ed. P. Günter), Springer-Verlag, Berlin (1987).
11. C. J. G. Kirby, in *Properties of $LiNbO_3$*, EMIS Datareview Series No. 5, INSPEC, London (1989), pp. 156–167.
12. O. G. Vlokh, *Ferroelectrics*, **75**, 119 (1987).
13. J. Fousek, in *Electrooptic and Photorefractive Materials* (ed. P. Günter), Springer-Verlag, Berlin (1987).
14. M. Cardona, *Modulation Spectroscopy*, Academic Press, London (1969).
15. D. E. Aspnes and N. Bottka, in *Semiconductors and Semimetals. Modulation Techniques*, Academic Press, New York (1972), pp. 457–543.
16. R. Guenther, *Modern Optics*, John Wiley, New York (1990).
17. Y. R. Shen, *The Principles of Nonlinear Optics*, John Wiley, New York (1984).
18. A. Alping and L. A. Coldren, *J. Appl. Phys.* **61**, 2430 (1987).
19. B. R. Bennet and R. A. Soref, *IEEE J. Quantum Electron.* **QE-23**, 2159 (1987).
20. J. I. Pankove, *Optical Processes in Semiconductors*, Dover, New York (1975).
21. I. P. Kaminow and E. H. Turner, *Phys. Rev.* **5**, 1564 (1972).
22. W. Hayes and R. Loudon, *Scattering of Light by Crystals*, John Wiley, New York (1978).
23. M. Cardona, in *Light Scattering in Solids*, (eds M. Cardona and G. Güntherodt), Springer-Verlag, Berlin (1982).
24. I. P. Kaminow and W. P. Johnston, *Phys. Rev.* **160**, 519 (1967).
25. R. C. Miller, *Appl. Phys. Lett.*, **5**, 17 (1964).
26. H. Benoit, *Ann. Phys.* **6**, 36 (1951).
27. J. F. Nye, *Physical Properties of Crystals*, Oxford University Press (1964).
28. S. Bhagavantam, *Crystal Symmetry and Physical Properties*, Academic Press, London (1966).
29. D. Eimerl, *IEEE J. Quantum Electron.* **QE-23**, 2104 (1987).
30. S. K. Kurtz and F. N. H. Robinson, *Appl. Phys. Lett.* **10**, 62 (1967).
31. P. Günter, *Optics Commun.* **11**, 285 (1974).
32. F. Lüty and A. Diaz-Gongora, *Phys. Stat. Sol. (b)* **85**, 693 (1978).
33. G. F. Lipscomb, R. S. Narang and A. F. Garito, *J. Chem. Phys.* **75**, 1509 (1981).
34. J. D. Beierlein and C. B. Arweiler, *Appl. Phys. Lett.* **49**, 917 (1986).
35. H. Nakatami, W. Bosenberg, L. K. Cheng and C. L. Tang, *Appl. Phys. Lett.* **52**, 1288 (1988).

36. M. Aillerie, M. D. Fontana, F. Abdi, C. Carabatos-Nedelec, N. Theofanous and G. Alexakis, *J. Appl. Phys.* **65**, 6 (1989).
37. R. M. A. Azzam and N. M. Bashara, *Ellipsometry and Polarized Light*, North-Holland, Amsterdam (1977).
38. C. C. Teng and H. T. Man, *Appl. Phys. Lett.* **56**, 1734 (1990).
39. J. S. Schildkraut, *Appl. Optics* **29**, 2839 (1990).
40. P. Rohl, B. Andress and J. Nordmann, *Appl. Phys. Lett.* **59**, 2793 (1991).
41. G. Gadret, F. Kajzar and P. Raimond, *Proc. SPIE* **1560**, 226 (1991).
42. Y. Tokura, A. Kurita and T. Koda, *Phys. Rev.* B **31**, 2588 (1985).
43. S. E. Acosta-Ortiz and A. Lastras-Martinez, *Phys. Rev.* B **40**, 1426 (1989).
44. S. E. Acosta-Ortiz, *J. Appl. Phys.* **70**, 3239 (1991).
45. M. Sigelle and R. Hierle, *J. Appl. Phys.* **52**, 4199 (1981).
46. K. D. Singer, M. G. Kuzyk, W. R. Holland, J. E. Sohn, S. J. Lalama, R. B. Comizzoli, H. E. Katz and M. L. Schilling, *Appl. Phys. Lett.* **53**, 1800 (1988).
47. M. G. Kuzyk and C. W. Dirk, *Appl. Phys. Lett.* **54**, 1628 (1989).
48. M. G. Kuzyk, J. E. Sohn and C. W. Dirk, *J. Opt. Soc. Am.* **B7**, 842 (1990).
49. M. Dumont, Y. Levy and D. Morichere, in *Organic Molecules for Nonlinear Optics and Photonics* (ed. J. Messier), NATO ASI Series, Kluwer Academic Publishers (1990).
50. N. Y. Zhang, X. H. He, Y. H. Shih and S. H. Tang, *Optics Commun.* **86**, 509 (1991).
51. H. Uchiki and T. Kobayashi, *J. Appl. Phys.* **64**, 2625 (1988).
52. T. Kobayashi, H. Uchibi and K. Minoshima, in *Nonlinear Optics of Organics and Semiconductors* (ed. T. Kobayashi), Springer Proceedings in Physics, Vol. 36, Springer-Verlag, Berlin (1989).
53. F. Meseguer, J. C. Merle and M. Cardona, *Ferroelectrics* **56**, 95 (1984).
54. W. L. Faust and C. H. Henry, *Phys. Rev. Lett.* **17**, 1265 (1966).

Appendix 3A. Pockels tensors for noncentrosymmetrical crystallographic point groups

Tensors will be given in contracted notation r_{ij}. X, Y, Z reference axes for all 32 crystallographic point groups and axial group ∞mm, have been chosen as follows:

- *Cubic groups:* X, Y, Z coincide with cubic axes for all groups 23, $m3$, $\bar{4}m3$, 432 and $m3m$.
- *Hexagonal groups:* Z along sixfold axis. X perpendicular to mirror plane for $6mm$, $6/mmm$ and $\bar{6}2m$; along binary axis for 622 and free for 6, $\bar{6}$ and $6/m$.
- *Tetragonal groups:* Z along fourfold axis. X perpendicular to mirror plane for $4mm$ and $4/mmm$; along binary axis for $42m$ and 422; free for 4, $\bar{4}$ and $4/m$.
- *Trigonal groups:* Z along trigonal axis. X perpendicular to mirror plane for $3m$ and $\bar{3}m$; along binary axis for 32; free for 3 and $\bar{3}$.
- *Orthorhombic groups:* Z along binary axis for 222 and $mm2$; perpendicular to mirror plane for mmm. X perpendicular to mirror plane for $mm2$ and mmm; along binary axis for 222.
- *Monoclinic groups:* Y along binary axis for 2 and $2/m$; perpendicular to mirror plane for m (and $2/m$). Other axes free.
- *Triclinic groups:* All axes free for 1 and $\bar{1}$.
- *Axial Symmetry* (∞mm): Z along symmetry axis (∞). Other axes free.

n indicates the number of nonzero independent components for each group.

Triclinic

1, $\bar{1}$

$$\begin{bmatrix} r_{11} & r_{12} & r_{13} \\ r_{21} & r_{22} & r_{23} \\ r_{31} & r_{32} & r_{33} \\ r_{41} & r_{42} & r_{43} \\ r_{51} & r_{52} & r_{53} \\ r_{61} & r_{62} & r_{63} \end{bmatrix}$$
$(n = 18)$

Monoclinic

2

$$\begin{bmatrix} 0 & r_{12} & 0 \\ 0 & r_{22} & 0 \\ 0 & r_{32} & 0 \\ r_{41} & 0 & r_{43} \\ 0 & r_{52} & 0 \\ r_{61} & 0 & r_{63} \end{bmatrix}$$
$(n = 8)$

m

$$\begin{bmatrix} r_{11} & 0 & r_{13} \\ r_{21} & 0 & r_{23} \\ r_{31} & 0 & r_{33} \\ 0 & r_{42} & 0 \\ r_{51} & 0 & r_{53} \\ 0 & r_{62} & 0 \end{bmatrix}$$
$(n = 10)$

Orthorhombic

222

$$\begin{bmatrix} 0 & 0 & 0 \\ 0 & 0 & 0 \\ 0 & 0 & 0 \\ r_{41} & 0 & 0 \\ 0 & r_{52} & 0 \\ 0 & 0 & r_{63} \end{bmatrix}$$
$(n = 3)$

$mm2$

$$\begin{bmatrix} 0 & 0 & r_{13} \\ 0 & 0 & r_{23} \\ 0 & 0 & r_{33} \\ 0 & r_{42} & 0 \\ r_{51} & 0 & 0 \\ 0 & 0 & 0 \end{bmatrix}$$
$(n = 5)$

Tetragonal

4

$$\begin{bmatrix} 0 & 0 & r_{13} \\ 0 & 0 & r_{13} \\ 0 & 0 & r_{33} \\ r_{41} & r_{42} & 0 \\ r_{42} & -r_{41} & 0 \\ 0 & 0 & 0 \end{bmatrix}$$
$(n = 4)$

Tetragonal

$\bar{4}$
$$\begin{bmatrix} 0 & 0 & r_{13} \\ 0 & 0 & -r_{13} \\ 0 & 0 & 0 \\ r_{41} & -r_{51} & 0 \\ r_{51} & r_{41} & 0 \\ 0 & 0 & r_{63} \end{bmatrix}$$
$(n = 4)$

422
$$\begin{bmatrix} 0 & 0 & 0 \\ 0 & 0 & 0 \\ 0 & 0 & 0 \\ r_{41} & 0 & 0 \\ 0 & -r_{41} & 0 \\ 0 & 0 & 0 \end{bmatrix}$$
$(n = 1)$

$4mm$
$$\begin{bmatrix} 0 & 0 & r_{13} \\ 0 & 0 & r_{13} \\ 0 & 0 & r_{33} \\ 0 & r_{51} & 0 \\ r_{51} & 0 & 0 \\ 0 & 0 & 0 \end{bmatrix}$$
$(n = 3)$

Tetragonal

$\bar{4}2m$
$$\begin{bmatrix} 0 & 0 & 0 \\ 0 & 0 & 0 \\ 0 & 0 & 0 \\ r_{41} & 0 & 0 \\ 0 & r_{41} & 0 \\ 0 & 0 & r_{63} \end{bmatrix}$$
$(n = 2)$

Trigonal

3
$$\begin{bmatrix} r_{11} & -r_{22} & r_{13} \\ -r_{11} & r_{22} & r_{13} \\ 0 & 0 & r_{33} \\ r_{41} & r_{51} & 0 \\ r_{51} & -r_{41} & 0 \\ -r_{22} & -r_{11} & 0 \end{bmatrix}$$
$(n = 6)$

32
$$\begin{bmatrix} r_{11} & 0 & 0 \\ -r_{11} & 0 & 0 \\ 0 & 0 & 0 \\ r_{41} & 0 & 0 \\ 0 & -r_{41} & 0 \\ 0 & -r_{11} & 0 \end{bmatrix}$$
$(n = 2)$

Trigonal

$3m$
$$\begin{bmatrix} 0 & -r_{22} & r_{13} \\ 0 & r_{22} & r_{13} \\ 0 & 0 & r_{33} \\ 0 & r_{51} & 0 \\ r_{51} & 0 & 0 \\ -r_{22} & 0 & 0 \end{bmatrix}$$
$(n = 4)$

Hexagonal

6
$$\begin{bmatrix} 0 & 0 & r_{13} \\ 0 & 0 & r_{13} \\ 0 & 0 & r_{33} \\ r_{41} & r_{51} & 0 \\ r_{51} & -r_{41} & 0 \\ 0 & 0 & 0 \end{bmatrix}$$
$(n = 4)$

$6mm$
$$\begin{bmatrix} 0 & 0 & r_{13} \\ 0 & 0 & r_{13} \\ 0 & 0 & r_{33} \\ 0 & r_{51} & 0 \\ r_{51} & 0 & 0 \\ 0 & 0 & 0 \end{bmatrix}$$
$(n = 3)$

Hexagonal

622
$$\begin{bmatrix} 0 & 0 & 0 \\ 0 & 0 & 0 \\ 0 & 0 & 0 \\ r_{41} & 0 & 0 \\ 0 & -r_{41} & 0 \\ 0 & 0 & 0 \end{bmatrix}$$
$(n = 1)$

$\bar{6}$
$$\begin{bmatrix} r_{11} & -r_{22} & 0 \\ -r_{11} & r_{22} & 0 \\ 0 & 0 & 0 \\ 0 & 0 & 0 \\ 0 & 0 & 0 \\ -r_{22} & -r_{11} & 0 \end{bmatrix}$$
$(n = 2)$

$\bar{6}m2$
$$\begin{bmatrix} 0 & -r_{22} & 0 \\ 0 & r_{22} & 0 \\ 0 & 0 & 0 \\ 0 & 0 & 0 \\ 0 & 0 & 0 \\ -r_{22} & 0 & 0 \end{bmatrix}$$
$(n = 1)$

Cubic *Axial*

432

$$\begin{bmatrix} 0 & 0 & 0 \\ 0 & 0 & 0 \\ 0 & 0 & 0 \\ 0 & 0 & 0 \\ 0 & 0 & 0 \\ 0 & 0 & 0 \end{bmatrix}$$

$(n = 0)$

$\bar{4}3m, 23$

$$\begin{bmatrix} 0 & 0 & 0 \\ 0 & 0 & 0 \\ 0 & 0 & 0 \\ r_{41} & 0 & 0 \\ 0 & r_{41} & 0 \\ 0 & 0 & r_{41} \end{bmatrix}$$

$(n = 1)$

∞mm

$$\begin{bmatrix} 0 & 0 & r_{13} \\ 0 & 0 & r_{13} \\ 0 & 0 & r_{33} \\ 0 & r_{51} & 0 \\ r_{51} & 0 & 0 \\ 0 & 0 & 0 \end{bmatrix}$$

$(n = 3)$

Appendix 3B. Kerr tensors for crystallographic point groups

Tensors for all 32 crystallographic point groups and isotropic symmetry are given in contracted notation s_{ij}. X, Y, Z axes have been chosen as described in Appendix 3A. n is the number of nonzero independent components.

Triclinic

$1, \bar{1}$

$$\begin{bmatrix} s_{11} & s_{12} & s_{13} & s_{14} & s_{15} & s_{16} \\ s_{21} & s_{22} & s_{23} & s_{24} & s_{25} & s_{26} \\ s_{31} & s_{32} & s_{33} & s_{34} & s_{35} & s_{36} \\ s_{41} & s_{42} & s_{43} & s_{44} & s_{45} & s_{46} \\ s_{51} & s_{52} & s_{53} & s_{54} & s_{55} & s_{56} \\ s_{61} & s_{62} & s_{63} & s_{64} & s_{65} & s_{66} \end{bmatrix}$$

$(n = 36)$

Monoclinic

$2, m, 2/m$

$$\begin{bmatrix} s_{11} & s_{12} & s_{13} & 0 & s_{15} & s_{16} \\ s_{21} & s_{22} & s_{23} & 0 & s_{25} & 0 \\ s_{31} & s_{32} & s_{33} & 0 & s_{35} & 0 \\ 0 & 0 & 0 & s_{44} & 0 & s_{46} \\ s_{51} & s_{52} & s_{53} & 0 & s_{55} & 0 \\ 0 & 0 & 0 & s_{64} & 0 & s_{66} \end{bmatrix}$$

$(n = 20)$

Orthorhombic

$222, mm2, mmm$

$$\begin{bmatrix} s_{11} & s_{12} & s_{13} & 0 & 0 & 0 \\ s_{21} & s_{22} & s_{23} & 0 & 0 & 0 \\ s_{31} & s_{32} & s_{33} & 0 & 0 & 0 \\ 0 & 0 & 0 & s_{44} & 0 & 0 \\ 0 & 0 & 0 & 0 & s_{55} & 0 \\ 0 & 0 & 0 & 0 & 0 & s_{66} \end{bmatrix}$$

$(n = 12)$

Tetragonal

$4, \bar{4}, 4/m$

$$\begin{bmatrix} s_{11} & s_{12} & s_{13} & 0 & 0 & s_{16} \\ s_{12} & s_{11} & s_{13} & 0 & 0 & -s_{16} \\ s_{31} & s_{31} & s_{33} & 0 & 0 & 0 \\ 0 & 0 & 0 & s_{44} & s_{45} & 0 \\ 0 & 0 & 0 & -s_{45} & s_{44} & 0 \\ s_{61} & -s_{61} & 0 & 0 & 0 & s_{66} \end{bmatrix}$$

$(n = 10)$

Tetragonal

$422, 4mm, \bar{4}2m, 4/mmm$

$$\begin{bmatrix} s_{11} & s_{12} & s_{13} & 0 & 0 & 0 \\ s_{12} & s_{11} & s_{13} & 0 & 0 & 0 \\ s_{31} & s_{31} & s_{33} & 0 & 0 & 0 \\ 0 & 0 & 0 & s_{44} & 0 & 0 \\ 0 & 0 & 0 & 0 & s_{44} & 0 \\ 0 & 0 & 0 & 0 & 0 & s_{66} \end{bmatrix}$$

$(n = 7)$

Trigonal

$3, \bar{3}$

$$\begin{bmatrix} s_{11} & s_{12} & s_{13} & s_{14} & -s_{25} & 2s_{62} \\ s_{12} & s_{11} & s_{13} & -s_{14} & s_{25} & -2s_{62} \\ s_{31} & s_{31} & s_{33} & 0 & 0 & 0 \\ s_{41} & -s_{41} & 0 & s_{44} & s_{45} & 2s_{52} \\ -s_{52} & s_{52} & 0 & -s_{45} & s_{44} & 2s_{41} \\ -s_{62} & s_{62} & 0 & s_{25} & s_{14} & s_{11} - s_{12} \end{bmatrix}$$

$(n = 12)$

Trigonal

$3m, 32, \bar{3}m$

$$\begin{bmatrix} s_{11} & s_{12} & s_{13} & s_{14} & 0 & 0 \\ s_{12} & s_{11} & s_{13} & -s_{14} & 0 & 0 \\ s_{31} & s_{31} & s_{33} & 0 & 0 & 0 \\ s_{41} & -s_{41} & 0 & s_{44} & 0 & 0 \\ 0 & 0 & 0 & 0 & s_{44} & 2s_{41} \\ 0 & 0 & 0 & 0 & s_{14} & s_{11}-s_{12} \end{bmatrix}$$

$(n = 8)$

Hexagonal

$6, \bar{6}, 6/m$

$$\begin{bmatrix} s_{11} & s_{12} & s_{13} & 0 & 0 & 2s_{61} \\ s_{12} & s_{11} & s_{13} & 0 & 0 & -2s_{61} \\ s_{31} & s_{31} & s_{33} & 0 & 0 & 0 \\ 0 & 0 & 0 & s_{44} & s_{45} & 0 \\ 0 & 0 & 0 & -s_{45} & s_{44} & 0 \\ s_{61} & -s_{61} & 0 & 0 & 0 & s_{11}-s_{12} \end{bmatrix}$$

$(n = 8)$

Hexagonal

$\bar{6}m2, 622, 6mm, 6/mmm$

$$\begin{bmatrix} s_{11} & s_{12} & s_{13} & 0 & 0 & 0 \\ s_{12} & s_{11} & s_{13} & 0 & 0 & 0 \\ s_{31} & s_{31} & s_{33} & 0 & 0 & 0 \\ 0 & 0 & 0 & s_{44} & 0 & 0 \\ 0 & 0 & 0 & 0 & s_{44} & 0 \\ 0 & 0 & 0 & 0 & 0 & s_{11}-s_{12} \end{bmatrix}$$

$(n = 6)$

Cubic

$23, m3$

$$\begin{bmatrix} s_{11} & s_{12} & s_{13} & 0 & 0 & 0 \\ s_{13} & s_{11} & s_{12} & 0 & 0 & 0 \\ s_{12} & s_{13} & s_{11} & 0 & 0 & 0 \\ 0 & 0 & 0 & s_{44} & 0 & 0 \\ 0 & 0 & 0 & 0 & s_{44} & 0 \\ 0 & 0 & 0 & 0 & 0 & s_{44} \end{bmatrix}$$

$(n = 4)$

Cubic

$\bar{4}3m, 432, m3m$

$$\begin{bmatrix} s_{11} & s_{12} & s_{12} & 0 & 0 & 0 \\ s_{12} & s_{11} & s_{12} & 0 & 0 & 0 \\ s_{12} & s_{12} & s_{11} & 0 & 0 & 0 \\ 0 & 0 & 0 & s_{44} & 0 & 0 \\ 0 & 0 & 0 & 0 & s_{44} & 0 \\ 0 & 0 & 0 & 0 & 0 & s_{44} \end{bmatrix}$$

$(n = 3)$

Isotropic

$$\begin{bmatrix} s_{11} & s_{12} & s_{12} & 0 & 0 & 0 \\ s_{12} & s_{11} & s_{12} & 0 & 0 & 0 \\ s_{12} & s_{12} & s_{11} & 0 & 0 & 0 \\ 0 & 0 & 0 & s_{11}-s_{12} & 0 & 0 \\ 0 & 0 & 0 & 0 & s_{11}-s_{12} & 0 \\ 0 & 0 & 0 & 0 & 0 & s_{11}-s_{12} \end{bmatrix}$$

$(n = 2)$

Chapter Four

Electrooptics and Nonlinear Optics

4.1. Introduction

Electrooptic phenomena involve a nonlinear response of the material and consequently fall outside the realm of *linear optics*. A linear medium, such as that described by an idealized Lorentz model with harmonic (proportional to elongation) forces, presents a dynamical response that is independent of additional excitation by a constant electric field. Therefore, it does not show any electrooptic behaviour. This is also inferred from the linear wave equation discussed in Chapter 2. The addition of a constant field E_0 to any propagating wave $E(\mathbf{r}, t)$ yields also a valid solution of the linear wave equation and so the propagation parameters (phase velocity or refractive index) are not modified. Consequently, it is useful, and formally more satisfactory, to study electrooptic effects within the general framework of nonlinear optics [1–4]. This will provide a deeper understanding of those processes and of their interrelations with other nonlinear phenomena. In fact, microscopic models accounting for the electrooptic response are also responsible for other nonlinear effects. Differences arise due to the wavelength dispersion behaviour and, more specifically, to vibrational (or, in general, relaxational) resonance effects influencing the electrooptic but not the (electronic) nonlinear response.

4.2. Nonlinear wave propagation

As discussed in Chapter 2, the propagation of optical waves in a material, free of charges and currents, is governed by the macroscopic Maxwell equations

$$\nabla \cdot \mathbf{D} = \nabla \cdot \mathbf{B} = 0 \tag{4.1a}$$

$$\nabla \times \mathbf{E} = -\frac{\delta \mathbf{B}}{\delta t} \tag{4.1b}$$

$$\nabla \times \mathbf{H} = \frac{\delta \mathbf{D}}{\delta t} \tag{4.1c}$$

where

$$\mathbf{D} = \epsilon_0 \mathbf{E} + \mathbf{P} \tag{4.2a}$$

$$\mathbf{B} = \mu_0 \mathbf{H} + \mathbf{M} = \mu_0 \mathbf{H} \tag{4.2b}$$

The induced magnetization $\mathbf{M}$ has been taken equal to zero for the high-frequency (optical) regime. Moreover, for *linear* optics it is assumed that $\mathbf{P}$ is linearly related to $\mathbf{E}$, i.e. $\mathbf{P} = \epsilon_0 \chi : \mathbf{E}$, χ being the susceptibility tensor. Then $\mathbf{D} = \epsilon_0 \epsilon : \mathbf{E}$, $\mathbf{B} = \mu_0 \mathbf{H}$, and one is led to the linear wave equation (2.22) to describe the propagation of an electromagnetic field $\mathbf{E}(t)$, $\mathbf{B}(t)$.

It is expected that for small enough wave fields, the linear approximation is appropriate. However, when they are not negligable with regard to the internal fields, responsible for the electron binding to atoms and molecules in matter ($\mathbf{E}_\mathrm{I} \approx 10^{11}$ V/m), a nonlinear response should emerge. Under these conditions, the linear constitutive relations between $\mathbf{D}$ and $\mathbf{E}$ must be modified to include a nonlinear term $\mathbf{P}_\mathrm{NL}$ in the polarization, i.e.

$$\mathbf{P} = \mathbf{P}_\mathrm{L} + \mathbf{P}_\mathrm{NL} \tag{4.3}$$

Applying the operator $\nabla\times$ to both sides of equation (4.1b) and using (4.1c) and (4.2a), one obtains

$$\nabla \times \nabla \times \mathbf{E} = -\mu_0 \frac{\partial(\nabla \times \mathbf{H})}{\partial t} = -\mu_0 \frac{\partial^2 \mathbf{D}}{\partial^2 t} = -\epsilon_0\mu_0 \frac{\partial^2(\epsilon : \mathbf{E})}{\partial^2 t} - \mu_0 \frac{\partial^2 \mathbf{P}}{\partial^2 t} \tag{4.4}$$

For a monochromatic complex wave $\mathbf{E} = \mathbf{E}(\mathbf{r}, \omega)\,\mathrm{e}^{-i\omega t}$, or one Fourier component of a given field, the source polarization appearing in (4.4) will oscillate also at a frequency ω, i.e. $\mathbf{P} = \mathbf{P}(\mathbf{r}, \omega)\,\mathrm{e}^{-i\omega t}$. Therefore, (4.4) can be written

$$\nabla \times \nabla \times \mathbf{E}(\mathbf{r}) = \left(\frac{\omega^2}{c^2}\right)\mathbf{E}(\mathbf{r}) + \mu_0\omega^2\mathbf{P}(\mathbf{r}) \tag{4.5}$$

In the nonlinear regime, one should use (4.3), which now becomes

$$\mathbf{P} = \epsilon_0\chi(\omega)\cdot\mathbf{E} + \mathbf{P}_\mathrm{NL} = \epsilon_0\varepsilon(\omega)\cdot\mathbf{E} - \epsilon_0\mathbf{E} + \mathbf{P}_\mathrm{NL} \tag{4.6}$$

$\chi(\omega)$ and $\epsilon(\omega)$ being, respectively, the susceptibility and the dielectric tensors for frequency ω. Taking (4.6) into account, equation (4.5) turns into

$$\nabla \times \nabla \times \mathbf{E}(\mathbf{r}) - \left(\frac{\omega^2}{c^2}\right)\varepsilon(\omega)\cdot\mathbf{E}(\mathbf{r}) = \mu_0\omega\mathbf{P}_\mathrm{NL}(\mathbf{r}) \tag{4.7}$$

which is the *wave equation for monochromatic waves* in a *nonlinear* medium. The nonlinear polarization $\mathbf{P}_\mathrm{NL}$ acts as a *source term* in the equation to generate the nonlinear solutions. For a linear medium, $\mathbf{P}_\mathrm{NL} = 0$, so that equation (4.7) reduces to (2.22), and monochromatic waves are adequate solutions for the propagating field. On the other hand, in the nonlinear case, monochromatic waves will no longer be propagating solutions and they will become mixed by the nonlinear polarization (*nonlinear wave-mixing*).

One can now go one step further assuming a plane wave solution

$$\mathbf{E}(\mathbf{r}) = E(z)\,\mathrm{e}^{ikz}\mathbf{u}_E \tag{4.8}$$

$\mathbf{u}_E$ being the unit vector in the field direction, and $\mathbf{k}$ the wavevector along the propagation direction Z. Dependence of the amplitude E on z has been explicitly included to account for energy exchange with other waves. Substitution of (4.8) into (4.7) leads, after some cumbersome manipulations (Appendix 4A), to

$$\frac{\mathrm{d}^2E}{\mathrm{d}z^2} + 2ik\frac{\mathrm{d}E}{\mathrm{d}z} = -\mu_0\omega^2\frac{(\mathbf{P}_\mathrm{NL}\cdot\mathbf{u}_E)\,\mathrm{e}^{-ikz}}{\cos^2\alpha} \tag{4.9}$$

which is the *nonlinear equation for plane waves.* α is the angle between $\mathbf{D}$ and $\mathbf{E}$ vectors (see Figure 2.6). The right-hand term couples the $E(z, \omega)$ wave to the other interacting waves in the nonlinear medium through the nonlinear polarization $\mathbf{P}_\mathrm{NL}$ (see Section 4.3).

For a simple common case, in which all waves are collinear along Z, it is expected that $\mathbf{P}_\mathrm{NL}(\mathbf{r}, \omega) = \mathbf{P}_\mathrm{NL}(\mathbf{z}, \omega)\,\mathrm{e}^{ik_\mathrm{p}z}$, k_p being polarization wavevector resulting from the effect of all other interacting plane-waves. Then, one finally arrives at

$$\frac{d^2E}{dz^2} + 2ik\frac{dE}{dz} = -\frac{\mu_0\omega^2 e^{i(k_p-k)z}(\mathbf{u}_E \cdot \mathbf{P}_{NL})}{\cos^2\alpha}, \tag{4.10}$$

which describes the attenuation of the wave due to the nonlinear coupling.

4.2.1. Slowly varying amplitude approximation

Equation (4.10) can now be further simplified by using the so-called *slowly-varying amplitude approximation*, which is very generally applicable (except for ultrashort pulses and strongly absorbing or extremely nonlinear media). It implies that the wave amplitude experiences a small relative change over a distance λ, i.e.

$$\left|\left(\frac{dE(z)}{dz}\right)\right|\frac{\lambda}{E} \ll 1 \tag{4.11}$$

which makes the second derivative in (4.10) negligible. Then, this equation turns into

$$\frac{dE}{dz} = \frac{i\omega}{2\epsilon_0 c n_\omega \cos^2\alpha}(\mathbf{u}_E \cdot \mathbf{P}_{NL})\, e^{-i(k-k_p)z} \tag{4.12}$$

which is first-order and simpler to handle. n_ω is the refractive index for the plane wave ($n_\omega^2 = \epsilon(\omega)$).

When dealing with several interacting plane waves, the energy coupling among them is described by a set of equations like (4.12), each applying to one of the waves.

4.3. Nonlinear polarization and nonlinear optical effects

For electromagnetic waves of large amplitude the induced polarization is no longer proportional to the electric field, but it should include higher-order terms. Assuming, for simplicity, a fully scalar formulation, one can write the expansion

$$P(E) = \epsilon_0\{\chi^{1)}E + \chi^{2)}E^2 + \chi^{3)}E^3 + \ldots\} \tag{4.13}$$

where $\chi^{1)}$ stands for the ordinary (linear) susceptibility, previously designated as χ, and $\chi^{n)}$ is the n-order susceptibility. It is equivalent to consider a field-dependent *linear* susceptibility $\chi(E)$, defined through

$$P(E) = \epsilon_0\chi(E)E \tag{4.14}$$

so that

$$\chi(E) = \chi(0) + \left\{\frac{d\chi(E)}{dE}\right\}_{E=0} E + \frac{1}{2!}\left\{\frac{d^2\chi(E)}{dE^2}\right\}_{E=0} E^2 + \ldots \tag{4.15}$$

Substituting (4.15) into (4.14) and comparing with (4.13), one obtains

$$\chi^{1)} = \chi(0) = \left(\frac{dP(E)}{dE}\right)_{E=0}\frac{1}{\epsilon_0} \tag{4.16a}$$

$$\chi^{2)} = \frac{1}{2!}\left(\frac{d\chi(E)}{dE}\right)_{E=0} = \left(\frac{d^2P(E)}{dE^2}\right)_{E=0}\frac{1}{(2!)\epsilon_0} \tag{4.16b}$$

$$\chi^{3)} = \frac{1}{3!}\left(\frac{d^2\chi(E)}{dE^2}\right)_{E=0} = \left(\frac{d^3P(E)}{dE^3}\right)_{E=0}\frac{1}{(3!)\epsilon_0} \tag{4.16c}$$

Table 4.1 *Units and conversion factors.*

Susceptibility	MKS unit	esu unit	Conversion factor
$\chi^{1)}$	—	—	$\chi^{1)}(\text{MKS}) = 12.56\,\chi^{1)}$ (esu)
$\chi^{2)}$	m/V	C/V^2	$\chi^{2)}(\text{MKS}) = 4.19 \times 10^{-4}\,\chi^{2)}$ (esu)
$\chi^{3)}$	m^2/V^2	Cm/V^3	$\chi^{3)}(\text{MKS}) = 1.40 \times 10^{-8}\,\chi^{3)}$ (esu)

These relations are the basis of a general theoretical approach to calculate nonlinear susceptibilities, *the finite-field method* (see Section 4.12.2).

One should note that expansion (4.13), defining the susceptibilities, implies a *spatially local* relation between **E** and **P**. There are interesting nonlinear mechanisms, involving charge transport, that do not fit into that scheme and *local* susceptibilities cannot be defined (see Chapter 10).

Remark concerning notation and units

MKS units will be used throughout this text. However, in many other books, particularly of a theoretical nature, the esu system, where $\epsilon_0 = 1$, is employed. Units and conversion factors are listed in Table 4.1. Some authors, even using MKS units, drop ϵ_0 from the nonlinear terms in (4.13). To obtain the values of $\chi^{2)}$ and $\chi^{3)}$ in this alternative system, one should multiply the corresponding MKS values used here by $\epsilon_0 = 8.85 \times 10^{-12}$.

4.3.1. Nonlinear optical effects

It is now easy to understand the occurrence of various nonlinear effects. Let us consider an incident monochromatic wave $E = E_0 \sin \omega t$. The induced polarization, up to third order in the field, becomes

$$\begin{aligned} P(t) &= \epsilon_0 \chi^{1)} \sin \omega t + \epsilon_0 \chi^{2)} E_0^2 \sin^2 \omega t + \epsilon_0 \chi^{3)} E_0^3 \sin^3 \omega t \\ &= \tfrac{1}{2}\epsilon_0 \chi^2 E_0^2 + \{\epsilon_0 \chi^{1)} E_0 + \tfrac{3}{4}\epsilon_0 \chi^{3)} E_0^3\} \sin \omega t - \tfrac{1}{2}\epsilon_0 \chi^{2)} E_0^2 \cos 2\omega t \\ &\quad - \tfrac{1}{4}\epsilon_0 \chi^{3)} E_0^3 \sin 3\omega t \end{aligned} \tag{4.17}$$

According to this expression, the polarization P, in the same direction of the field, is made up of four contributions oscillating in time at frequencies 0, ω, 2ω and 3ω, i.e.

$$P_0 = \epsilon_0 \chi^{2)} \frac{E_0^2}{2} \tag{4.18a}$$

$$P_\omega = \epsilon_0 \left(\chi^{1)} E_0 + 3\chi^{3)} \frac{E_0^3}{4} \right) \sin \omega t \tag{4.18b}$$

$$P_{2\omega} = \left(-\epsilon_0 \chi^{2)} \frac{E_0^2}{2} \right) \cos 2\omega t \tag{4.18c}$$

$$P_{3\omega} = \left(-\epsilon_0 \chi^{3)} \frac{E_0^3}{4} \right) \sin 3\omega t \tag{4.18d}$$

The static contribution P_0 implies a constant induced voltage in the material and the effect can be considered as an *optical rectification* (or an *inverse Pockels effect*). The terms $P_{2\omega}$ and $P_{3\omega}$ are the sources that generate the *second* and *third* harmonic, respectively, of the incident wave. Since the radiated power is proportional to P^2, the intensity of those harmonics is respectively proportional to the square and cube of the incident intensity, in accordance with experimental results. Finally, the P_ω term includes the ordinary linear response $\chi^{1)}\epsilon_0 E_0 \sin \omega t$, modified by an additional term proportional to $\chi^{3)} E_0^3$. To examine this modification in P_ω, and so in the dielectric constant or refractive index, one can write, recalling (4.18b)

$$D(\omega) = \epsilon_0 \epsilon(\omega) E(\omega) = \epsilon_0 E(\omega) + P(\omega) = \epsilon_0 \epsilon(\omega) + P_{\mathrm{L}}(\omega) + P_{\mathrm{NL}}(\omega)$$
$$= \epsilon_0 \{\epsilon_{\mathrm{L}}(\omega) + \tfrac{3}{4}\chi^{3)} E_0^2\}(\omega) \tag{4.19}$$

the subscript L being added here to stress the linear character of the dielectric constant.

So, the nonlinear field-induced change in the (linear) dielectric constant is

$$\Delta\epsilon = \epsilon_{\mathrm{NL}} - \epsilon_{\mathrm{L}} = \tfrac{3}{4}\chi^{3)} E_0^2 \tag{4.20}$$

i.e. proportional to $\chi^{3)}$ and to the incident intensity. The associated refractive index change is

$$\Delta n = n_{\mathrm{NL}} - n_{\mathrm{L}} = \frac{\Delta\epsilon}{2n_{\mathrm{L}}} = \frac{3}{8n_{\mathrm{L}}}\chi^{3)} E_0^2 \tag{4.21}$$

Dropping the subscripts L and NL and recalling that for a plane wave the intensity is $I = (\epsilon_0 \epsilon c/2n) E_0^2$, one can finally write

$$n(I) = n + \Delta n = n + n_2 I \tag{4.22}$$

with

$$n_2 = \frac{3}{4\epsilon_0 \epsilon c}\chi^{3)} \tag{4.23}$$

Formula (4.22) expresses the linear variation of the refractive index with the light intensity of the exciting plane wave (*optical Kerr effect*).

In summary, the second-order susceptibility $\chi^{2)}$ is responsible for:

- optical rectification;
- second harmonic generation;

and the cubic susceptibility $\chi^{3)}$ induces

- generation of the third harmonic;
- light-induced change in refractive index (Optical Kerr effect).

4.3.2. Anisotropic media

A more rigorous formulation has to take into account the tensorial character of the susceptibilities in an anisotropic medium. Then, expansion (4.13) becomes

$$P_i = \epsilon_0 \{\chi_{ij}^{1)} E_j + \chi_{ijk}^{2)} E_j E_k + \chi_{ijkl}^{3)} E_j E_k E_l + \ldots\} \tag{4.24}$$

where the n-order susceptibility $\chi_{ij\ldots s}^{n)}$ is an $(n+1)$-rank tensor having 3^{n+1} components. Equations (4.18) should also be accordingly modified.

So far, frequency dispersion effects have been ignored (*instantaneous response*) and consequently the tensor $\chi_{ij\ldots r}^{n)}$ is invariant *vs* any permutation of $(j \ldots r)$. In particular, $\chi_{ijk}^{2)} = \chi_{ikj}^{2)}$, and a contracted notation $\chi_{is}^{2)}$ $(i = 1, 2, 3;\ s = 1, 2, 3, 4, 5, 6)$ can be employed, as for the electrooptic coefficients r_{is} (see section 3.2.1).

Very often, the tensor

$$d_{ijk} = \chi_{ijk}^{2)} \tag{4.25a}$$

or, in contracted notation,

$$d_{is} = \frac{\chi_{is}^{2)}}{2} \tag{4.25b}$$

is used instead of $\chi_{is}^{2)}$, particularly when dealing with second-harmonic experiments.

Dispersion effects and their influence on the nonlinear susceptibilities will be considered in Section 4.6.

4.4. Second harmonic generation: phenomenological approach

The first and more thoroughly investigated nonlinear phenomenon is second-harmonic generation [5].

4.4.1. Efficiency

The second-harmonic power output from any local region of the material is, ignoring anisotropy

$$W(2\omega) \approx \frac{|P(2\omega)|^2}{n(2\omega)} \approx \frac{|\chi^{2)}|^2 |E(\omega)|^2}{n(2\omega)} \tag{4.26}$$

Taking into account that $I(\omega) \approx n(\omega)|E(\omega)|^2$, we find that

$$\frac{W(2\omega)}{I^2(\omega)} \approx \frac{|\chi^{2)}|^2}{n^2(\omega)n(2\omega)} \tag{4.27}$$

Therefore, an appropriate *figure of merit* to evaluate the efficiency for second-harmonic generation is the parameter

$$M_{2\omega} = \frac{|\chi^{2)}|^2}{n^3} \tag{4.28}$$

For a medium with symmetry lower than cubic, one uses for $\chi^{2)}$ in (4.27) an effective susceptibility $\chi_{\text{eff}}^{2)}$, which is dependent on incidence direction and light polarization. Assuming that $\mathbf{u}_\omega$ and $\mathbf{u}_{2\omega}$ are the polarization unit vectors for the fundamental and harmonic waves, respectively, the effective susceptibility is defined through

$$\mathbf{u}_{2\omega} \cdot \mathbf{P}_{2\omega} = \mathbf{u}_{2\omega} \cdot \boldsymbol{\chi}^{2)} \cdot \mathbf{EE} = \mathbf{u}_{2\omega} \cdot \chi^{2)} \cdot \mathbf{u}_\omega \mathbf{u}_\omega E^2 = \chi_{\text{eff}}^{2)} E^2 \tag{4.29}$$

with

$$\chi_{\text{eff}}^{2)} = \mathbf{u}_{2\omega} \cdot \boldsymbol{\chi}^{2)} \cdot \mathbf{u}_\omega \mathbf{u}_\omega \tag{4.30}$$

Assuming collinear propagation of the fundamental and harmonic waves, $\chi_{\text{eff}}^{2)}$ (or d_{eff}) is univocally determined as long as one fixes the incidence direction and the two polarization states (ordinary or extraordinary for an uniaxial crystal). Expressions for d_{eff} as a function of the incidence angle θ (with regard to the Z optic axis) and azimuthal angle ϕ (with regard to the X axis) are given in Appendix 4B for the relevant noncentrosymmetrical point groups. Kleinmann symmetry is assumed (see Section 4.7.2).

4.4.2. Phase (index) matching

When considering a macroscopic piece of material, the total power output is obtained after coherently adding all emerging harmonic electric fields. For a plate of isotropic material with thickness l (Figure 4.1), illuminated by the fundamental wave $E = E \sin(kx - \omega t)$, the emerging collinear harmonic field coming from a slice of thickness dx at X is (section 4.3.1),

$$dE_{2\omega}(l,t) \approx \cos\{2kx - 2\omega(t - \Delta t)\}\,dx \tag{4.31}$$

where the time Δt needed by the harmonic wave to travel from x to l is

$$\Delta t = \frac{l-x}{v_{2\omega}} = \frac{(l-x)k_{2\omega}}{2\omega} \tag{4.32}$$

the subscript 2ω referring to the harmonic wave. At the end face of the plate the total harmonic field is

$$\begin{aligned} E_{2\omega}(t) &= \int dE_{2\omega}(x, t - \Delta t)\,dx \\ &= \frac{1}{2k - k_{2\omega}} \sin\left[\frac{l(2k - k_{2\omega})}{2}\right] \cos\left[kl + \frac{k_{2\omega}}{2} - 2\omega t\right] \end{aligned} \tag{4.33}$$

and the associated intensity

$$I_{2\omega} \approx E_{2\omega}^2 \approx \frac{1}{n_\omega - n_{2\omega}} \sin^2\left[\frac{l\omega(n - n_{2\omega})}{c}\right] \tag{4.34}$$

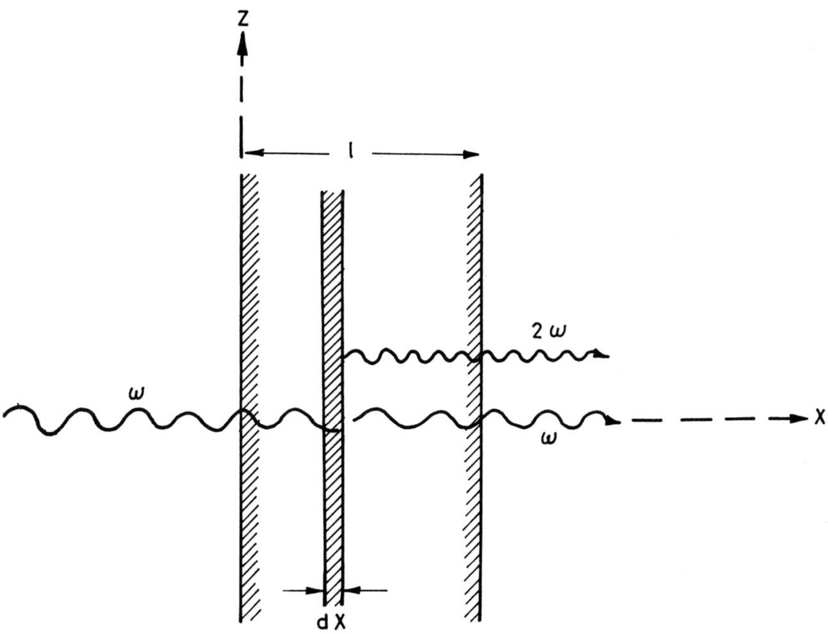

Figure 4.1 *Second-harmonic generation by a thin slice inside a slab. Harmonic waves generated by different slices are coherently added.*

This simple treatment has assumed that the intensity of the fundamental wave remains constant when propagating through the material (*undepleted approximation*). A more rigorous analysis, taking into account the *depletion* of the fundamental wave, will be given in Section 4.17.

Equation (4.33) emphasizes the key role of the index or wavevector matching, most often called *phase matching*, between the fundamental and the harmonic waves.

For $n \neq n_{2\omega}$, i.e. in the absence of matching, the outgoing intensity (4.34) experiences periodic variations with l and reaches maxima for thicknesses l_n

$$l_m = \frac{m\lambda}{4(n_\omega - n_{2\omega})} = ml_c \tag{4.35}$$

which are multiples of the so-called *coherence length* $l_0 = \lambda/4(n - n_{2\omega})$ (Figure 4.2). Typically, $n_\omega - n_{2\omega} \approx 10^{-2}$ and $l_c \approx 10$–$100\ \mu$m, leading to very low efficiencies for second-harmonic generation and so justifying the undepleted approximation. This behaviour is clearly illustrated in experiments such as those initially carried out by Maker and coworkers [6], where $I_{2\omega}$ is measured as a function of the orientation of the plate with regard to the incident fundamental beam (Figure 4.3).

For $n_\omega = n_{2\omega}$, i.e. perfect phase matching, the outgoing intensity is proportional to the square of the plate thickness:

$$I_{2\omega} \approx \frac{l^2\omega^2}{c^2} \tag{4.36}$$

allowing very high conversion efficiencies from the fundamental to the harmonic wave. This behaviour is expected from the fact that the harmonic and the fundamental waves propagate at the same speed and so the harmonic wavelets generated at different points of the material are in phase. As a consequence of this coherence, the emitted harmonic intensity is proportional to the *square* of the emitting centre concentration. For this

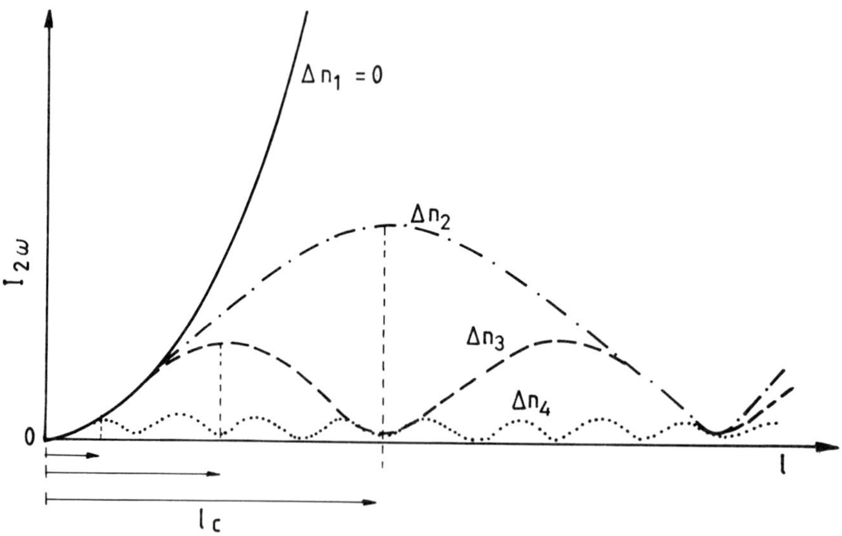

Figure 4.2 *Intensity of the second-harmonic wave as a function of slab thickness l for different index mismatches* $\Delta n_1 = 0 < \Delta n_2 < \Delta n_3 < \Delta n_4$. $\Delta n = n_\omega - n_{2\omega}$ *(see text).*

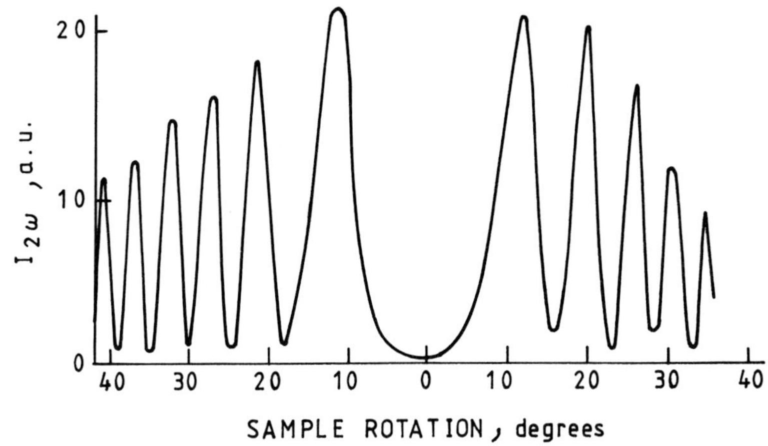

Figure 4.3 *Variation of the second-harmonic intensity with slab orientation* (Maker fringes) *[6].*

situation the undepleted approximation is no longer valid, except for low enough efficiencies.

4.4.3. Index matching: corpuscular view

From a corpuscular point of view, second-harmonic generation involves the coalescence of two photons $\hbar\omega$ to yield a single photon of energy $2\hbar\omega$ (Figure 4.4). The principle of momentum conservation requires

$$\mathbf{k}_1(\omega) + \mathbf{k}_2(\omega) = \mathbf{k}(2\omega) \tag{4.37}$$

For the usual case of collinear propagation of the three waves,

$$2\mathbf{k}(\omega) = \mathbf{k}(2\omega) \tag{4.38a}$$

or in terms of the refractive indices

$$n_\omega = n_{2\omega} \tag{4.38b}$$

providing additional insight into the result previously obtained with the wave model. This corpuscular image can be generalized to the mixing of two photons $\hbar\omega_1$ and $\hbar\omega_2$ to give a new photon with energy $\hbar(\omega_1 + \omega_2)$.

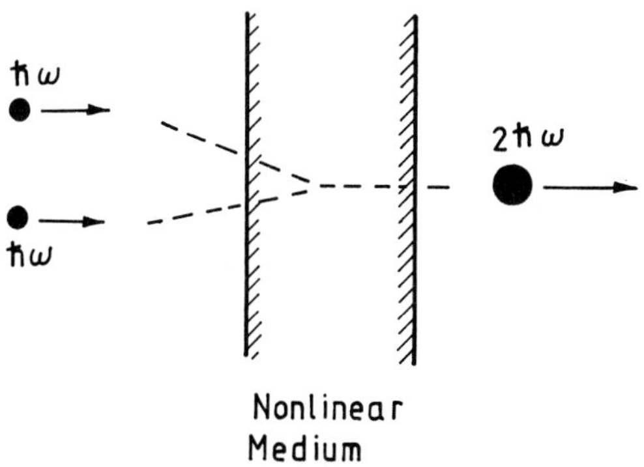

Figure 4.4 *Corpuscular view of second-harmonic generation.*

4.5. Experimental methods for phase matching

The occurrence of wavelength dispersion makes it difficult to achieve the condition $n = n_{2\omega}$, required for collinear phase matching. However, it is possible to use the natural birefringence of many materials to match the ordinary index for the fundamental wave with the extraordinary index for the harmonic wave or *vice versa*. The method is illustrated in Figure 4.5, where sections of the extraordinary index ellipsoid and ordinary index sphere are drawn for the fundamental and harmonic frequencies. Normal dispersion and negative birefringence ($n_e < n_o$), have been assumed in this example. It is observed that for a certain incidence direction defined by the angle θ_c with the optic axis $n_e(\theta_c, 2\omega) = n_o(\omega)$. The critical angle θ_c is given by the formula

$$\sin^2 \theta_c = \frac{n_{\omega,o}^{-2} - n_{2\omega,o}^{-2}}{n_{2\omega,e}^{-2} - n_{2\omega,o}^{-2}} \tag{4.39}$$

The efficiency for second-harmonic generation is very critically dependent on the fulfilment of the phase-matching condition. Figure 4.6 shows this behaviour for KDP; an angular deviation $<0.1^\circ$ from the critical angle θ_c reduces the efficiency in more than one order of magnitude. This illustrates the very severe alignment conditions demanded for efficient optical-mixing experiments.

The possibility of birefringence phase matching and the critical angle θ_c depend on light wavelength and crystal temperature. So, for each temperature one may define a wavelength range, where matching is achievable. When $\theta_c = \pi/2$, the matching is called *noncritical*. Its advantage is that fundamental (ordinary) and harmonic (extraordinary) waves travel in the

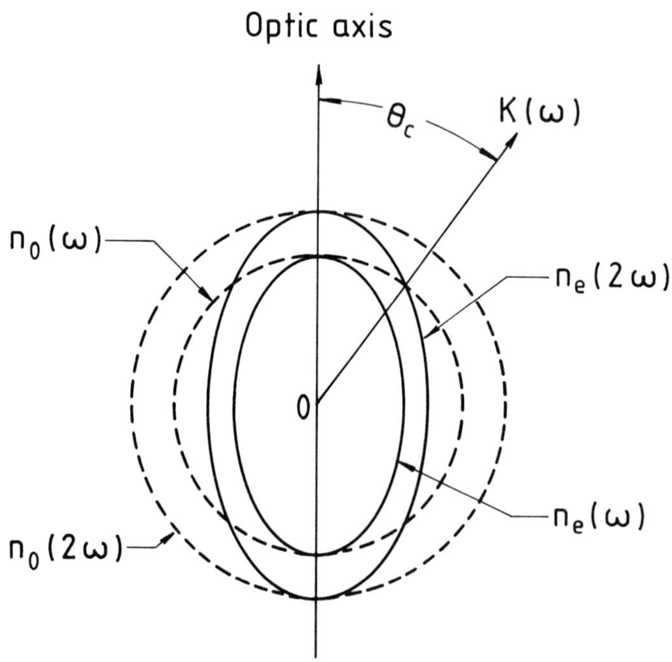

Figure 4.5 *Index ellipsoids for the fundamental and harmonic frequencies, showing the incidence direction at which phase matching is achieved.*

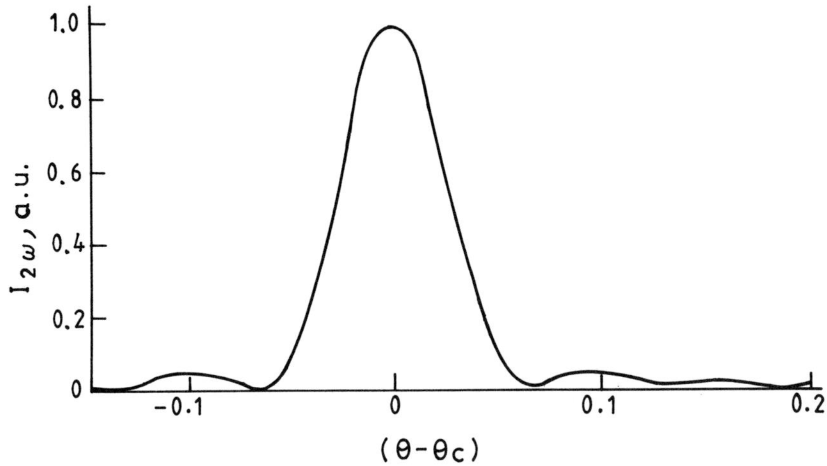

Figure 4.6 *Variation of the second-harmonic intensity in KDP with the angular deviation from the phase-matching direction [7].*

same direction, keeping the matching condition during propagation. This is not the case for $\theta_c \neq \pi/2$, *critical* phase matching, where *walk-off* of the waves occurs.

So far, we have been considering a particular example of the so-called *type I* phase matching, where the two fundamental waves at frequency ω have the same polarization (either ordinary or extraordinary). It can be achieved in two relevant materials, $LiNbO_3$ and KDP, for Nd : YAG laser light at 1.06 μm. One may also use *type II* phase matching, in which the two fundamental waves have orthogonal polarizations. It has been achieved at 1.06 μm in KDP but not in $LiNbO_3$. Neither type of birefringent phase matching is possible for quartz.

An interesting method for phase matching [8] induces in the material a periodic variation of $\chi^{2)}$, whose wavevector **K** matches the difference between the optical $2\mathbf{k}(\omega)$ and $\mathbf{k}(2\omega)$ wavevectors. Using this method, efficient blue light generation has been recently achieved in periodically poled $LiNbO_3$ [9]. Another phase-matching technique, using alternative segments of positive and negative index mismatch, has also been proposed and checked in KTP waveguides [10].

4.6. Dispersive behaviour: formal definition of $\chi^{n)}$

In expansions (4.13) and (4.24) for the polarization, it has been assumed that the material response is instantaneous. In real media there is a finite delay between the electrical excitation and the polarization, leading to frequency-dependent susceptibilities (*dispersive behaviour*).

4.6.1. Second-order susceptibilities

In order to simplify the formalism, let us consider the second-order susceptibility corresponding to the nonlinear mixing of two monochromatic waves of frequencies ω_1 and ω_2, which in complex notation can be written

$$\mathbf{E}_1(t) = \tfrac{1}{2}\{\mathbf{E}_1\,e^{i\omega_1 t} + \mathbf{E}_1^*\,e^{-i\omega_1 t}\} \tag{4.40a}$$

$$\mathbf{E}_2(t) = \tfrac{1}{2}\{\mathbf{E}_2\,e^{i\omega_2 t} + \mathbf{E}_2^*\,e^{-i\omega_2 t}\} \tag{4.40b}$$

Similarly, the induced polarization at a frequency ω can be expressed as

$$\mathbf{P}(t) = \tfrac{1}{2}\{\mathbf{P}\,\mathrm{e}^{i\omega t} + \mathbf{P}^*\,\mathrm{e}^{-i\omega t}\} \tag{4.41}$$

Taking into account the finite response time of the material, the *second-order* polarization should be written in the form

$$P_i^{2)}(t) = \int_0^\infty \int_0^\infty \kappa_{ijk}^{2)}(\tau_1, \tau_2) E_j(t-\tau_1) E_k(t-\tau_2)\,\mathrm{d}\tau_1\,\mathrm{d}\tau_2 \tag{4.42}$$

where $E = E_1 + E_2$, and the sum over j and k should be performed. $\kappa_{ijk}^{2)}$ is a real *time-response function* and use has been made of the causality principle $(\tau_1, \tau_2 > 0)$. By substituting expression (4.40) into (4.42), one obtains

$$\begin{aligned} P_i^{2)}(t) &= P_i^{2)}(\omega_1+\omega_2)\,\mathrm{e}^{i(\omega_1+\omega_2)t} + P_i(\omega_1-\omega_2)\,\mathrm{e}^{i(\omega_1-\omega_2)t} \\ &\quad + P_i^{2)}(0) + P_i^{2)}(2\omega_1)\,\mathrm{e}^{i2\omega_1 t} + P_i^{2)}(2\omega_2)\,\mathrm{e}^{i\omega_2 t} + \text{c.c.} \end{aligned} \tag{4.43}$$

where c.c. is the complex conjugate of the preceding expression. The polarization is made up of five contributions oscillating at frequencies $\omega_1 + \omega_2$, $\omega_1 - \omega_2$, 0, $2\omega_1$ and $2\omega_2$. In detail, one has for the complex polarization oscillating at the sum frequency $\omega_1 + \omega_2$,

$$\begin{aligned} P_i^{2)}(\omega_1+\omega_2) &= \tfrac{1}{2}\{\chi_{ijk}^{2)}(\omega_1+\omega_2 : \omega_1, \omega_2) E_{1j}(\omega_1) E_{2k}(\omega_2) \\ &\quad + \chi_{ijk}^{2)}(\omega_1+\omega_2 : \omega_2, \omega_1) E_{2j}(\omega_2) E_{lk}(\omega_1)\} \end{aligned} \tag{4.44}$$

where the *frequency-dependent complex susceptibilities* are defined through

$$\chi_{ijk}^{2)}(\omega_1+\omega_2 : \omega_1, \omega_2) = \int \kappa_{ijk}^{2)}(\tau_1, \tau_2)\,\mathrm{e}^{i\omega_1\tau_1}\,\mathrm{e}^{i\omega_2\tau_2}\,\mathrm{d}\tau_1\,\mathrm{d}\tau_2 \tag{4.45a}$$

$$\chi_{ijk}^{2)}(\omega_1+\omega_2 : \omega_2, \omega_1) = \int \kappa_{ijk}^{2)}(\tau_1, \tau_2)\,\mathrm{e}^{i\omega_2\tau_1}\,\mathrm{e}^{i\omega_1\tau_2}\,\mathrm{d}\tau_1\,\mathrm{d}\tau_2 \tag{4.45b}$$

Similar expressions to (4.44) are obtained for the other polarization components $P_i^{2)}(\omega_1 - \omega_2)$, $P_i^{2)}(2\omega_1)$, $P_i^{2)}(2\omega_2)$, and $P_i^{2)}(0)$.

It is convenient to redefine the susceptibility $\chi_{ijk}^{2)}$ in the form [11],

$$\begin{aligned} \chi_{ijk}^{2)}(\omega_1+\omega_2 : \omega_1, \omega_2) &= \tfrac{1}{2}\{\chi_{ijk}^{2)}(\omega_1+\omega_2 : \omega_1, \omega_2) + \chi_{ijk}^{2)}(\omega_1+\omega_2 : \omega_2, \omega_1)\} \\ &= \int K_{ijk}^{2)}(\tau_1, \tau_2)\,\mathrm{e}^{i\omega_1\tau_1}\,\mathrm{e}^{i\omega_2\tau_2}\,\mathrm{d}\tau_1\,\mathrm{d}\tau_2 \end{aligned} \tag{4.46}$$

This is a transform of the symmetrized response function

$$K_{ijk}^{2)} = \tfrac{1}{2}\{\kappa_{ijk}^{2)}(\tau_1, \tau_2) + \kappa_{ikj}^{2)}(\tau_2, \tau_1)\} \tag{4.47}$$

that takes into account the irrelevance of the ordering of the fields in (4.42), whenever each field becomes associated to the same tensor subindex. Then, by using (4.46) in (4.43), the complex polarization at $\omega_1 + \omega_2$ becomes (skipping the indices for the various fields)

$$P_i^{2)}(\omega_1+\omega_2) = \chi_{ijk}^{2)}(\omega_1+\omega_2 : \omega_1, \omega_2) E_j(\omega_1) E_k(\omega_2) \tag{4.48a}$$

By following the same formal procedure one obtains the expressions for the other contributions to the nonlinear polarization,

$$P_i^{2)}(\omega_1-\omega_2)=\chi_{ijk}^{2)}(\omega_1-\omega_2:\omega_1,-\omega_2)E_j(\omega_1)E_k^*(\omega_2) \tag{4.48b}$$

$$P_i^{2)}(2\omega)=\tfrac{1}{2}\chi_{ijk}^{2)}(2\omega:\omega,\omega)E_j(\omega)E_k(\omega) \qquad (\omega=\omega_1,\omega_2) \tag{4.48c}$$

$$P_i^{2)}(0)=\tfrac{1}{2}\chi_{ijk}^{2)}(0:\omega,-\omega)E_j(\omega)E_k^*(\omega) \tag{4.48d}$$

with susceptibilities defined as natural extensions of (4.46).

One should note the factor $\frac{1}{2}$ appearing when one moves from the polarization at $\omega_1+\omega_2$ (or $\omega_1-\omega_2$) to that for the second harmonic 2ω (or 0). This is a consequence of the fact that in the second case one is using the total field amplitude of the fundamental wave for each mixing field in expressions (4.48c) and (4.48d).

From (4.46) we find that

$$\chi_{ijk}^{2)}(\omega_1+\omega_2:\omega_1,\omega_2)=\chi_{ikj}^{2)}(\omega_1+\omega_2:\omega_2,\omega_1) \tag{4.49}$$

Assuming that one is operating in a dispersion-free spectral region (i.e. ω_1 and ω_2 are exchangeable), the tensor $\chi_{ijk}^{2)}$ is symmetrical with regard to the j,k indices and one may recourse to the contracted notation $\chi_{is}^{2)}$ $(i=1,2,3; s=1,\ldots,6)$. Alternatively, one often uses the coefficients $d_{is}=\frac{1}{2}\chi_{is}^{2)}$, defined in Section 4.3.2. In terms of these latter coefficients, equations (4.49a) and (4.49c) can be written in explicit matrix form as

$$\begin{bmatrix} P_x(\omega_1+\omega_2) \\ P_y(\omega_1+\omega_2) \\ P_z(\omega_1+\omega_2) \end{bmatrix} = 2\begin{bmatrix} d_{11}\, d_{12}\, d_{13}\, d_{14}\, d_{15}\, d_{16} \\ d_{21}\, d_{22}\, d_{23}\, d_{24}\, d_{25}\, d_{26} \\ d_{31}\, d_{32}\, d_{33}\, d_{34}\, d_{35}\, d_{36} \end{bmatrix} \begin{bmatrix} E_x(\omega_1)E_x(\omega_2) \\ E_y(\omega_1)E_y(\omega_2) \\ E_z(\omega_1)E_z(\omega_2) \\ E_y(\omega_1)E_z(\omega_2)+E_z(\omega_1)E_y(\omega_2) \\ E_x(\omega_1)E_z(\omega_2)+E_z(\omega_1)E_x(\omega_2) \\ E_x(\omega_1)E_y(\omega_2)+E_y(\omega_1)E_x(\omega_2) \end{bmatrix} \tag{4.50}$$

$$\begin{bmatrix} P_x(2\omega) \\ P_y(2\omega) \\ P_z(2\omega) \end{bmatrix} = \begin{bmatrix} d_{11}\, d_{12}\, d_{13}\, d_{14}\, d_{15}\, d_{16} \\ d_{21}\, d_{22}\, d_{23}\, d_{24}\, d_{25}\, d_{26} \\ d_{31}\, d_{32}\, d_{33}\, d_{34}\, d_{35}\, d_{36} \end{bmatrix} \begin{bmatrix} E_x^2(\omega) \\ E_y^2(\omega) \\ E_z^2(\omega) \\ 2E_y(\omega)E_z(\omega) \\ 2E_x(\omega)E_z(\omega) \\ 2E_x(\omega)E_y(\omega) \end{bmatrix} \tag{4.51}$$

Electrooptic susceptibility

When one of the two mixing frequencies is zero, i.e. in electrooptic experiments, a similar analysis leads to

$$P_i^{2)}(\omega)=2\chi_{ijk}^{2)}(\omega:\omega,0)E_j(\omega)E_k(0)=4d_{ijk}(\omega:\omega,0)E_j(\omega)E_k(0) \tag{4.52}$$

where an additional factor 2 has appeared.

4.6.2. *n*-order susceptibilities

The discussion given above for nonlinear two-wave mixing can be generalized to the mixing of n waves and define the n-order susceptibility $\chi^{n)}(\omega : \omega_1, \ldots, \omega_p, -\omega_{p+1}, \ldots, -\omega_n)$, with $\omega = \omega_1 + \ldots + \omega_p - \omega_{p+1} - \ldots - \omega_n$, through

$$\chi^{n)}_{ij\ldots s}(\omega : \omega_1, \ldots, \omega_p, -\omega_{p+1}, \ldots, -\omega_n)$$

$$= \int K^{n)}_{ij\ldots s}(\tau_1, \ldots, \tau_p, \tau_{p+1}, \ldots, \tau_n)\, \mathrm{e}^{i\omega_1\tau} \ldots \mathrm{e}^{i\omega_p\tau_p}\, \mathrm{e}^{-i\omega_{p+1}\tau_{p+1}} \ldots \mathrm{e}^{-i\omega_n\tau_n} \tag{4.53}$$

The real response function $K^{n)}$ is defined as

$$K^{n)}_{ij\ldots s} = \frac{1}{n!} \sum_{P} \kappa^{n)}_{ij\ldots s}(\tau_1, \ldots, \tau_p, \tau_{p+1}, \ldots, \tau_n) \tag{4.54}$$

where the sum extends over all permutations of the pairs $(\tau_{\mathrm{m}}, \omega_{\mathrm{m}})$. This susceptibility connects the interacting fields at frequencies ω_{i} to the induced polarization at frequency ω by the expression

$$P^{n)}_i(\omega) = \chi^{n)}_{ij\ldots s}(\omega : \omega_1, \ldots, -\omega_p, \omega_{p+1}, \ldots, -\omega_n) E_1(\omega_1) \ldots E_p(\omega_p) E^*_{p+1}(\omega_{p+1}) \ldots E^*_n(\omega_n) \tag{4.55}$$

Note that for any frequency preceded by the minus sign, the corresponding complex field amplitudes have to be conjugated.

Remark. Other authors (e.g. [3]) use an alternative expression for the fields, namely

$$E = 2E\cos(\omega t + \phi) = E\,\mathrm{e}^{i\omega t} + E^*\,\mathrm{e}^{-i\omega t} \tag{4.56}$$

leading to an extra factor 2 in all expressions (4.48) for the nonlinear polarizations.

Miller's parameter [12]

The frequency dependence of the second-order susceptibilities is smoothed out by using Miller's parameter

$$\Delta_{ijk} = \frac{\chi^{2)}_{ijk}(\omega : \omega_1, \omega_2)}{\chi^{1)}_{ii}(\omega)\chi^{1)}_{jj}(\omega_1)\chi^{1)}_{kk}(\omega_2)} \tag{4.57}$$

already mentioned in connection with electrooptic phenomena. Values of Δ_{ijk}, which are also weakly dependent on material, are around 3 pm/V (1.5×10^{-5} esu). The formal equivalence between (4.57) and (3.20) can be confirmed by using the results of Section 4.9.

4.7. Symmetry properties of the susceptibilities

The tensor $\chi^{n)}_{ij\ldots s}(\omega)$ has a number of useful symmetry properties. Some of them are independent of the structure of the material and are named *intrinsic*. Others are of a *geometrical* or *structural* nature and are related to the *point group symmetry* of the medium. They determine the number of *nonzero independent* components of the tensors.

4.7.1. Point-group symmetry relations

As for any n-rank tensor, the various symmetry operations of a given point group establish linear relations among the components of $\chi^{n)}$. This yields a simplified structure of the tensor, having a reduced number of independent nonzero components [13,14]. In particular, the third-rank tensor d_{ijk} (or $\chi^{2)}_{ijk}(2\omega : \omega, \omega)$), which is symmetrical in the jk indices (see (4.60)), has the same symmetry restrictions as the Pockels tensor r_{ijk} (invariant *vs* ij permutation). In the contracted notation the structure of d_{ij} $(i = 1, 2, 3;\ j = 1, \ldots, 6)$ is just the transposed of that of r_{ij}. Therefore, it can be immediately inferred from Appendix 3A for all crystallographic point groups.

A very important consequence of this symmetry is that d_{ij} (and any other even-rank susceptibility) is identically zero for a centrosymmetric material and, in particular, for isotropic solids, liquids and gases.

4.7.2. Intrinsic symmetry properties

There are, also, a number of intrinsic symmetry properties that are going to be enumerated below without proof.

(a) Reality

$$\chi^{n)}_{ri\ldots s}(\omega : \omega_1, \omega_2, \ldots, \omega_n) = \chi^{n)}_{ri\ldots s}(-\omega : -\omega_1, -\omega_2, \ldots, -\omega_n) \tag{4.58}$$

(b) Intrinsic permutation symmetry

$\chi^{n)}_{rij\ldots s}(\omega : \omega_1, \ldots, \omega_n)$ is invariant under any permutation of the pairs $i, \omega_1; \ldots; s, \omega_n$

(4.59)

For the second-order susceptibility one obtains the relation (4.49) already mentioned in Section 4.6.1.

For the particular and relevant case of second-harmonic generation, $\omega_1 = \omega_2 = \omega$,

$$\chi^{2)}_{rij}(2\omega : \omega, \omega) = \chi^{2)}_{rji}(2\omega : \omega, \omega) \tag{4.60}$$

indicating that the corresponding susceptibility tensor is symmetrical in the two last indices.

(c) Overall permutation symmetry

For lossless media, property (b) is generalized to state that $\chi^{n)}_{ri\ldots s}(\omega : \omega_1, \ldots, \omega_n)$ is invariant *vs* any permutation among all pairs r, ω; i, ω_1; $\ldots$; s, ω_n. In particular

$$\begin{aligned}\chi^{2)}_{rij}(\omega : \omega_1, \omega_2) &= \chi^{2)}_{rji}(\omega : \omega_2, \omega_1) = \chi^{2)}_{irj}(\omega_1 : \omega, -\omega_2) = \chi^{2)}_{ijr}(\omega_1 : -\omega_2, \omega)\\ &= \chi^{2)}_{jir}(\omega_2 : -\omega_1, \omega) = \chi^{2)}_{jri}(\omega_2 : \omega, -\omega_1)\end{aligned} \tag{4.61}$$

Note that the sign of the frequencies should be always consistent with the algebraic relationship $\omega = \omega_1 + \omega_2$ (or in general $\omega = \sum_i \omega_i$).

(d) Kleinmann symmetry [15]

In the absence of wavelength dispersion, relations (c) lead to the invariance of $\chi^{n)}_{ij\ldots s}$ *vs* any permutation of cartesian indices, regardless of frequency ordering. For the second-order susceptibility the additional relations among the reduced-notation components, imposed by Kleinmann symmetry, are

$$d_{ijk} = d_{ikj} = d_{jik} = d_{kij} = d_{jki} = d_{kji} \tag{4.62}$$

and so the general structure of the d_{ij} matrix (contracted notation) can be written

$$\begin{bmatrix} a & b & c & d & f & g \\ g & h & i & j & d & b \\ f & j & k & i & c & d \end{bmatrix} \tag{4.63}$$

with 10 independent nonzero components. Using the information given in Appendix 3A and relations (4.62), the general structures of d_{ij} tensor, allowed by *Kleinmann symmetry*, can be obtained for all noncentrosymmetrical point groups. The results are tabulated in Appendix 4C.

A similar analysis can be carried out for $\chi^{3)}_{ijkl}$, where only 15 independent tensor components remain when Kleinmann symmetry applies.

The Kleinmann relations are expected to apply when all involved frequencies are located in a dispersion-free spectral region, i.e. away from any resonances. Consequently, the electrooptic susceptibilities, where frequencies 0 and ω are separated by the acoustical and optical vibrational resonances, do not satisfy them (see Section 4.9).

(e) Generalized Kramers–Krönig relations

As a consequence of the causality principle, $\chi^{n)}(\omega : \omega_1, \omega_2, \ldots, \omega_n)$ is an analytical function over the upper halves of the complex planes $\omega_1, \ldots, \omega_n$. This analyticity implies the existence of integral relations between the real and imaginary parts of $\chi^{n)}$, generalizing the Kramers–Krönig relations introduced in Chapter 2 for the linear susceptibility [1].

4.8. Nonlinear processes and susceptibilities

Table 4.2 lists various nonlinear processes and the corresponding frequency-dependent susceptibilities. One should note from this table that the Pockels and Kerr electrooptic effects are nonlinear processes respectively governed by the susceptibilities $\chi^{2)}(\omega : 0, \omega)$ and $\chi^{3)}(\omega : 0, 0, \omega)$. They correspond to curves obtained by cutting the general surfaces $\chi^{2)}(\omega : \omega_1, \omega_2)$ and $\chi^{3)}(\omega : \omega_1, \omega_2, \omega_3)$ by the coordinate planes $\omega_1 = 0$ and $\omega_1 = 0, \omega_2 = 0$, respectively.

Another remarkable electrooptic phenomenon is the *electric-field-induced second-harmonic generation* (EFISH) [16,17], where second-harmonic generation is estimulated by an applied static (or low-frequency field) in a centrosymmetrical medium. This effect can be understood as a breaking of the inversion symmetry due to the field. The phenomenon is particularly interesting in solutions containing second-order nonlinear molecules, and it has been often used to measure their hyperpolarizabilities. One may also consider that a large static field converts a Kerr material into a Pockels material as sensed by an additional probe field.

Table 4.2 *Susceptibilities and nonlinear processes.*

Susceptibility	Process
$\chi^{2)}(2\omega:\omega,\omega)$	Second-harmonic generation
$\chi^{2)}(\omega_1\pm\omega_2:\omega_1,\pm\omega_2)$	Sum or difference of frequencies (Parametric processes)
$\chi^{2)}(\omega:0,\omega)$	Pockels electrooptic effect
$\chi^{2)}(0:\omega,-\omega)$	Optical rectification (Inverse Pockels effect)
$\chi^{3)}(3\omega:\omega,\omega,\omega)$	Third-harmonic generation
$\chi^{3)}(\omega_1\pm\omega_2\pm\omega_3:\omega_1,\pm\omega_2,\pm\omega_3)$	Frequency mixing
$\chi^{3)}(\omega:\omega,\omega,-\omega)$	Optically induced birefringence (optical Kerr effect)
$\chi^{3)}(\omega:0,0,\omega)$	Kerr electrooptic effect
$\chi^{3)}(2\omega:0,\omega,\omega)$	Electric-field-induced second-harmonic generation (EFISH)

4.9. Electrooptic coefficients and susceptibilities

The electronic electrooptic coefficients r^{e}_{ijk} and s^{e}_{ijkl} defined in Chapter 3 can be now related to the corresponding Pockels $\chi^{2)}_{ijk}(\omega:0,\omega)$ and Kerr $\chi^{3)}_{ijkl}(\omega:0,0,\omega)$ susceptibilities (assumed to be real).

In the presence of a static electric field, the dielectric tensor and its inverse can be written

$$\epsilon_{ij}(E)=\epsilon_{ij}(0)+(\delta\epsilon)_{ij} \tag{4.64a}$$

$$\epsilon^{-1}_{ij}(E)=\epsilon^{-1}_{ij}(0)+(\delta\epsilon^{-1})_{ij} \tag{4.64b}$$

By using the matrix inversion rules, one obtains

(a) For $i\neq j\,(\neq k)$:

$$(\delta\epsilon)_{ij}=\frac{-(\delta\epsilon^{-1})_{ij}\{\epsilon^{-1}_{kk}+(\delta\epsilon^{-1})_{kk}\}}{\{\epsilon^{-1}_{11}+(\delta\epsilon^{-1})_{11}\}\{\epsilon^{-1}_{22}+(\delta\epsilon^{-1})_{22}\}\{\epsilon^{-1}_{33}+(\delta\epsilon^{-1})_{33}\}} \tag{4.65}$$

and assuming $(\delta\epsilon)^{-1}\ll\epsilon^{-1}$

$$(\delta\epsilon)_{ij}\approx-\epsilon_{ii}\epsilon_{jj}(\delta\epsilon^{-1})_{ij} \tag{4.66}$$

(b) For $i=j\,(\neq l\neq k)$:

$$\begin{aligned}\epsilon_{ii}+(\delta\epsilon)_{ii}&=\frac{\{\epsilon^{-1}_{ll}+(\delta\epsilon^{-1})_{ll}\}\{\epsilon^{-1}_{kk}+(\delta\epsilon^{-1})_{kk}\}}{\{\epsilon^{-1}_{11}+(\delta\epsilon^{-1})_{11}\}\{\epsilon^{-1}_{22}+(\delta\epsilon^{-1})_{22}\}\{\epsilon^{-1}_{33}+(\delta\epsilon^{-1})_{33}\}}\\&=\frac{1}{\epsilon^{-1}_{ii}+(\delta\epsilon^{-1})_{ii}}\approx\epsilon_{ii}-(\delta\epsilon^{-1})_{ii}\epsilon^{2}_{ii}\end{aligned} \tag{4.67}$$

and so

$$(\delta\epsilon)_{ii}=-(\epsilon_{ii})^2(\delta\epsilon^{-1})_{ii} \tag{4.68}$$

formally identical to (4.66) which, therefore, has general validity for any i,j pair

Remembering that $\delta\chi^{1)}_{ij}=\delta\epsilon_{ij}$, one should write for the polarization at frequency ω in the presence of a static field

$$\begin{aligned}P_{\omega,i}(E)&=P_{\omega,i}(0)+\epsilon_0\{\delta\chi^{1)}_{ij}(\omega)\}E_j(\omega)\\&=P_{\omega,i}(0)-\epsilon_0\epsilon_{ii}\epsilon_{jj}\{r^{e}_{ijk}E_k(0)+s^{e}_{ijkl}E_k(0)E_l(0)\}E_j(\omega)\end{aligned} \tag{4.69}$$

and comparing with expansion (4.24) in terms of the susceptibilities, one obtains

$$r^{\mathrm{e}}_{ijk} = -\frac{2\chi^{2)}_{ijk}(\omega : 0, \omega)}{\epsilon_{ii}\epsilon_{jj}} \tag{4.70a}$$

$$s^{\mathrm{e}}_{ijkl} = -\frac{3\chi^{3)}_{ijkl}(\omega : 0, 0, \omega)}{\epsilon_{ii}\epsilon_{jj}} \tag{4.70b}$$

which can be written approximately as

$$r^{\mathrm{e}}_{ijk} = -\frac{2\chi^{2)}_{ijk}(\omega : 0, \omega)}{n^4(\omega)} \tag{4.70c}$$

$$s^{\mathrm{e}}_{ijkl} = -\frac{3\chi^{3)}_{ijkl}(\omega : 0, 0, \omega)}{n^4(\omega)} \tag{4.70d}$$

Very often, experimental values of the susceptibilities are available in the optical regime, i.e. $\chi^{2)}(\omega : 0, \omega)$ and $\chi^{3)}(\omega : 0, 0, \omega)$. From these data and using (4.70), one may obtain the purely *electronic* or direct contribution to the electrooptic coefficients. Comparison with the full coefficients (*unclamped*) determined from electrooptic experiments, allows the contribution of ionic relaxation to the electrooptic behaviour to be evaluated. As an illustrative example, Table 5.3 (next chapter) compares values r^{e} inferred from $\chi^{2)}(2\omega : \omega, w)$ with r^{T} for several relevant materials. In order to separate the optical mode and acoustic mode (piezoelectric) contributions, *clamped* data are also required.

Comparison between electrooptic and optical wave-mixing experiments

Electrooptic experiments allow for the determination of the nonlinear $\chi^{2)}(\omega : 0, \omega)$ and $\chi^{3)}(\omega : 0, 0, \omega)$ susceptibilities with some advantages over optical wave-mixing experiments measuring $\chi^{2)}(2\omega : \omega, \omega)$ and $\chi^{3)}(3\omega : \omega, \omega, \omega)$. They do not require high-power lasers and do not have phase-matching limitations. In fact, they can be routinely performed with low-power conventional incoherent sources. The reasons accounting for such apparently different behaviour are:

(1) Static (or low-frequency) electric fields required in electrooptic experiments are $\gtrsim 10^4$ V/cm, i.e. of the same order of magnitude as those involved in optical beams of $\gtrsim$MW/cm^2, generated by high-power pulsed lasers.
(2) Since $\mathbf{k} = 0$ for static fields, the phase-matching condition $\mathbf{k}(\omega = 0) + \mathbf{k}(\omega) = \mathbf{k}(\omega)$ for Pockels (or $\mathbf{k}(\omega = 0) + \mathbf{k}(\omega = 0) + \mathbf{k}(\omega) = \mathbf{k}(\omega)$ for Kerr) experiments is automatically fulfilled.

4.10. Molecular formulation: hyperpolarizabilities

For organic materials, binding energy is mostly associated with weak Van der Waals forces and hydrogen bonds. Consequently, individual molecules keep their identity to a large extent. Therefore, it is very often convenient to describe the macroscopic nonlinear behaviour as resulting from the addition of the individual molecular responses. The same description could be applied, in principle, to an inorganic solid in terms of the constitutive bonds or microscopic units.

Table 4.3 *Typical values for the molecular hyperpolarizabilities.*

Polarizability	MKS (SI)	esu
α	$10^{-39}\,C^2\,m^2/J$	$10^{-23}\,cm^3$
β	$10^{-50}\,C^3\,m^3/J^2$	$10^{-30}\,cm^5/sC$
γ	$10^{-61}\,C^4\,m^4/J^3$	$10^{-37}\,cm^7/sC^2$

For the molecular polarization one may write a similar expansion to (4.24), i.e.

$$p_i(\mathbf{E}) = p_i(0) + \alpha_{ij} E_j + \beta_{ijk} E_j E_k + \gamma_{ijkl} E_j E_k E_l + \ldots \tag{4.71}$$

where $\mathbf{p}(0)$ stands for a permanent electric dipole moment occurring in polar molecules. $\mathbf{E}$ is now the local field acting on the molecule, α the linear polarizability and β and γ the quadratic and cubic *hyperpolarizabilities*, respectively. Higher-order terms are usually neglected. Typical values for α, β and γ are given in Table 4.3.

For molecules with axial symmetry around the Z axis, β_{zzz} is, generally, the dominant second-order coefficient, allowing a remarkable simplification of the corresponding tensorial term in (4.71). When considering wavelength dispersion, hyperpolarizabilities depend on all active frequencies and are expressed with the same nomenclature used for the susceptibilities, e.g. $\beta(2\omega : \omega, \omega)$ is the second-harmonic hyperpolarizability.

In a naive picture, assuming that all molecules contributing to the macroscopic polarization are parallel, one may expect to achieve the optimum macroscopic susceptibilities:

$$\left.\begin{aligned} \chi^{1)} &= \frac{N\alpha}{\epsilon_0} \approx 1 \\ \chi^{2)} &= \frac{N\beta}{\epsilon_0} \approx 10^{-11}\,\mathrm{m/V} \\ \chi^{3)} &= \frac{N\gamma}{\epsilon_0} \approx 10^{-22}\,\mathrm{m^2/V^2} \end{aligned}\right\} \tag{4.72}$$

which are in the range of measured values. A more rigorous analysis, taking into account partial molecular ordering, is deferred to Section 4.14.

For dense media, the local field acting on the molecules differs from the macroscopic field used in Maxwell and wave equations. Consequently, some correction factors are required to express the hyperpolarizabilities in terms of the macroscopic fields (see 4.11).

4.10.1. Estimate of the harmonic generation efficiencies

One can now estimate the second- and third-harmonic power radiated by an individual molecule, using the classical result $W_1(\omega) = p^2\omega^4/12\pi\epsilon_0 c^3$. In SI units, it is

$$W_1(2\omega) \approx 10^{-60} I^2 \tag{4.73a}$$

$$W_1(3\omega) = 10^{-140} I^3 \tag{4.73b}$$

The role of coherence (*phase matching*) in harmonic generation by a molecular medium can be also illustrated. Let us consider a light-excited volume of 1 mm length and 1 cm^2 cross-section. If all molecules ($N \sim 10^{22}$) radiate *incoherently*, one obtains for the total second-harmonic emitted power,

$$W_N^{\mathrm{ih}}(2\omega) = N W_1(2\omega) \sim 10^{-38} I^2 \tag{4.74a}$$

and for the efficiency $\eta_N^{\text{ih}} = W_N^{\text{ih}}(2\omega)/I(\omega)$

$$\eta_N^{\text{ih}}(2\omega) = 10^{-38} I \tag{4.74b}$$

This has to be compared with the value for fully *coherent* emission,

$$W_N^{\text{ch}}(2\omega) = N^2 W_1(2\omega) \sim 10^{-16} I^2 \tag{4.74c}$$

leading to

$$\eta_N^{\text{ch}} = 10^{-16} I \tag{4.74d}$$

As noted in Section 4.4.2 there is always coherence in the second-harmonic emission along a certain distance (*coherence length*). Consequently, the efficiencies should reach values between (4.47b) and (4.47d).

4.11. Local field corrections

When establishing the connection between the microscopic and macroscopic formulations of the polarization in a dense medium, local field corrections, accounting for the reaction of the polarization on the field, have to be included. In fact, local field effects may play a key role in enhancing the nonlinear response of a material [18]. The calculation of these effects constitutes a very difficult problem in condensed matter physics, particularly in the nonlinear regime [19]. A number of simplifying approaches are generally used, although some more rigorous analyses have been tried [20–22]. There are two relevant situations to be considered next.

4.11.1. Lorentz correction factor [23]

For a dielectric medium consisting of well-localized (bound) electrons, one follows the Lorentz treatment that takes into account the electronic and ionic polarization. For isotropic or cubic symmetry, it is shown that

$$\mathbf{E}_{\text{L}} = \mathbf{E} + \frac{\mathbf{P}}{3\epsilon_0} \tag{4.75}$$

$\mathbf{E}_{\text{L}}$ being the local field and $\mathbf{E}$ the macroscopic field. In the *macroscopic formulation* $\mathbf{P}$ includes a part derived from the linear constitutive relations and an additional term associated with the nonlinear response, i.e.

$$\mathbf{P} = \mathbf{P}_{\text{L}}^{(\text{ma})} + \mathbf{P}_{\text{NL}}^{(\text{ma})} = \epsilon_0(\epsilon - 1)\mathbf{E} + \mathbf{P}_{\text{NL}}^{(\text{ma})} \tag{4.76}$$

where the macroscopic character for the polarization has been explicitly indicated. ϵ stands for the (linear) dielectric constant. Substituting (4.76) into (4.75), we find that

$$\mathbf{E}_{\text{L}} = \frac{(\epsilon + 2)\mathbf{E}}{3} + \frac{\mathbf{P}_{\text{NL}}^{(\text{ma})}}{3\epsilon_0} \tag{4.77}$$

that (neglecting $\mathbf{P}_{\text{NL}}^{(\text{ma})}$ *vs* $\mathbf{P}_{\text{L}}^{(\text{ma})}$) can be well approximated by

$$\mathbf{E}_{\text{L}} = \frac{\epsilon + 2}{3}\mathbf{E} \tag{4.78}$$

relating the local to the macroscopic field, through the *Lorentz correction factor*

$$f_{LZ} = \frac{\epsilon + 2}{3} \tag{4.79}$$

On the other hand, at a *molecular* (or, in general, *microscopic*) level, one writes

$$\mathbf{P} = \mathbf{P}_{L}^{mo} + \mathbf{P}_{NL}^{mo} = \mathbf{N}\alpha\mathbf{E}_{L} + \mathbf{P}_{NL}^{mo} \tag{4.80}$$

where $\mathbf{P}_{NL}^{mo}$ includes the contribution of higher-order molecular hyperpolarizabilities. On the other hand, one should note that the linear term in (4.80) already includes a part of the macroscopic nonlinear response through the local field (see formula (4.77)). Substituting (4.78) into (4.80) and comparing with (4.76), one obtains from the equality of linear terms in $\mathbf{E}$

$$1 - \frac{N\alpha}{3\epsilon_0} = \frac{3}{\epsilon + 2} \tag{4.81}$$

Then, the nonlinear polarization terms lead to

$$\mathbf{P}_{NL}^{ma} = \frac{\mathbf{P}_{NL}^{mo}}{1 - (N\alpha/3\epsilon_0)} = \frac{(\epsilon + 2)\mathbf{P}_{NL}^{mo}}{3} = f_{LZ}\mathbf{P}_{NL}^{mo} \tag{4.82}$$

Equation (4.81) relates the dielectric constant to the linear polarizability. Equation (4.82) expresses $\mathbf{P}_{NL}^{ma}$, which acts as a *source term* in the macroscopic nonlinear wave equation, in terms of the nonlinear polarization in the molecular formalism. This implies that, in addition to the field, the nonlinear polarization also has to be corrected when going from the microscopic to the macroscopic formulation. One should note that the difference between $\mathbf{P}_{NL}^{ma}$ and $\mathbf{P}_{NL}^{mo}$ arises from the fact that in the macroscopic formulation the effect of the nonlinear polarization on the linear response is fully ignored.

For materials with symmetry lower than cubic, expression (4.75) is generalized to

$$\mathbf{E}_{L} = \mathbf{E} + \mathbf{L}\mathbf{P} \tag{4.83}$$

$\mathbf{L}$ now being a local field tensor.

For free electrons in a medium [24], the local field factor becomes $f = 1$.

4.11.2. Onsager correction factor

A different situation occurs when the medium contains dipoles that can reorient under the applied electric field, if the oscillating frequency is lower than or comparable to the reorientation rates (e.g. in electrooptic experiments). Under these conditions, the correction factor should include the contribution of oriented dipoles in addition to that corresponding to the electronically induced dipoles. Then, according to Onsager [25],

$$f_{ON}(0) = \frac{\epsilon(0)\{\epsilon^{e}(0) + 2\}}{2\epsilon(0) + \epsilon^{e}(0)} \tag{4.84}$$

where $\epsilon(0)$ stands for the static dielectric constant and $\epsilon^{e}(0)$ refers to the electronic contribution to it. In the absence of reorientational motions, $\epsilon(0) = \epsilon^{e}(0)$, and $f_{ON} = \{\epsilon(0) + 2\}/3$, in accordance with the Lorentz result (4.79). Also, $f_{ON} = f_{LZ}$ for optical frequencies where molecular reorientation is quenched.

4.12. Calculation of susceptibilities: theoretical approaches

There are two main general quantum-mechanical approaches to calculate nonlinear susceptibilities and electrooptic coefficients [1,3,4,11,26,27]: a perturbation method, often called sum-over-states (SOS) method and the finite-field (or derivative) method. Because of their generality in the field of nonlinear optics, including electrooptics, they were skipped in Chapter 3 and are summarized here.

The SOS method involves the use of a matrix density formalism and a perturbation expansion. It permits us to express the susceptibilities as closed analytical expressions containing the eigenvalues of the electron system and matrix elements of the dipolar electrical interaction between pairs of eigenstates. In a way, it provides a quantum-mechanical basis to justify and extend the simple classical anharmonic oscillator model introduced in Section 4.3.

The finite-field method expresses the induced polarization as a function of the applied field as a parameter. Subsequently, the higher-order susceptibilities are obtained by successive derivatives of the polarization with regard to the field.

The success of both methods relies critically on a judicious formulation of the hamiltonians and an adequate evaluation of their eigenvalues and eigenfunctions. Therefore, there are extensive opportunities to apply the abundant paraphernalia of quantum-mechanical methods: *ab initio*, Hartree–Fock, pseudopotential etc. The two methods will be now schematically reviewed and followed by a succinct introduction to more phenomenological models. Deeper and more detailed discussions should be consulted in the quoted references and original papers.

4.12.1. SOS method

In order to calculate the susceptibilities, one has to solve the time-dependent equation for the density operator ρ

$$i\hbar\frac{\mathrm{d}\rho}{\mathrm{d}t} = [\mathcal{H}, \rho] \tag{4.85}$$

with the perturbed hamiltonian $H = H_0 + H'$, H_0 being the unperturbed hamiltonian and H' the term associated with the energy coupling to the electric field

$$\mathcal{H}' = -\mathbf{P}\cdot\mathbf{E} \tag{4.86}$$

Then, one uses a perturbation expansion for ρ and takes a basis formed by the eigenstates of $\mathcal{H}_0$. Relaxation processes are treated phenomenologically by introducing a *relaxation time* or a *damping coefficient* $\Gamma_{nn'}$ associated with each $n - n'$ transition. After some long and tedious mathematical derivation, one arrives at final expressions for the susceptibilities. It is often convenient for a more physical insight to use the additivity principle referred to N equivalent bound electrons per unit volume. Moreover, it can generally be assumed that in the absence of the field, all bound electrons are in the ground (g) state, i.e. $\rho_{gg} = 1$, $\rho_{nn'} = 0$, for n or $n' = g$. Then, first- and second-order susceptibilities can be written [3]

$$\chi_{ij}^{1)}(\omega) = \frac{N}{\epsilon_0\hbar}\sum \frac{\langle p_i\rangle_{ng}\langle p_j\rangle_{gn}}{\omega + \omega_{ng} + i\Gamma_{ng}} - \frac{\langle p_j\rangle_{ng}\langle p_i\rangle_{gn}}{\omega - \omega_{ng} + i\Gamma_{ng}} \tag{4.87a}$$

$$\chi_{ijk}^{2)}(\omega : \omega_1, \omega_2) = -\frac{N}{\epsilon_0\hbar}\sum \frac{\langle p_i\rangle_{gn}\langle p_j\rangle_{nm}\langle p_k\rangle_{mg}}{(\omega - \omega_{ng} + i\Gamma_{ng})(\omega_2 - \omega_{mg} + i\Gamma_{mg})} \tag{4.87b}$$

where $\langle \mathbf{p} \rangle_{ab}$ is the matrix element of the dipole moment operator $q\mathbf{r}$ for each electron between states a and b. For $\chi_{ijk}^{2)}$ only one out of the various terms contributing to the sum in (4.87b) have been included. More cumbersome expressions apply to $\chi_{ijkl}^{3)}$, while keeping the same formal structure. Useful diagramatic techniques to write all terms in the sums have been given by Yee and Gustafson [28]. Anyhow, formulae (4.87) illustrate the results obtained with the perturbative method and emphasize the relevant features of the susceptibilities (e.g. the resonant enhancement near the material eigenfrequencies).

A rather drastic but useful simplification to formulae (4.87), in the transparency domain of the material, is provided by the Unsold approximation [29]. It assumes that an average transition energy can be defined, which applies to all terms in the summation over the excited states.

4.12.2. Finite-field method

Under an applied field, either static or oscillatory, the total hamiltonian $\mathcal{H}$ includes the term $\mathcal{H}'$, (4.86), describing the Stark coupling. In the dipolar approximation H is written

$$\begin{aligned}\mathcal{H}' &= -\mathbf{P}\cdot\mathbf{E} \\ &= -\epsilon_0\{P_{0i}E_i + \chi_{ij}^{1)}E_iE_j + \chi_{ijk}^{2)}E_iE_jE_k + \ldots + \chi_{ij\ldots p}^{n)}E_iE_j\ldots E_p + \ldots\}\end{aligned} \tag{4.88}$$

By considering E_i as parameters in the hamiltonian one may then obtain the eigenvalues $\mathscr{E}_n$ (corresponding to the eigenstates n), which will be a function of E_i. Let us assume the material in the ground state with energy $\mathscr{E}_g(E_i)$. In accordance with expression (4.88), the susceptibilities can be calculated through

$$\chi_{ij\ldots p}^{n)} = \frac{\partial^{n+1)}\mathscr{E}_g}{\partial E_i\,\partial E_j\ldots\partial E_p} \tag{4.89}$$

Therefore, one is left with the quantum-mechanical problem of determining $\mathscr{E}_g(E_i)$.

4.12.3. Phenomenological models

In many cases, phenomenological models, appropriate to specific materials, are successfully used. They involve some drastic approximations on the general theoretical schemes and often have recourse to empirical parameters.

Most phenomenological models use an *additivity principle*, i.e. it is assumed that nonlinear susceptibilities are the tensorial sum of the corresponding hyperpolarizabilities of some elementary building units making up the material structure. They can be either interatomic bonds, molecules or molecular clusters. The additivity principle is well adapted to deal with covalent semiconductor materials (in terms of bonds) and weakly bonded organic materials (in terms of individual molecules).

In other cases, one selects for the calculation the part of the total electron system which is considered to play a dominant role in the electrooptic response. Examples are provided by *free electrons* in low-gap semiconductors or *π-electrons* in conjugated molecules.

A different example is the so-called *two-level model* [30], which derives directly from the SOS method, after assuming that only the ground and one excited state are relevant in the expressions for the susceptibilities. It provides a useful description of *charge-transfer organic molecules*.

All these phenomenological models will be discussed in Chapters 5 and 6, when dealing with inorganic and organic materials, respectively. One should also include here the anharmonic oscillator model already described in Section 3.9.

4.13. Resonances

Both the general quantum mechanical formulation of Section 4.12, and the phenomenological models (e.g. anharmonic oscillator), reveal a key feature of the susceptibilities: the occurrence of *resonances*. In fact, $\chi^{n)}(\omega : \omega_1, \omega_2, \ldots, \omega_n)$ reach very high values whenever any of the intervening frequencies or their algebraic sums coincide with one of the eigenfrequencies of the medium. As an example, resonances of $\chi^{2)}(\omega : \omega_1, \omega_2)$ occur at ω_1, ω_2 and $\omega_1 + \omega_2$ and for $\chi^{3)}(\omega : \omega_1, \omega_2, \omega_3)$ at ω_1, ω_2, ω_3, $\omega_1 + \omega_2$, $\omega_1 + \omega_3$, $\omega_2 + \omega_3$, and ω. Nonlinear effects will be strongly enhanced at or near those resonances, at the cost (in many cases) of a reduced transparency.

On the other hand, as a generalization of the linear case, the resonances in $\chi^{n)}$ provide a very interesting spectroscopic tool [31] to determine the energy level structure of the material. In this respect, the third-order susceptibility presents definite advantages over $\chi^{1)}$ and $\chi^{2)}$, since resonances may occur at frequencies where none of the propagating waves is absorbed.

4.14. Relation between the molecular and the bulk nonlinear response

Taking into account the weakness of molecular interactions in an organic solid, one may write the bulk nonlinear susceptibility as a sum of all molecular contributions inside the unit volume. Let us first consider the second-order susceptibility. In order to write each molecular contribution, it is first necessary to relate the hyperpolarizability β_{ijk} (referred to molecular axes) for an arbitrarily oriented molecule, to the appropriate components b_{ijk} in the common laboratory reference frame. For a generic molecule u.

$$b^u_{lmn} = \cos\theta_{li}\cos\theta_{mj}\cos\theta_{nk}\beta_{ijk} \tag{4.90}$$

θ_{ab} being the angle between the x_a molecular axis and the X_b laboratory axis.

On the other hand, for a situation where Lorentz field corrections apply, one writes for the second-order polarization in the microscopic formulation

$$\begin{aligned} P_l^{2)(\mathrm{mo})}(\omega : \omega_1, \omega_2) &= \sum_{u=1}^{N} b^u_{lmn} E_{\mathrm{L},m}(\omega_1) E_{\mathrm{L},n}(\omega_2) \\ &= \sum_{u=1}^{N} b^u_{lmn} f_{\mathrm{LZ}}(\omega_1) f_{\mathrm{LZ}}(\omega_2) E_m(\omega_1) E_n(\omega_2) \end{aligned} \tag{4.91}$$

$f_{\mathrm{LZ}}(\omega_1) = (\epsilon + 2)/3$ being the Lorentz factor and N the density of molecules. Taking (4.82) into account, the macroscopic polarization is then expressed as

$$\begin{aligned} P_l^{2)(\mathrm{ma})}(\omega : \omega_1, \omega_2) &= \sum_{u=1}^{N} b^u_{lmn} f_{\mathrm{LZ}}(\omega) f_{\mathrm{LZ}}(\omega_1) f_{\mathrm{LZ}}(\omega_2) E_m E_n \\ &= N f_{\mathrm{LZ}}(\omega) f_{\mathrm{LZ}}(\omega_1) f_{\mathrm{LZ}}(\omega_2) \langle b_{lmn} \rangle E_m E_n \end{aligned} \tag{4.92}$$

where $\langle\ \rangle$ refers to an orientational average.

From the definition of the second-order susceptibility one finally obtains

$$\epsilon_0 \chi^{2)}_{lmn} = \frac{N}{9}\{\epsilon(\omega)+2\}\{\epsilon(\omega_1)+2\}\{\epsilon(\omega_2)+2\}\langle\cos\theta_{li}\cos\theta_{mj}\cos\theta_{nk}\rangle\beta_{ijk} \tag{4.93}$$

relating $\chi^{2)}_{lmn}$ to the molecular hyperpolarizability β_{ijk} and to a matrix $\langle\,\rangle_{lmn,ijk} = \langle\cos\theta_{li}\cos\theta_{mj}\cos\theta_{nk}\rangle$, characterizing the distribution of molecular orientations.

Expression (4.93) can be immediately generalized for the third-order susceptibility

$$\epsilon_0 \chi^{3)}_{ijkl}(\omega:\omega_1,\omega_2,\omega_3) = f_{\mathrm{LZ}}(\omega) f_{\mathrm{LZ}}(\omega_1) f(\omega_2) f(\omega_3) N \gamma^{3)}_{ijkl}(\omega:\omega_1,\omega_2,\omega_3) \tag{4.94}$$

In the case where any of the acting frequencies is zero, as in electrooptic phenomena, equations (4.93) and (4.94) are still valid except that one should substitute the appropriate correction factor (either Lorentz or Onsager) in the expression for the local static field.

4.15. Example: oriented thin film

One simple and useful example of this procedure is that of an oriented thin film, having ∞mm ($C_{\infty v}$) symmetry. According to Appendix 3A, the $\chi^{2)}_{rs}$ (or d_{rs}) tensor has only three nonzero components:

$$\chi^{2)}_{15} = \chi^{2)}_{xxz} = \chi^{2)}_{yyz} = \chi^{2)}_{xzx} = \chi^{2)}_{yzy} \tag{4.95a}$$

$$\chi^{2)}_{31} = \chi^{2)}_{zxx} = \chi^{2)}_{zyy} \tag{4.95b}$$

$$\chi^{2)}_{33} = \chi^{2)}_{zzz} \tag{4.95c}$$

In a dispersion-free region, where Kleinmann symmetry applies, the two first coefficients become equal and only two independent nonzero coefficients $\chi^{2)}_{31}$ and $\chi^{2)}_{33}$ remain (see Appendix 4C). Similarly, the two nonzero coefficients of the hyperpolarizability tensor for axial symmetry molecules, are β_{31} and β_{33}.

Molecular orientation in the laboratory frame can be now conveniently characterized by the angles $\theta = \theta_{zZ}$, formed by the molecular axis z with the laboratory axis Z, and ϕ, between planes OzZ and OZX (Figure 4.7). The calculation of the bulk susceptibility is markedly simplified if we make the reasonable assumption that β_{31} is negligible with regard to β_{33} and set equal to zero. Then the only angular factors appearing in formula (4.90) for the nonzero b_{ijk} terms are

$$\cos\theta_{zZ} = \cos\theta \tag{4.96a}$$

$$\cos\theta_{zX} = \sin\theta\cos\phi \tag{4.96b}$$

$$\cos\theta_{zY} = \sin\theta\sin\phi \tag{4.96c}$$

Therefore, one obtains in contracted notation

$$\langle b_{33}\rangle = \langle\cos^3\theta\rangle\beta_{33} \tag{4.97a}$$

$$\langle b_{31}\rangle = \langle\cos\theta\sin^2\theta\cos^2\phi\rangle\beta_{33} = \tfrac{1}{2}\langle\cos\theta\sin^2\theta\rangle\beta_{33} \tag{4.97b}$$

where the averages over θ and ϕ have been performed sequentially. The corresponding expressions (4.93) for the susceptibilities become

$$\epsilon_0 \chi^{2)}_{33} = NF\langle\cos^3\theta\rangle\beta_{33} \tag{4.98a}$$

$$\epsilon_0 \chi^{2)}_{31} = \tfrac{1}{2} NF\langle\cos\theta\sin^2\theta\rangle\beta_{31} \tag{4.98b}$$

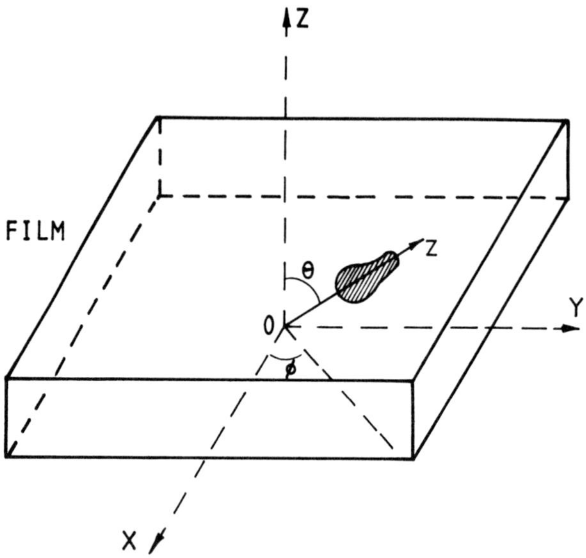

Figure 4.7 *Relation between molecular and laboratory frames in an oriented thin film.*

F being an overall local field factor. For *perfect molecular alignment* along Z, $\theta = 0$ and then

$$\chi_{31}^{2)} = 0 \tag{4.99}$$

When the molecular ordering is caused by a field $\boldsymbol{E}$ applied along Z and acting on polar molecules with dipole moment p, then, mechanical–statistical considerations [4] lead to

$$\langle \cos^3 \theta \rangle = L_3\left(\frac{pE}{kT}\right) \tag{4.100a}$$

$$\langle \cos\theta \sin^2 \theta \rangle = L_1\left(\frac{pE}{kT}\right) - L_3\left(\frac{pE}{kT}\right) \tag{4.100b}$$

where L_1 and L_2 are Langevin functions. For $pE \ll kT$, $L_3 \cong pE/5kT$ and $L_3 - L_1 \cong 2pE/15kT$, and one obtains to a good approximation

$$\epsilon_0 \chi_{33}^{2)} = \frac{NFpE\beta_{33}}{5kT} \tag{4.101a}$$

$$\epsilon_0 \chi_{31}^{2)} = \frac{NFpE\beta_{33}}{15kT} \tag{4.101b}$$

and consequently,

$$\chi_{33}^{2)} = 3\chi_{31}^{2)} \tag{4.102}$$

which has been already mentioned in Section 3.14.1.

For other macroscopic structures such as single crystals, the calculations become more involved and have been carried out by Zyss and Oudar [32] for all point-group symmetries.

4.16. Second-order processes: three-wave mixing

A rigorous analysis of these processes, taking into account the exchange of energy among interacting waves, requires the use of the nonlinear wave equation, as described in Section 4.2. Let us consider three collinear plane waves of frequencies ω_1, ω_2 and $\omega_3 = \omega_1 + \omega_2$, which propagate along Z and are nonlinearly coupled through the second-order susceptibility. The generated second-order polarizations are

$$\mathbf{P}^{2)}(z, \omega_1) = \epsilon_0 \chi^{2)}(\omega_1 : -\omega_2, \omega_3) : \mathbf{u}_2\mathbf{u}_3 E_2^*(z) E_3(z)\, e^{i(k_3 - k_2)z} \tag{4.103a}$$

$$\mathbf{P}^{2)}(z, \omega_2) = \epsilon_0 \chi^{2)}(\omega_2 : -\omega_1, \omega_3) : \mathbf{u}_1\mathbf{u}_3 E_1^*(z) E_3(z)\, e^{i(k_3 - k_1)z} \tag{4.103b}$$

$$\mathbf{P}^{2)}(z, \omega_3) = \epsilon_0 \chi^{2)}(\omega_3 : \omega_1, \omega_2) : \mathbf{u}_1\mathbf{u}_2 E_1(z) E_2(z)\, e^{i(k_1 + k_2)z} \tag{4.103c}$$

$\mathbf{u}_i$ being the polarization unit vector for wave i. On the other hand, for a loseless medium conditions (4.61) impose

$$\begin{aligned} \mathbf{u}_1 \cdot \chi^{2)}(\omega_1 : -\omega_2, \omega_3) : \mathbf{u}_2\mathbf{u}_3 &= \mathbf{u}_2 \cdot \chi^{2)}(\omega_2 : -\omega_1, \omega_3) : \mathbf{u}_1\mathbf{u}_3 \\ &= \mathbf{u}_3 \cdot \chi^{2)}(\omega_3 : \omega_1, \omega_2) : \mathbf{u}_1\mathbf{u}_2 = \chi \end{aligned} \tag{4.104}$$

χ being an effective scalar second-order susceptibility. Introducing (4.104) and (4.103) into (4.12), with $\cos\alpha = 1$ one obtains the following set of coupled equations:

$$\frac{dE_1(z)}{dz} = i\frac{\omega_1 \chi}{2cn_1} E_2^*(z) E_3(z)\, e^{i\Delta k z} \tag{4.105a}$$

$$\frac{dE_2(z)}{dz} = i\frac{\omega_2 \chi}{2cn_2} E_1^*(z) E_3(z)\, e^{i\Delta k z} \tag{4.105b}$$

$$\frac{dE_3(z)}{dz} = i\frac{\omega_3 \chi}{2cn_3} E_1(z) E_2(z)\, e^{-i\Delta k z} \tag{4.105c}$$

with $\Delta \mathbf{k} = \mathbf{k}_3 - \mathbf{k}_1 - \mathbf{k}_2$ being the *momentum mismatch*. They describe the exchange of energy among the three coupled waves.

Manley–Rowe relations

Designating by $S_i(z)$ the energy flux along z of the i interacting wave (frequency ω_i), the following relations can be inferred from equations (4.105):

$$\frac{d(S_1 + S_2 + S_3)}{dz} = 0 \tag{4.106a}$$

$$\begin{aligned} \frac{d[(S_1/\omega_1) - (S_2/\omega_2)]}{dz} &= \frac{d[(S_1/\omega_1) + (S_3/\omega_3)]}{dz} \\ &= \frac{d[(S_2/\omega_2) + (S_3/\omega_3)]}{dz} = 0 \end{aligned} \tag{4.106b}$$

The first one simply expresses energy conservation. On the other hand, (4.106b) permits us to interpret the wave-mixing process in terms of individual photons. The gain (loss) of one photon by wave 3, is associated to the simultaneous loss (gain) of one photon by each one of the waves 1 and 2. They are known as the *Manley–Rowe relations*.

4.17. Reanalysis of second-harmonic generation

A particularly relevant example of three-wave mixing is second harmonic generation, where $\omega_1 = \omega_2 = \omega$, $\omega_3 = 2\omega$. For this case one should use $\chi/2$ instead of χ and equations (4.105) become

$$\frac{dE_\omega}{dz} = i\frac{\omega\chi}{4cn_\omega}E^*_\omega E_{2\omega}\,e^{i\Delta k z} \tag{4.107a}$$

$$\frac{dE_{2\omega}}{dz} = i\frac{\omega\chi}{2cn_{2\omega}}E^2_\omega\,e^{-i\Delta k z} \tag{4.107b}$$

If $\Delta k \neq k_{2\omega} - 2k_\omega \neq 0$, i.e. $n_{2\omega} \neq n_\omega$ (no phase matching) the oscillatory term $e^{i\Delta k z}$ makes the energy transfer between the two waves very inefficient and so one may use the ***undepleted approximation*** in which E_ω and I_ω remain constant. Then equation (4.107b) has the solution

$$E_{2\omega}(z) = -\frac{\omega\chi}{2cn_{2\omega}}E^2_\omega\frac{(e^{-i\Delta k Z} - 1)}{\Delta k} \tag{4.108a}$$

with the boundary condition $E_{2\omega}(0) = 0$. The second-harmonic intensity is then

$$I_{2\omega}(z) = \frac{4\omega^2\chi^2}{\epsilon_0 n_{2\omega} n^2_\omega c^3}\left(\frac{\sin(\Delta k z/2)}{\Delta k}\right)^2 I^2_\omega \tag{4.108b}$$

in accordance with expression (4.33) derived from the phenomenological approach.

When the phase-matching condition $\Delta k = 0$ ($n_\omega = n_{2\omega} = n$) is obeyed, equations have to be solved taking into account energy depletion of the fundamental beam. The intensities of the fundamental and harmonic wave become [2,3]

$$I_\omega(z) = I_\omega(0)\,\mathrm{sech}^2\left(\frac{E_\omega(0)\chi\omega z}{2cn}\right) \tag{4.109a}$$

$$I_{2\omega}(z) = I_\omega(0)\tanh^2\left(\frac{E_\omega(0)\chi\omega z}{2cn}\right) \tag{4.109b}$$

The conversion efficiency is defined as $\eta = I_{2\omega}/I_\omega$. For small penetration depths z or small incident fundamental intensities, η is proportional to $I^{1/2}_\omega(0)$ and $z^{1/2}$. On the other hand, for thick enough samples or high enough incident intensities, η approaches 1. Figure 4.8 shows the experimental dependence of η on the incident intensity for KDP. The thickness $z_{1/2}$ required to reduce the fundamental intensity to a half of its initial value is given by

$$z_{1/2} = \frac{2cn}{\chi\omega E_\omega(0)}\tanh^{-1}2^{-1/2} \tag{4.110}$$

For KDP, operating with a fundamental He–Ne laser beam of $I_\omega(0) = 100\,\mathrm{MW/cm^2}$, $z_{1/2} = 3.3\,\mathrm{cm}$. Along this distance the crystal must keep an excellent refractive index homogeneity to ensure the phase-matching condition. This point illustrates the stringent requirements imposed by this condition on the optical quality of nonlinear crystals.

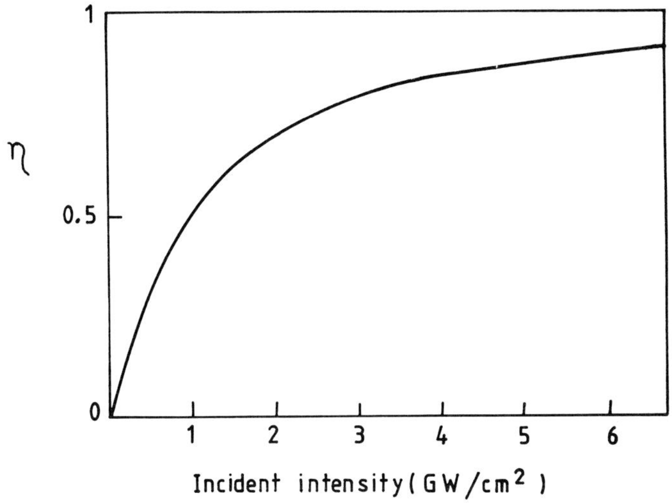

Figure 4.8 *Dependence of the second-harmonic conversion efficiency on incident energy for KDP [6].*

4.18 Third-order processes: four-wave mixing

The propagation of four coupled waves in a nonlinear medium is described by the third-order susceptibility $\chi^{3)}_{ijkl}$. Rate equations governing the exchange of energy among the waves are a trivial extension of those (4.105) explicitly written for three-wave mixing.

A well documented four-wave mixing process is third-harmonic generation governed by $\chi^{3)}_{ijkl}(3\omega : \omega, \omega, \omega)$. Efficient generation requires that the phase-matching condition $3\mathbf{k}_\omega = \mathbf{k}_{3\omega}$ be obeyed by recoursing to the crystalline anisotropy. Although the process has an efficiency much smaller than that corresponding to second-order effects, it has no symmetry restrictions and can be observed in all media including amorphous solids, liquids and gases.

Another interesting process is the *optical Kerr effect* or *optically induced birefringence*, where a light beam of frequency ω modifies the refractive index for a light frequency ω'. It can be considered as an electrooptic Kerr effect, in which the modulation frequency Ω of the applied field is within the optical range $\Omega = \omega$. It is governed by $\chi^{3)}(\omega : \omega, -\omega, \omega')$.

A particular application of the optical Kerr effect is the so-called *degenerated four-wave mixing* (DFWM) where $\omega' = \omega$ [33–35]. Let us consider three plane waves with complex amplitudes $E_1(\omega)$, $E_2(\omega)$ and $E_3(\omega)$, incident on a slab of material with the geometry of Figure 4.9. Waves 1 and 2, with opposite wavevectors ($\mathbf{k}_1 = -\mathbf{k}_2$), are the so-called *pump waves* and are, generally, much more intense than the *signal wave* 3. For simplicity, a scalar formation will be used. The nonlinear interaction induces a nonlinear polarization at the same frequency ω, given by

$$P_4(\omega) = \chi^{3)}(\omega : \omega, -\omega, \omega)E_1 E_2^* E_3 \tag{4.111}$$

which acts as a source term in the wave equation describing the generation of a fourth plane wave $E_4(\omega)$. The phase-matching condition requires that $\mathbf{k}_4 = -\mathbf{k}_3$. On the other hand, the *undepleted approximation* can be used if the pump waves are much more intense than the signal wave. Then, the complex amplitude of the pump waves can be taken as constant and from (4.111) one may write

$$P_4(\omega) \approx E_3^*(\omega) \tag{4.112}$$

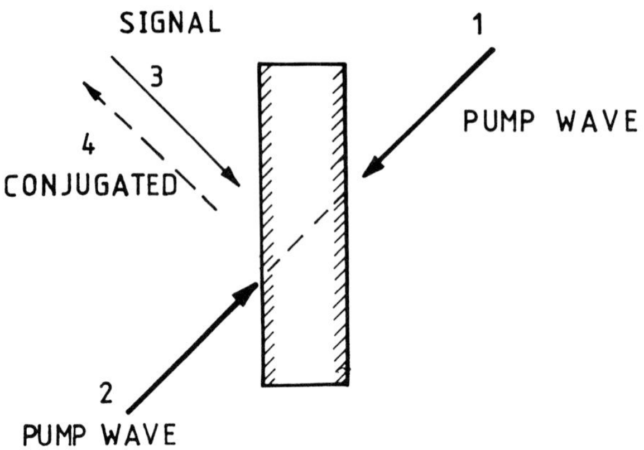

Figure 4.9 *Geometrical arrangement of beams in a degenerated four-wave mixing experiment.* ***k*** *conservation is ensured.*

Then, from a comparison between the nonlinear wave equation as applied to waves 3 and 4, with source terms $P_3 \approx E_3$ and $P_4 \approx E_3^*$, respectively, one arrives at

$$E_4(\omega) \approx E_3^*(\omega) \tag{4.113}$$

i.e. the generated wave 4 has a complex amplitude that is (except for a constant factor) *conjugated* of that of the *signal wave* 3. One may write in detail

$$E_3 = A_3\,\mathrm{e}^{i(k_3 r+\phi)}\,\mathrm{e}^{-i\omega t} \tag{4.114a}$$

$$E_4 = A_4\,\mathrm{e}^{-i(k_3 r+\phi)}\,\mathrm{e}^{-i\omega t} \tag{4.114b}$$

explicitly indicating that the phase factor of wave 4 is just the conjugate of that of wave 3. Alternatively, one can think of wave 4 as being the *time reverse* of 3. This permits us to consider the nonlinear material as a peculiar *phase conjugated mirror* in which incident and reflected waves travel along the same direction but with opposite senses (Figure 4.10).

The process of DFWM can be physically understood as a real-time holography procedure [36]. The interference of the pump wave 1 and the signal beam 3 writes a refractive index grating in the material through the optical Kerr effect. Then, the other

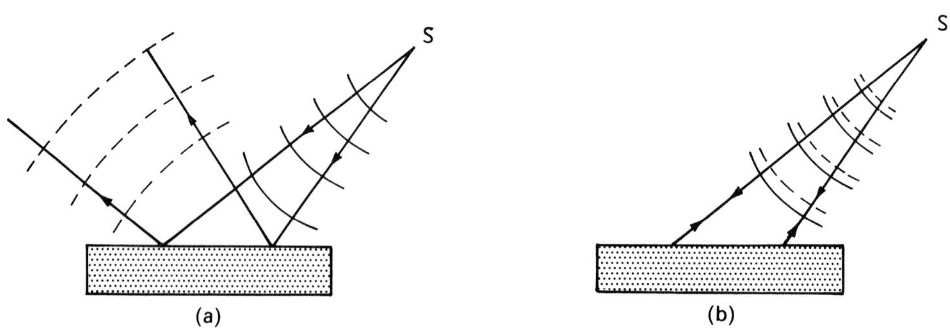

Figure 4.10 *Comparison between the operation of (a) a normal and (b) a phase-conjugated mirror.*

Table 4.4 *Nonresonant second-order susceptibilities* $\chi_{ij}^{2)}(2\omega : \omega, \omega)$ *for some inorganic and organic crystals.*

Crystal	Symmetry	$\chi_{ij}^{2)}$ (pm/V)	Reference
α-SiO_2	32	$\chi_{11}^{2)} = 0.8$ $\chi_{14}^{2)} = 0.018$	[42]
KH_2PO_4 (KDP)	$\bar{4}2m$	$\chi_{14}^{2)} = 1.00$ $\chi_{36}^{2)} = 0.92$	[42]
$LiNbO_3$	$3m$	$\chi_{22}^{2)} = 6.14$ $\chi_{31}^{2)} = -11.7$ $\chi_{33}^{2)} = 81.4$	[42]
$BaTiO_3$	$4mm$	$\chi_{15}^{2)} = -34.4$ $\chi_{31}^{2)} = -36$ $\chi_{33}^{2)} = -13.4$	[42]
β-BaB_2O_4 (BBO)	$3m$	$\chi_{31}^{2)} = 0.16$ $\chi_{22}^{2)} = 3.2$	[43]
LiB_3O_5 (LBO)	$mm2$	$\chi_{33}^{2)} = 0.15$ $\chi_{32}^{2)} = 2.97$ $\chi_{31}^{2)} = 2.75$	[44]
$KTiOP_4$ (KTP)	$mm2$	$\chi_{33}^{2)} = 27.4$ $\chi_{32}^{2)} = 10$ $\chi_{31}^{2)} = 11$ $\chi_{24}^{2)} = 15.2$ $\chi_{15}^{2)} = 12.2$	[45]
$LiIO_3$	$\bar{6}$	$\chi_{33}^{2)} = -10.9$ $\chi_{36}^{2)} = -8.4$	[42]
GaAs	$43m$	$\chi_{14}^{2)} = 280$	[46]
Ag_3AsS_3 (proustite)	$3m$	$\chi_{22}^{2)} = 57$ $\chi_{31}^{2)} = 30$	[42]
Urea	$\bar{4}2m$	$\chi_{14}^{2)} = 4.5$	[47]
mNA	$mm2$	$\chi_{31}^{2)} = 68$ $\chi_{32}^{2)} = 1.7$ $\chi_{33}^{2)} = 17.6$	[47]
MNA	m	$\chi_{11}^{2)} = 210$ $\chi_{12}^{2)} = 32$	[47]

Table 4.4 *Continued.*

Crystal	Symmetry	$\chi_{ij}^{2)}$ (pm/V)	Reference
MAP	2	$\chi_{22}^{2)} = 36.8$	[47]
POM	222	$\chi_{14}^{2)} = 18.4$	[47]
NPP	2	$\chi_{21}^{2)} = 168$	[4]

mNA: meta-nitroaniline.
MNA: 2-methyl-4-nitroaniline.
MAP: methyl-(2,4-dinitro-phenyl)-aminopropanoate.
POM: 3-methyl-4-nitropyridine-1-oxide.
NPP: N-(4-nitrophenyl)-(L)-prolinol.

pump beam 3 is diffracted by this grating at the Bragg (or phase-matching) condition and generates the phase-conjugated wave 4.

Phase-conjugation has very exciting applications [33–35,37,38] in a variety of fields, e.g. adaptive optics, coherent optical processing and computing, and laser technology. A more detailed analysis of the process in some peculiar nonlinear materials, known as *photorefractive materials*, will be presented in Chapter 10 and some applications discussed in Chapter 11.

4.19. Experimental methods

There are a variety of methods [3,4,39–41] to determine the elements of the susceptibility tensor. If one is interested in every component, single crystals and several experiments with different light polarizations and crystal orientations are generally required. Absolute measurements are very difficult because the analysis of experiments requires a precise knowledge of laser mode profile, beam focusing and birefringence *walk-off*. In many cases, absolute values are not the aim and one looks for a comparative evaluation with regard to some well-characterized standards, such as quartz, KDP, GaAs, $LiIO_3$ or urea. This simplifies the experimental procedures and may allow rapid classification of available materials according to their nonlinear performances. Measurements of the nonlinear susceptibilities yield the values of the corresponding electronic electrooptic coefficients through the application of relations (4.70). Therefore it is opportune for the purposes of this book to review summarily some important techniques to measure $\chi_{ijk}^{2)}$ and $\chi_{ijkl}^{3)}$.

4.19.1. Second-order susceptibilities

The $\chi_{ijk}^{2)}(2\omega : \omega, \omega)$ or d_{ij} susceptibilities, applying to second-harmonic generation, are most often measured. For single crystals, the usual method involves monitoring of the second-harmonic intensity oscillations either on rotating a crystal slab [6] or shifting a wedged sample in the fundamental laser beam [41,42]. These methods can be adapted to determine the relative sign of the various components of the susceptibility tensor. Measured values of $\chi_{rs}^{2)}$ components for a few relevant electrooptic crystals are listed in Table 4.4. Data from references [42–47] have been used. Other data are given in refs. [48–51]. They should be consulted for more extensive information.

Very often, absolute measurements are not required. Then, an easy and useful technique for a *first* nonlinear test on a novel material can be applied and is described next.

4.19.2. Powder method

A method for a qualitative evaluation of the second-order nonlinear response of a powder sample has been developed by Kurz and Perry [52]. It is very useful to provide a ranking of materials according to their nonlinear efficiency, without requiring a bulk crystalline sample, which is often difficult or even impossible to obtain. In particular, the method permits us to differentiate between phase-matchable (for some incident direction) and non-phase-matchable materials. For the first case, the second-harmonic intensity $I_{2\omega}$ is given approximately by the formula

$$I_{2\omega} = C\langle d_{ij}\rangle I_{\omega}^{2} L \frac{\bar{l}_c^2}{\bar{r}} \sin^2\left(\frac{\pi\bar{r}}{2\bar{l}_c}\right) \tag{4.115}$$

where $\langle d_{ijk}\rangle$ stands for an angular average of the effective d-tensor component along the incidence direction, and C a coefficient including geometrical constants and refractive indices at ω and 2ω. $\bar{l}_c$ and $\bar{r}$ are the average coherence length and particle size, respectively. The dependence (4.115) is illustrated in Figure 4.11. For $\bar{r} \ll \bar{l}_c$, $I_{2\omega} \approx \bar{r}L$, yielding the initial linearly increasing region of the curve. On the other hand, for $\bar{r} \gg \bar{l}_c$ the $1/\bar{r}$ dependence dominates and leads to $I_{2\omega}$ approaching 0 asymptotically (the oscillations associated to the $\sin^2$ factor are, generally, averaged out due to the size distribution).

For phase-matchable materials, the shape of the $I_{2\omega}(\bar{r})$ curve is markedly different from that corresponding to the above case (see Figure 4.11). In particular, a constant $I_{2\omega}$ is predicted for $\bar{r} \gg \bar{l}_c$. This marked difference allows us to sort out phase-matchable materials if the full dependence on $\bar{r}$ is experimentally determined.

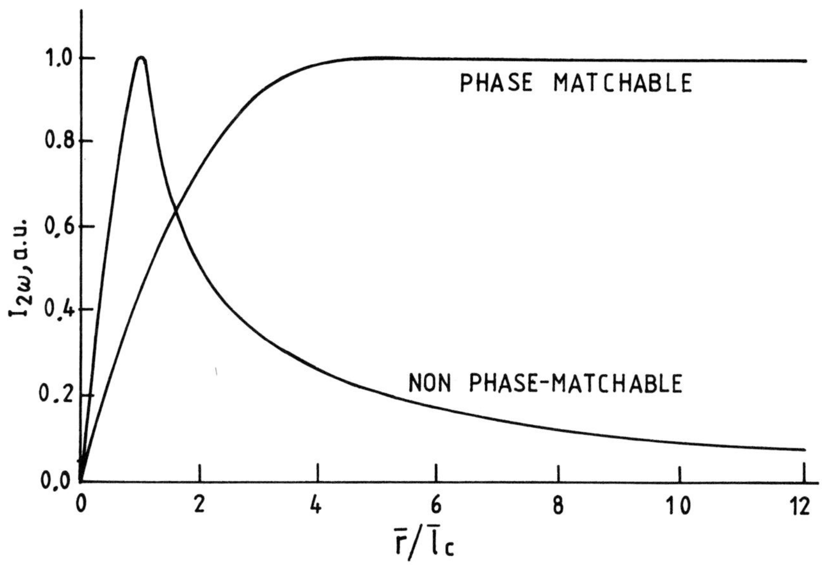

Figure 4.11 ***Second-harmonic intensity from a powder as a function of the average grain radius $\bar{r}$ (normalized to the average coherence length $\bar{l}_c$) [36].***

Table 4.5 *Second-harmonic efficiencies of powder samples relative to quartz [27,51,52].*

Material	Efficiency
$LiIO_3$	300
$LiNbO_3$	600
$KNbO_3$	2000
Urea	20
POM	280
MAP	200
mNA	800
MNA	1600
DAN	2300
NPP	3000

The technique has been generally used by comparing $I_{2\omega}$, generated by the material, to that obtained from a quartz or urea powdered sample taken as a reference. The region $\bar{r} \approx 100\,\mu m$ is typically explored for evaluation and ranking. Second-harmonic efficiencies, relative to quartz, are listed in Table 4.5 for a number of powdered materials ($\lambda = 1.06\,\mu m$). Data quoted in references [27,51,52] have been used.

4.19.3. Third-order susceptibilities

A number of techniques are available to measure third-order susceptibilities [4]. They are associated with the different nonlinear phenomena mentioned in previous sections: third-harmonic generation, EFISH, optical Kerr effect and degenerated four-wave mixing (DFWM). The third-harmonic generation exclusively probes the electronic nonlinearity. On the other hand, DFWM senses both the electronic and dynamical nonlinearities. EFISH has a particular interest for liquid samples. In fact, the analysis of the EFISH data includes, in addition to the third-order hyperpolarizability γ of the molecules, a contribution associated with the second-order polarizability β [16,17] which is often used for its determination.

Recently, a rather simple single-beam method, called *Z-scan*, has been put forward [53]. It is based on the *self-focusing* effect caused by a third-order nonlinearity. However, the method demands a very high gaussian quality of the input beam and does not yield any information on the time response of the nonlinear mechanisms.

Experimental data on third-order nonlinear susceptibilities are very dependent on the measuring technique and the physical conditions. The interested reader should consult research papers for specific information. However, $\chi^{3)}_{ijkl}$ values range from $10^{-23}\,m^2/V^2$ (10^{-15} esu) for ordinary glass and $10^{-21}\,m^2/V^2$ (10^{-13} esu) for CS_2, to $10^{-18}\,m^2/V^2$ (10^{-10} esu) for polyacetylene and up to $10^{-16}\,m^2/V^2$ (10^{-8} esu) for semiconductor-doped glasses and even $10^{-10}\,m^2/V^2$ (10^{-2} esu) for semiconductor multiple quantum wells (operating at the lowest exciton peak).

References

1. N. Bloembergen, *Nonlinear Optics*, W. A. Benjamin, New York (1965).
2. Harper, P. G. and Wherrett, B. S. (eds), *Nonlinear Optics*, Academic Press, New York (1977).

3. Y. R. Shen, *The Principles of Nonlinear Optics*, John Wiley, New York (1984).
4. P. N. Prasad and D. J. Williams, *Introduction to Nonlinear Optical Effects in Molecules and Polymers*, John Wiley, New York (1991).
5. P. A. Franklen, A. E. Hill, C. W. Peters and G. Weinreich, *Phys. Rev.* **7**, 118 (1961).
6. P. D. Maker, R. W. Terhune, M. Nisenoff and C. M. Savage, *Phys. Rev. Lett.* **8**, 21 (1962).
7. A. Ashkin, G. D. Boyd and J. M. Dziedzic, *Phys. Rev. Lett.* **11**, 14 (1963).
8. S. Somekh and A. Yariv, *Opt. Commun.* **6**, 301 (1972).
9. G. A. Magle, M. M. Fejer and R. L. Byer, *Appl. Phys. Lett.* **56**, 108 (1990).
10. J. D. Bierlein, D. B. Laubacher, J. B. Brown and R. J. van der Poel, *Appl. Phys. Lett.* **56**, 1725 (1990).
11. D. S. Chemla, *Rep. Prog. Phys.* **43**, 1191 (1980).
12. R. C. Miller, *Appl. Phys. Lett.* **5**, 17 (1964).
13. J. F. Nye, *Physical Properties of Crystals*, Oxford University Press (1964).
14. S. Bhagavantam, *Crystal Symmetry and Physical Properties*, Academic Press, London (1966).
15. D. A. Kleinman, *Phys. Rev.* **126**, 1761 (1962).
16. B. F. Levine and C. G. Bethea, *J. Chem. Phys.* **63**, 2666 (1975).
17. J. L. Oudar, *J. Chem. Phys.* **67**, 446 (1977).
18. R. W. Boyd, J. J. Maki and J. E. Sipe, in *Nonlinear Optics. Fundamentals, Materials and Devices* (ed. S. Miyata), Elsevier, Amsterdam (1992).
19. D. Bedeaux and N. Bloembergen, *Physica* **69**, 57 (1973).
20. S. Adler, *Phys. Rev.* **126**, 413 (1962).
21. A. S. Pine, *Phys. Rev.* **139**, A901 (1966).
22. R. W. Munn, *Mol. Phys.* **64**, 1 (1988).
23. H. A. Lorentz, *The Theory of Electrons*, Teubner, Leipzig (1916).
24. C. G. Darwin, *Proc. R. Soc. (Lond.)* **146A**, 17 (1934).
25. J. Onsager, *J. Am. Chem. Soc.* **58**, 1456 (1936).
26. J. Ducuing, in *Nonlinear Optics* (eds P. G. Harper and B. S. Wherret), Academic Press, New York (1977).
27. Chemla, D. S. and Zyss, J. (eds), *Nonlinear Optical Properties of Organic Molecules and Crystals*, Vols 1 and 2, Academic Press, London (1987).
28. T. K. Yee and T. K. Gustafson, *Phys. Rev.* A**18**, 1597 (1978).
29. A. Unsold, *Z. Phys.* **43**, 388 (1927).
30. J. L. Oudar and D. S. Chemla, *J. Chem. Phys.* **66**, 2664 (1977).
31. N. Bloembergen, *Rev. Mod. Phys.* **54**, 685 (1982).
32. J. Zyss and J. L. Oudar, *Phys. Rev.* A**26**, 2025 (1982).
33. R. A. Fisher (ed), *Optical Phase Conjugation*, Academic Press, New York (1983).
34. B. Y. Zeldovich, N. F. Pilipetsky and V. V. Shkunov, *Principles of Phase Conjugation*, Optical Science Series, Vol. 42, Springer-Verlag, Berlin (1985).
35. D. M. Pepper, *Opt. Engng* **21**, 156 (1982).
36. A. Yariv, *Opt. Commun.* **25**, 23 (1978); *IEEE J. Quantum Electron.*, **QE-14**, 650 (1978).
37. D. M. Pepper, D. A. Rockwell and H. W. Bruesselbach, *Photon. Spectra* **95**, August (1986).
38. A. D. McAulay, *Optical Computer Architectures*, John Wiley, New York (1991).
39. F. Zernike and J. E. Midwinter, *Applied Nonlinear Optics*, Academic Press, New York (1973).
40. J. Jerphagnon and S. K. Kurtz, *J. Appl. Phys.* **41**, 1667 (1970).
41. J. E. Bjorkholm, *IEEE J. Quantum Electron.* **QE-5**, 260 (1969).
42. R. W. Boyd, *Nonlinear Optics*, Academic Press, San Diego (1992).
43. C. L. Tang, W. R. Bosenberg, T. Ukachi, R. J. Lang and C. K. Cheng, *Laser Focus Wld* Sept. 87 (1990).
44. C. Chen, Y. Wu, G. You, R. Li and S. Lin, *J. Opt. Soc. Am.* **B6**, 616 (1989).
45. J. D. Bierlein and H. Vanherzeele, *J. Opt. Soc. Am.* **B6**, 622 (1989).
46. P. Günter, *Ferroelectrics* **75**, 5 (1987).
47. B. K. Nayar and C. S. Winter, *Opt. Quantum Electron.* **22**, 297 (1990).

48. R. L. Byer, in *Nonlinear Optics* (eds P. G. Harper and B. S. Wherret), Academic Press, New York (1977), pp. 47–160.
49. D. N. Nikogosyan, *Sov. J. Quantum Electron.* **7**, 1 (1977).
50. A. Yariv and P. Yeh, *Optical Waves in Crystals*, John Wiley, New York (1984).
51. R. T. Bailey, F. R. Cruickshank, D. Pugh and J. N. Sherwood, *Int. J. Optoelectron.*, **5**, 89 (1990).
52. S. K. Kurtz and T. T. Perry, *J. Appl. Phys.* **39**, 3798 (1968).
53. M. A. Sheik-bahae, A. A. Said and E. W. VanStryland, *Opt. Lett.* **14**, 955 (1989).

Appendix 4A. Derivation of the nonlinear equation for plane waves

One starts from the wave equation for a monochromatic wave $\mathbf{E}(\mathbf{r}, t) = \mathbf{E}(\mathbf{r})\,\mathrm{e}^{-i\omega t}$:

$$\nabla \times \nabla \times \mathbf{E}(\mathbf{r}) = \frac{\omega^2}{c^2}\boldsymbol{\varepsilon}(\omega)\cdot\mathbf{E} + \mu_0\omega^2\mathbf{P}_{\mathrm{NL}} \tag{4A.1}$$

Let us now consider that $\mathbf{E}(\mathbf{r}, t)$ is a plane wave propagating along Z, and polarized in the XZ plane, i.e.

$$\mathbf{E}(\mathbf{r}) = E(z)\,\mathrm{e}^{ikz}\,\mathbf{u}_{\mathrm{E}} = E\,\mathrm{e}^{ikz}(\cos\alpha)\mathbf{u}_X + E\,\mathrm{e}^{ikz}(\sin\alpha)\mathbf{u}_Z \tag{4A.2}$$

α being the angle between $\mathbf{E}$ and the X axis. Wave attenuation is accounted for by the $E(z)$ dependence.

One first uses

$$\nabla \times \nabla \times \mathbf{E} = \nabla(\nabla\cdot\mathbf{E}) - \mathbf{E} \tag{4A.3}$$

Substituting (4A.2) into the right-hand side terms of (4A.3) leads to

$$\nabla(\nabla\cdot\mathbf{E}) = \sin\alpha\,\mathrm{e}^{ikz}\left\{-k^2E + 2ik\frac{\mathrm{d}E}{\mathrm{d}z} + \frac{\mathrm{d}^2E}{\mathrm{d}z^2}\right\}\mathbf{u}_Z \tag{4A.4a}$$

and

$$\begin{aligned}\mathbf{E} = {} & (\cos\alpha)\,\mathrm{e}^{ikz}\left\{-k^2E + 2ik\frac{\mathrm{d}E}{\mathrm{d}z} + \frac{\mathrm{d}^2E}{\mathrm{d}z^2}\right\}\mathbf{u}_X \\ & + (\sin\alpha)\,\mathrm{e}^{ikz}\left\{-k^2E + 2ik\frac{\mathrm{d}E}{\mathrm{d}z} + \frac{\mathrm{d}^2E}{\mathrm{d}z^2}\right\}\mathbf{u}_Z\end{aligned} \tag{4A.4b}$$

and so to

$$\nabla \times \nabla \times \mathbf{E} = -(\cos^2\alpha)\,\mathrm{e}^{ikz}\left\{-k^2E + 2ik\frac{\mathrm{d}E}{\mathrm{d}z} + \frac{\mathrm{d}^2E}{\mathrm{d}z^2}\right\}\mathbf{u}_X \tag{4A.5}$$

Substituting into (4A.1) and multiplying the resulting equation by $\mathbf{u}_{\mathrm{E}}$, one obtains

$$-(\cos^2\alpha)\,\mathrm{e}^{ikz}\left\{-k^2E + 2ik\frac{\mathrm{d}E}{\mathrm{d}z} + \frac{\mathrm{d}^2E}{\mathrm{d}z^2}\right\} = \frac{\omega^2}{c^2}\mathrm{e}\,E\mathbf{u}_{\mathrm{E}}\cdot\boldsymbol{\varepsilon}\cdot\mathbf{u}_{\mathrm{E}} + \mu_0\omega^2\mathbf{P}_{\mathrm{NL}}\cdot\mathbf{u}_{\mathrm{E}} \tag{4A.6}$$

In the linear regime, $\mathbf{P}_{\mathrm{NL}} = 0$ and $\mathrm{d}E/\mathrm{d}z = 0$ (no wave attenuation in a non-lossy medium) and equation (4A.6) reduces to

$$k^2\cos^2\alpha = \frac{\omega^2}{c^2}\mathbf{u}_{\mathrm{E}}\cdot\boldsymbol{\varepsilon}\cdot\mathbf{u}_{\mathrm{E}} \tag{4A.7}$$

Using (4A.7) in (4A.6), this takes the form

$$-\mathrm{e}^{ikz}\cos^2\alpha\left\{2ik\frac{\mathrm{d}E}{\mathrm{d}z} + \frac{\mathrm{d}^2E}{\mathrm{d}z^2}\right\} = \mu_0\omega^2\mathbf{P}_{\mathrm{NL}}\cdot\mathbf{u}_{\mathrm{E}} \tag{4A.8}$$

and finally

$$\frac{\mathrm{d}^2E}{\mathrm{d}z^2} + 2ik\frac{\mathrm{d}E}{\mathrm{d}z} = -\frac{\mu_0\omega^2}{\cos^2\alpha}(\mathbf{P}_{\mathrm{NL}}\cdot\mathbf{u}_{\mathrm{E}})\,\mathrm{e}^{-ikz} \tag{4A.9}$$

which is the nonlinear equation for plane waves.

Appendix 4B. Effective nonlinear coefficients d_{eff} for crystallographic point groups

Crystal class	Wave polarizations*	
	$E_\omega(e)E_{2\omega}(o)$	$E_\omega(e)E_{2\omega}(e)$
4, 4*mm*, 6, 6*mm*	0	$d_{15}\sin\theta$
422, 622	0	0
$\bar{4}$	$\sin 2\theta(d_{14}\cos 2\phi - d_{15}\sin 2\phi)$	$-\sin\theta(d_{14}\sin 2\phi + d_{15}\cos 2\phi)$
$\bar{4}2m$	$d_{14}(\sin 2\theta\cos 2\phi)$	$-d_{14}\sin\theta\sin 2\phi$
$\bar{6}$	$\cos^2\theta(d_{11}\sin 3\phi + d_{22}\cos 3\phi)$	$\cos\theta(d_{11}\cos 3\phi + d_{22}\sin 3\phi)$
$\bar{6}m2$	$d_{22}\cos^2\theta\cos 3\phi$	$d_{22}\cos\theta\sin 3\phi$
3	$\cos^2\theta(d_{11}\sin 3\phi + d_{22}\cos 3\phi)$	$d_{15}\sin\theta + \cos\theta(d_{11}\cos 3\phi + d_{22}\sin 3\phi)$
3*m*	$d_{22}\cos^2\theta\cos 3\phi$	$d_{15}\sin\theta - d_{22}\cos\theta\sin 3\phi$
32	$d_{11}\cos^2\theta\sin 3\phi$	$d_{11}\cos\theta\cos 3\phi$

* e: extraordinary; o: ordinary.

Appendix 4C. Structure of the d_{ij} tensor assuming Kleinmann symmetry

Coordinate axes are taken as indicated in Appendix 3A. n is the number of nonzero independent components.

Triclinic

$1, \bar{1}$

$$\begin{bmatrix} d_{11} & d_{12} & d_{13} & d_{14} & d_{15} & d_{16} \\ d_{21} & d_{22} & d_{23} & d_{24} & d_{25} & d_{26} \\ d_{31} & d_{32} & d_{33} & d_{34} & d_{35} & d_{36} \end{bmatrix}$$

$(n = 10)$

Monoclinic

2

$$\begin{bmatrix} 0 & 0 & 0 & d_{14} & 0 & d_{16} \\ d_{16} & d_{22} & d_{23} & 0 & d_{14} & 0 \\ 0 & 0 & 0 & d_{23} & 0 & d_{14} \end{bmatrix}$$

$(n = 4)$

Monoclinic

m

$$\begin{bmatrix} d_{11} & d_{12} & d_{13} & 0 & d_{15} & 0 \\ 0 & 0 & 0 & d_{24} & 0 & d_{12} \\ d_{15} & d_{24} & d_{33} & 0 & d_{13} & 0 \end{bmatrix}$$

$(n = 6)$

Orthorhombic

222

$$\begin{bmatrix} 0 & 0 & 0 & d_{14} & 0 & 0 \\ 0 & 0 & 0 & 0 & d_{14} & 0 \\ 0 & 0 & 0 & 0 & 0 & d_{14} \end{bmatrix}$$

$(n = 1)$

Orthorhombic

$mm2$

$$\begin{bmatrix} 0 & 0 & 0 & 0 & d_{15} & 0 \\ 0 & 0 & 0 & d_{24} & 0 & 0 \\ d_{15} & d_{24} & d_{33} & 0 & 0 & 0 \end{bmatrix}$$

$(n = 3)$

Tetragonal

4

$$\begin{bmatrix} 0 & 0 & 0 & 0 & d_{15} & 0 \\ 0 & 0 & 0 & d_{15} & 0 & 0 \\ d_{15} & d_{15} & d_{33} & 0 & 0 & 0 \end{bmatrix}$$

$(n = 2)$

Tetragonal

$\bar{4}$

$$\begin{bmatrix} 0 & 0 & 0 & d_{14} & d_{15} & 0 \\ 0 & 0 & 0 & -d_{15} & d_{14} & 0 \\ d_{15} & -d_{15} & 0 & 0 & 0 & d_{14} \end{bmatrix}$$

$(n = 2)$

422

$$\begin{bmatrix} 0 & 0 & 0 & 0 & 0 & 0 \\ 0 & 0 & 0 & 0 & 0 & 0 \\ 0 & 0 & 0 & 0 & 0 & 0 \end{bmatrix}$$

$(n = 0)$

Tetragonal

$4mm$

$$\begin{bmatrix} 0 & 0 & 0 & 0 & d_{15} & 0 \\ 0 & 0 & 0 & d_{15} & 0 & 0 \\ d_{15} & d_{15} & d_{33} & 0 & 0 & 0 \end{bmatrix}$$

$(n = 2)$

$\bar{4}2m$

$$\begin{bmatrix} 0 & 0 & 0 & d_{14} & 0 & 0 \\ 0 & 0 & 0 & 0 & d_{14} & 0 \\ 0 & 0 & 0 & 0 & 0 & d_{14} \end{bmatrix}$$

$(n = 1)$

Trigonal

3

$$\begin{bmatrix} d_{11} & -d_{11} & 0 & 0 & d_{15} & d_{16} \\ -d_{16} & -d_{16} & 0 & d_{15} & 0 & -d_{11} \\ d_{15} & d_{15} & d_{33} & 0 & 0 & 0 \end{bmatrix}$$

$(n = 4)$

32

$$\begin{bmatrix} d_{11} & -d_{11} & 0 & 0 & 0 & 0 \\ 0 & 0 & 0 & 0 & 0 & -d_{11} \\ 0 & 0 & 0 & 0 & 0 & 0 \end{bmatrix}$$

$(n = 1)$

Trigonal

$3m$

$$\begin{bmatrix} 0 & 0 & 0 & 0 & d_{15} & d_{16} \\ d_{16} & -d_{16} & 0 & d_{15} & 0 & 0 \\ d_{15} & d_{15} & d_{33} & 0 & 0 & 0 \end{bmatrix}$$

$(n = 3)$

Hexagonal

6

$$\begin{bmatrix} 0 & 0 & 0 & 0 & d_{15} & 0 \\ 0 & 0 & 0 & d_{15} & 0 & 0 \\ d_{15} & d_{15} & d_{33} & 0 & 0 & 0 \end{bmatrix}$$

$(n = 2)$

Hexagonal

$\bar{6}$

$$\begin{bmatrix} d_{11} & -d_{11} & 0 & 0 & 0 & d_{16} \\ d_{16} & -d_{16} & 0 & 0 & 0 & -d_{11} \\ 0 & 0 & 0 & 0 & 0 & 0 \end{bmatrix}$$

$(n = 2)$

622

$$\begin{bmatrix} 0 & 0 & 0 & 0 & 0 & 0 \\ 0 & 0 & 0 & 0 & 0 & 0 \\ 0 & 0 & 0 & 0 & 0 & 0 \end{bmatrix}$$

$(n = 0)$

Hexagonal

$6mm$

$$\begin{bmatrix} 0 & 0 & 0 & 0 & d_{15} & 0 \\ 0 & 0 & 0 & d_{15} & 0 & 0 \\ d_{15} & d_{15} & d_{33} & 0 & 0 & 0 \end{bmatrix}$$

$(n = 2)$

$\bar{6}m2$

$$\begin{bmatrix} 0 & 0 & 0 & 0 & 0 & d_{16} \\ d_{16} & -d_{16} & 0 & 0 & 0 & 0 \\ 0 & 0 & 0 & 0 & 0 & 0 \end{bmatrix}$$

$(n = 1)$

Cubic

432

$$\begin{bmatrix} 0 & 0 & 0 & 0 & 0 & 0 \\ 0 & 0 & 0 & 0 & 0 & 0 \\ 0 & 0 & 0 & 0 & 0 & 0 \end{bmatrix}$$

$(n = 0)$

$\bar{4}3m, 23$

$$\begin{bmatrix} 0 & 0 & 0 & d_{14} & 0 & 0 \\ 0 & 0 & 0 & 0 & d_{14} & 0 \\ 0 & 0 & 0 & 0 & 0 & d_{14} \end{bmatrix}$$

$(n = 1)$

Axial

∞mm

$$\begin{bmatrix} 0 & 0 & 0 & 0 & d_{15} & 0 \\ 0 & 0 & 0 & d_{15} & 0 & 0 \\ d_{15} & d_{15} & d_{33} & 0 & 0 & 0 \end{bmatrix}$$

$(n = 2)$

Chapter Five

Inorganic Electrooptic Materials

5.1. Introduction

There is a very large variety of useful electrooptic materials, covering a wide range of values for the response parameters: spectral range, electrooptic coefficients, refractive index tensor, response time etc. Therefore, each of them may find specific applications for particular technological functions. However, most commercial devices only use a few common materials, including KDP, $LiNbO_3$, PLZT ceramics and some liquid crystals. This is a consequence of the very stringent requirements for commercial applications. Unfortunately, the size, structural and chemical perfection and homogeneity of many single crystals (particularly organics) do not yet reach the standards demanded. For polycrystalline, ceramic and amorphous materials, a major problem is associated with the lack of adequate reproducibility and reliability. Cost considerations also limit the practical use of some interesting materials. It is, however, expected that recent advances in materials design, preparation and characterization, will increase the above list in the near future offering a much wider range of useful options.

The large span in electrooptic properties derivates from the diversity of operative physical mechanisms. Although several of them may be active simultaneously, it is possible to classify materials in broad groups, wherein a given mechanism is dominant for typical operating conditions. This is mostly determined by the type of bonding and electronic structure. A major division can thus be established between *inorganic materials* with predominantly ionic and covalent bonds, and *organic materials*, presenting weak Van der Waals and hydrogen bonding interactions. In this first chapter on the subject the main features and dominant physical mechanisms for various types of inorganic materials are reviewed and their relevant parameters listed. The following chapter will be concerned with the rapidly developing field of electrooptic organic materials, including liquid crystals already having a substantial market quota. Chapter 9 is devoted to semiconductor nanostructures that can be fabricated as a consequence of the progress in atomic and quantum engineering.

The organization of the chapter is as follows: Sections 5.2 to 5.5 deal with dielectric and ferroelectric single crystals. Sections 5.6 and 5.7 are concerned with electrooptic ceramics, mostly PLZT (Section 5.7). Semiconductors are discussed in Sections 5.8 to 5.11. Finally, Section 5.12 briefly addresses the problem of glasses as Kerr materials. Measured values for the Pockels electrooptic coefficients of relevant dielectric, ceramic and semiconductor

Table 5.1 *Pockels coefficients for dielectric and ferroelectric crystals (pm/V).*

Crystal	Point symmetry	Refractive index	Transparency range (μm)	r^T_{ij}	ϵ^T_{ij}
KH_2PO_4 (KDP) [5]	$\bar{4}2m$	$n_o = 1.5074$ $n_e = 1.4669$	0.2–1.5	$r_{41} = 8.77$ $r_{63} = -10.3$	$\epsilon_{33} = 21$ $\epsilon_{11} = 42$
KD_2PO_4 (KD*P) [5]	$\bar{4}2m$ (F)	$n_o = 1.4931$ $n_e = 1.4582$	0.2–2.1	$r_{41} = 8.8$ $r_{63} = 26.4$	$\epsilon_{33} = 50$
$(NH_4)H_2PO_4$ (ADP) [5]	$4\bar{2}m$ (F)	$n_o = 1.5087$ $n_e = 1.4680$	0.2–1.4	$r_{41} = 25$ $r_{63} = 8.6$	$\epsilon_{33} = 15$ $\epsilon_{11} = 56$
$BaTiO_3$ [6]	$4mm$ (F)	$n_o = 2.437$ $n_e = 2.365$	0.4–4	$r_{33} = 103$ $r_{13} = 14.5$ $r_{42} = 1700$	$\epsilon_{33} = 135$ $\epsilon_{11} = 3700$
$LiNbO_3$ [7]	$3m$ (F)	$n_o = 2.286$ $n_e = 2.200$	0.4–5	$r_{13} = +10.0$ $r_{33} = +32.6$ $r_{22} = +6.8$ $r_{51} = +32.6$	$\epsilon_{33} = 29$ $\epsilon_{11} = 85$
$LiTaO_3$ [2]	$3m$ (F)	$n_o = 2.176$ $n_e = 2.180$	0.4–5	$r_{13} = +8$ $r_{33} = +33$ $r_{51} = +20$ $r_{22} \approx 1$	$\epsilon_{33} = 45$ $\epsilon_{11} = 51$
$KNbO_3$ [8]	$mm2$ (F)	$n_a = 2.279$ $n_b = 2.329$ $n_c = 2.167$	0.35–4	$r_{13} = +28$ $r_{33} = +64$ $r_{42} = 380$ $r_{23} = +1.3$ $r_{51} = 105$	$\epsilon_{33} = 55$ $\epsilon_{22} = 1000$ $\epsilon_{11} = 160$
$KTiOPO_4$ [9]	$mm2$ (F)	$n_a = 1.7367$ $n_b = 1.7395$ $n_c = 1.8305$	0.35–4.4	$r_{13} = +9.5$ $r_{23} = +15.7$ $r_{33} = +36.3$ $r_{51} = 7.3$ $r_{42} = 9.3$	$\epsilon_{33} > 18$ $\epsilon_{22} = 11$ $\epsilon_{11} = 12$
$Sr_{0.6}Ba_{0.4}Nb_2O_6$ (SBN: 60) [10]	$4mm$ (F)	$n_o = 2.3116$ $n_e = 2.2953$	0.37–6	$r_{13} = 37$ $r_{33} = 237$	$\epsilon_{33} = 1000$
$Sr_{0.75}Ba_{0.25}Nb_2O_6$ (SBN: 75) [1]	$4mm$ (F)	$n_o = 2.299$ $n_e = 3.32$	0.35–6	$r_{13} = 67$ $r_{33} = 1340$	$\epsilon_{33} = 3400$
$Ba_2NaNb_5O_{15}$ [2]	$mm2$ (F)	$n_x = 2.326$ $n_y = 2.324$ $n_z = 2.221$	0.45–4.5	$r_{13} = 15$ $r_{23} = 13$ $r_{33} = 48$ $r_{42} = 92$ $r_{51} = 90$	$\epsilon_{33} = 51$ $\epsilon_{22} = 242$ $\epsilon_{11} = 246$
$Bi_{12}SiO_{20}$ (BSO) [11]	$m3$	$n = 2.530$	0.4–6	$r_{41} = 4.1$	$\epsilon = 56$
$Bi_{12}GeO_{20}$ (BGO) [11]	$m3$	$n = 2.530$	0.4–6	$r_{41} = 3.8$	$\epsilon = 39$
$Bi_{12}TiO_{20}$ (BTO) [11]	$m3$	$n = 2.58$	0.4–6	$r_{41} = 5.2$	$\epsilon = 47$
β-BaB_2O_4 (BBO) [12]	$3m$	$n_o = 1.655$ $n_e = 1.5425$	0.19–3.5	$r_{22} = 2.7$	$\epsilon_{11} = 6.7$ $\epsilon_{33} = 8.1$

Table 5.1 *Continued.*

Crystal	Point symmetry	Refractive index	Transparency range (μm)	r_{ij}^{T}	ϵ_{ij}^{T}
α-SiO2 (quartz) [1]	32	$n_o = 1.546$ $n_e = 1.555$	0.18–3.5	$r_{11} = -0.5$ $r_{41} = 0.2$	$\epsilon_{33} = 4$ $\epsilon_{11} = 4$
$LiIO_3$ [1]	6	$n_o = 1.85$ $n_e = 1.72$	0.3–5.5	$r_{33}^{S} = +6.4$ $r_{13}^{S} = +4.1$ $r_{51}^{S} = +3.3$ $r_{42}^{S} = 1.4$	$\epsilon_{33} = 554$ $\epsilon_{11} = 65$

Electrooptic data have been taken from the references indicated against each crystal. Absolute values are quoted except when the sign is explicitly shown. All data refer to *unclamped* values, except for $LiIO_3$ where they correspond to *clamped* experiments. The probing wavelength is 0.633.
(F): ferroelectric phase.

materials are listed in Tables 5.1, 5.6 and 5.8 respectively. Kerr coefficients are given in Tables 5.6 (ceramics) and 5.9 (several isotropic and anisotropic materials). More extensive data can be found in references [1–4].

5.2. Dielectric and ferroelectric single crystals

5.2.1. General considerations

Dielectric or insulating crystals are very interesting for optical applications because of their wide transparency window, covering the visible region (and often part of the near ultraviolet). Moreover, they generally have high cohesive energy and good thermal and mechanical stability. For Pockels activity crystals must be accentric (lacking an inversion centre) and consequently they often present ferroelectric phases. In fact, most useful electrooptic materials are ferroelectric and so they will constitute the major part of this section. Their electrooptic coefficients are typically one order of magnitude higher than those for nonferroelectric materials and show a marked anisotropy related to the presence of a ferroelectric axis. This behaviour can be clearly appreciated from Table 5.1, where r^{T} coefficients for a number of dielectric crystals are listed. A more complete description of physical and electrooptic properties for a few of them will be given in Sections 5.3 to 5.5.

5.2.2. Bond anharmonic polarizability descriptions

The simplest models to understand electrooptic behaviour in dielectric crystals use anharmonic bond polarizabilities. Assuming that the relevant electronic structure can be described in terms of bonds connecting pairs of atoms, one may use the additivity principle to relate the nonlinear susceptibilities and so the electrooptic coefficients to the bond hyperpolarizabilities. Ignoring local field factors, one should write for the second-order susceptibility, in accordance with Section 4.14,

$$\chi_{ijk}^{2)} = \sum_{u} \cos\theta_{il}^{u} \cos\theta_{jm}^{u} \cos\theta_{kn}^{u} \beta_{lmn} \tag{5.1}$$

where the sum extends over all (equivalent) u bonds in the unit volume and θ_{ab} is the angle formed by axis X_a of the laboratory frame with axis x_b of the intrinsic bond frame. β_{lmn}

Table 5.2 *Second-order polarizabilities β of some bonds [13].*

Bond	$\beta_{\parallel}/\epsilon_0$ (m^4/V)	$\beta_{\perp}/\epsilon_0$ (m^4/V)	Crystals
P–O	4.1×10^{-12}	1.2×10^{-12}	KDP
I–O	13.5×10^{-12}	3.8×10^{-12}	$LiIO_3$, HIO_3
Ta–O	19.9×10^{-12}	1.8×10^{-12}	$LiTaO_3$
Nb–O	32×10^{-12}	4.7×10^{-12}	$LiNbO_3$
Ti–O	82.7×10^{-12}	6.2×10^{-12}	$BaTiO_3$

stands for the second-order bond hyperpolarizability tensor referred to the latter frame. This description has been successfully used for organic materials (Chapter 6) and covalent semiconductors (Section 5.4). For the wide gap dielectric and ferroelectric materials considered here (e.g. oxides), the model is not as useful and well supported. However, the description has been sometimes used to find relations among nonlinear susceptibilities, as illustrated when discussing the behaviour of $LiNbO_3$ (Section 5.4.2). On the other hand, one may obtain the bond polarizabilities β_{ijk} from the measured values of the nonlinear susceptibilities (or electronic Pockels coefficients), and so establish some connection between the electronic structure and the electrooptic response. As an example, Table 5.2 shows the second-order polarizabilities $\beta_{\parallel}$ and $\beta_{\perp}$ for several oxide bonds [13], inferred from $\chi^{(2)}_{ijk}$ (or d_{ij}) by using (5.1). The superior performance of niobates and titanates is clearly apparent.

In principle, the model could be further exploited if one is able to compute the bond polarizabilities from first principles or one uses some empirical relationships between bond parameters (e.g. length) and polarizabilities.

5.2.3. Effects related to the ferroelectric transition

The enhanced electrooptic response of ferroelectric crystals is associated with the paraelectric–ferroelectric transition and can be inferred from expression (3.9),

$$r_{ijk} = \epsilon_0 f_{ijl}(\epsilon_{lk} - \delta_{lk}) \tag{5.2}$$

relating the electrooptic coefficients r_{ijk} to the polarization-optic coefficients f_{ijk}. Since f_{ijk} depend smoothly on temperature, it is expected that r_{ijk} should roughly follow the behaviour of the low-frequency dielectric constant ϵ_{ij} and reach high values near the Curie temperature T_c. This singular temperature behaviour for r_{ijk} has been, indeed, experimentally observed.

For displacive ferroelectrics, the rise in the dielectric constants ϵ_{ij} near the phase transition is caused by the strong lattice polarization associated with the ferroelectric soft mode [13]. Therefore, it is expected that the coupling between this polar mode and the electron dynamics will provide a dominant contribution to the r^T_{ijk} and r^S_{ijk} values well above the purely electronic term r^e_{ijk}. Moreover, the high piezoelectric coefficients of the ferroelectric phase will also induce a strong additional enhancement of r^T_{ijk} with regard to r^S_{ijk}.

Experimental values of r^e_{ij}, r^S_{ij} and r^T_{ij} (contracted notation) for several relevant ferroelectric crystals are given in Table 5.3. Some nonferroelectric crystals have been also included for comparison. It appears that r^e_{ij} does not show any large systematic dependence on the material. On the other hand, r^T_{ij} (and r^S_{ij}) is much higher for ferroelectric crystals. As an example, in the oxide crystals r^T_{ij} increases on going from $LiNbO_3$ to $BaTiO_3$, roughly in accordance with the behaviour of the low-frequency dielectric constant. One should also note the influence of P_s and geometrical factors on the measured coefficients.

Table 5.3 *Electrooptic coefficients r^T, r^S and r^e (pm/V).*

Crystal	r^T	r^S	r^e
$LiNbO_3$ [7]	$r_{33} = +32.6$	+30.8	5.34
	$r_{22} = 6.8$	34	0.62
	$r_{13} = +10.0$	+8.6	0.92
	$r_{51} = 32.6$	+28	—
$BaTiO_3$ [6]	$r_{42} = 1640$	820	+2.3
	$r_{33} = 80$	28	+0.85
	$r_{13} = 24$	8	+2.3
$KNbO_3$ [8]	$r_{42} = 380$	—	2.7
	$r_{33} = 64$	25	+5.0
	$r_{13} = 28$	10	+2.7
KDP [2]	$r_{63} = 10.5$	8.8	−0.38
	$r_{41} = +8.6$	—	−0.38
SiO_2 [2]	$r_{11} = -0.5$	0.23	−0.26
	$r_{41} = 0.20$	—	0
ZnS ($\bar{4}3m$) [2]	$r_{41} = +1.5$	−2	−4.0
GaAs [14]	$r_{41} = -1.17$	−1.33	−4.2

More insight into the electrooptic response of ferroelectric crystals can be obtained from a simple approach relating the polarization-optic response in the ferroelectric and paraelectric phases. In the centrosymmetrical paraelectric phase, the second-order (Kerr) effect is written

$$\Delta\left(\frac{1}{n^2}\right)_{ij} = g_{ijkl} P_k P_l \tag{5.3}$$

P_k and P_l being the field-induced polarizations in such a phase. The ferroelectric accentric phase, obtained below T_c, can be considered equivalent to the paraelectric phase after imposing a bias polarization $\mathbf{P}_s$ (a single ferroelectric domain is assumed). One may then apply formula (5.3) in the form

$$\begin{aligned}\Delta\left(\frac{1}{n^2}\right) &= g_{ijkl}(P_{s,k} + P_k)(P_{s,l} + P_l) \\ &= (2g_{ijkl} P_s \cos\theta_l) P_k + g_{ijkl} P_k P_l + g_{ijkl} P_s^2 \cos\theta_k \cos\theta_l\end{aligned} \tag{5.4}$$

where P_q are now the field-induced polarizations in the ferroelectric phase and θ_q the angles formed by $\mathbf{P}_s$ with the X_q coordinate axis. Use has been made of the symmetry property $g_{ijkl} = g_{ijlk}$. The first term on the right-side of equation (5.4) can be reinterpreted as a Pockels effect governed by the polarization-optic coefficients f_{ijk},

$$f_{ijk} = g_{ijk}^{\text{eff}} P_s \tag{5.5}$$

with

$$g_{ijk}^{\text{eff}} = 2g_{ijkl} \cos\theta_l \tag{5.6}$$

which are determined by the g-tensor coefficients in the paraelectric phase and geometrical factors dependent on the direction of the spontaneous polarization. For the particular case of $\mathbf{P}_s$ pointing in the Z direction

$$g_{ijk}^{\text{eff}} = 2g_{ij3k} \tag{5.7}$$

and

$$f_{ijk} = 2g_{ij3k}P_s \tag{5.8}$$

Using (5.2) one finally obtains

$$r_{ijk} = 2\epsilon_0 g_{ij3l}P_s(\epsilon_{lk} - \delta_{lk}) \tag{5.9}$$

Expressions (5.8) and (5.9) establish a proportionality relationship between r_{ijk} (and f_{ijk}) and the spontaneous polarization P_s. Another relevant feature of these formulae is that they account for the anisotropy in the f and r coefficients, which is related to the specific orientation of $\mathbf{P}_s$.

5.3. KDP-family crystals

KH_2PO_4 (KDP) is a hydrogen-bonded ferroelectric material with interesting electrooptic applications [5,15]. Large homogeneous crystals, with good optical quality, can be easily grown at a low cost. Also, they have a large forbidden gap, providing a very wide window for optical work. However, they are soluble in water and fragile. In addition to KDP, much research attention has been paid to other members of the same structural family, e.g. $(NH_4)H_2PO_4$ (ADP), RbH_2PO_4 (RDP) and KH_2AsO_4 (KDA). Their deuterated versions (where H is replaced by D, generally designated with an asterisk, e.g. KD*P, AD*P) also present interesting properties.

Structure

The crystal symmetry of KDP (KH_2PO_4) is tetragonal (point group $\bar{4}2m$) at room temperature [15]. Therefore it lacks inversion symmetry although it is not ferroelectric. At $T = 123\,\text{K}$ it experiences a transition into an orthorhombic ferroelectric phase (point group mm^2), due to a shear along the c-axis. The structure contains octahedral phosphate groups PO_4 connected by hydrogen bonds, as illustrated in Figure 5.1. Protons are in the lines connecting pairs of oxygen atoms on neighbouring octahedra, shifted from the centre towards one of the oxygens. In the ferroelectric phase only two protons are closer to each phosphate group, either to the lower or upper oxygens, giving rise to a dipole moment pointing in the $+c$ or $-c$ direction respectively.

Electrooptic behaviour

Since the KDP structure lacks inversion symmetry it presents a Pockels effect in both the ferroelectric (low temperature) and the paraelectric (RT) phases [5]. The electrooptic and nonlinear coefficients of the KDP-family members show a smaller dispersion of experimental values in comparison to other electrooptic crystals, such as $LiNbO_3$ and $LiIO_3$. This reproducibility has made them useful reference materials for the characterization of electrooptic materials.

Electrooptic coefficients are very similar for all crystals of the family. However, r_{36}^{T} is markedly enhanced on deuteration, as illustrated in Table 5.1 for KDP. Between room

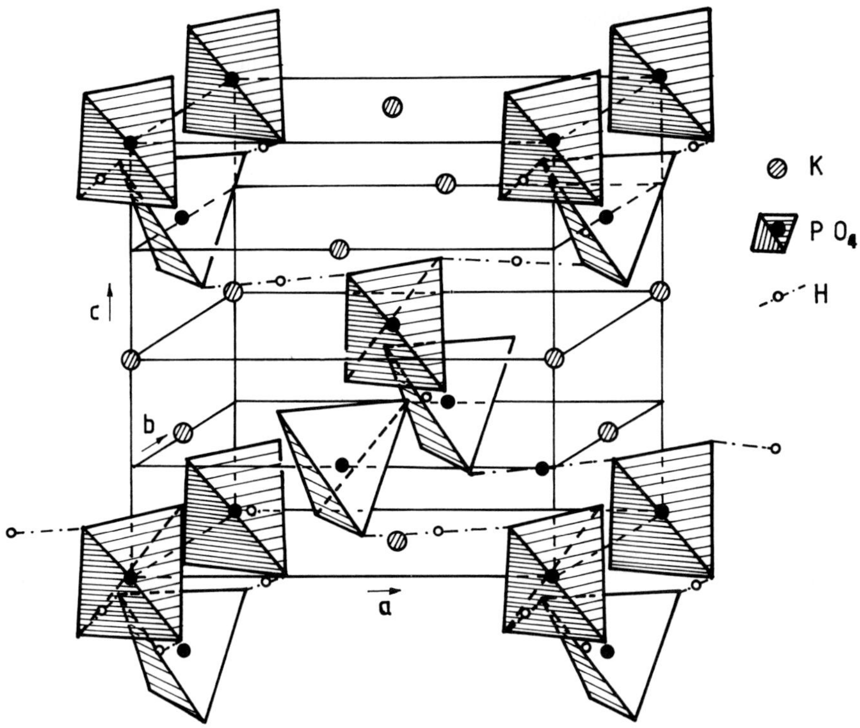

Figure 5.1 *Structure of KDP showing the PO_4 octahedra and H bonds [16].*

temperature and the Curie point, r_{36}^{T} follows the behaviour of the static dielectric constant and obeys the Curie–Weiss law.

5.4. Oxide ferroelectrics

Oxide ferroelectrics include some of the most efficient electrooptic materials, such as $BaTiO_3$, $LiNbO_3$ and KTP. Crystal structure is in all cases related to the basic perovskite structure. We will exclusively consider in some detail the above quoted materials due to their particular relevance. Some physical properties of those and other useful ferroelectric oxides are given in Table 5.4.

In addition to pure compounds, interesting mixed oxides are grown with properties that can be controlled by adjusting the composition. One should mention KTN ($KTaO_3 + KNbO_3$) and SBN ($SrNb_2O_6 + BaNb_2O_6$). For KTN [17], the Curie temperature T_c lies very close to room temperature, and shifts from 20 °C for [Nb]/[K] = 0.37 to 40 °C for [Nb]/[K] = 0.40. As to BSN, T_c goes from 60 °C ([Ba]/[Sr] = 1/3) to 205 °C ([Ba]/[Sr] = 3) [18].

Although most work has been carried out on bulk single crystals, there is a growing interest in thin-film deposition by a variety of techniques, stimulated by the enormous effort devoted to high T_c superconductors [19]. This would permit integration of the electrooptic devices with the microelectronics technologies. On the other hand, an active research is also being carried out on the preparation of single crystal fibres, particularly by *laser-heated pedestal growth* (LHPG) [20].

Table 5.4 *Physical parameters of ferroelectric oxides.*

Material	Phases and transition temperatures (°C)	Melting point (°C)	$\mathscr{E}_g$ (eV)
$BaTiO_3$	Rho.—Ort.—Tet.—Cub.—Hex. −90 5 120 1460	1.618	3.26
$KNbO_3$	Rho.—Ort.—Tet.—Cub. −10 225 435	1.050	3.2
$LiNbO_3$	Tri.—Tri. 1210	1.260	3.8
$LiTaO_3$	Tri.—Tri. 665	1.650	3.8
KTP	Ort. (*mm*2)—Ort. (*mmm*) 936	1.150	3.6
$Ba_2NaNb_3O_{15}$	Ort.—Tet.—Cub. 300 560	1.440	2.8

$\mathscr{E}_g$ is the energy gap at room temperature.
Rho.: rhombic; ort.: orthorhombic; tet.: tetragonal; cub.: cubic; hex.: hexagonal; tri.: trigonal.

5.4.1. $BaTiO_3$

Structure

This is the most thoroughly studied ferroelectric material [5–15,17–22]. Above the Curie temperature ($T_c = 120°C$) it presents the cubic *perovskite* structure, with the unit cell shown in Figure 5.2. Note the TiO_6 octahedron inside the cube of Ba atoms. Below T_c the crystal is tetragonal down to 5 °C, where it becomes monoclinic. At temperatures lower than −90 °C the structure is rhombohedral. These ferroelectric phases can be derived from the paraelectric cubic structure by successively deforming it along the <100>, <110> and

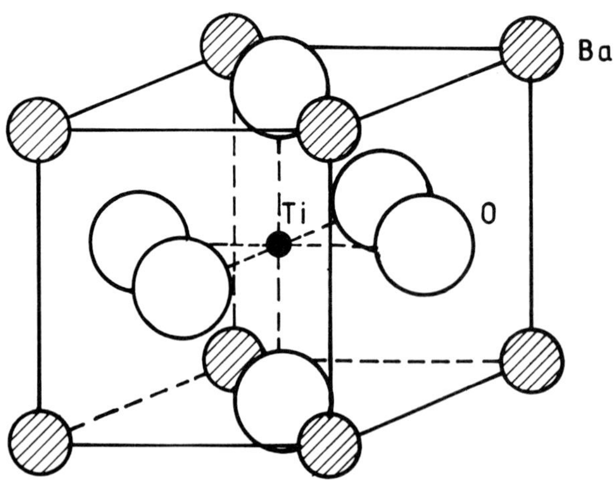

Figure 5.2 *Perovskite structure for the high temperature phase of* $BaTiO_3$.

<111> axes, on lowering the temperature. These deformation axes determine the direction of the spontaneous polarization vector $\mathbf{P}_s$.

At first, crystals were grown by the flux method (Remeika), using $BaTiO_3$ powder dissolved in molten KF. Much better and larger crystals are now grown by the top-seeded solution growth (TSSG).

The dielectric tensor is strongly anisotropic below the paraelectric–ferroelectric transition at 120 °C (4*mm* phase) and $\epsilon_{11} = \epsilon_{22} \gg \epsilon_{33}$. This difference can be understood taking into account that in the ferroelectric phase the Ti^{4+} ion has been trapped at one of the two potential wells along the *c*-axis. Therefore, it cannot efficiently respond to an electric field along the *c* direction, whereas its motion along the perpendicular direction is not restricted. On the other hand, ϵ_{11} grows markedly near the tetragonal–orthorhombic transition at 9 °C.

Electrooptic properties

Similarly to the dielectric constant, electrooptic coefficients are strongly anisotropic. This behaviour derives from the relationship (5.2) between electrooptic coefficients and dielectric constant which can be now formulated in detail for $BaTiO_3$. Let us consider a single domain of the tetragonal phase with ferroelectric axis along Z. From (5.8) one obtains for the nonzero coefficients

$$f_{113} = f_{13}\,(\text{contracted notation}) = 2g_{13}P_s \tag{5.10a}$$

$$f_{333} = f_{33} = 2g_{33}P_s \tag{5.10b}$$

$$f_{322} = f_{42} = 2g_{44}P_s \tag{5.10c}$$

and then

$$r_{13} = 2g_{13}\epsilon_0 P_s(\epsilon_{33} - 1) \tag{5.11a}$$

$$r_{33} = 2g_{33}\epsilon_0 P_s(\epsilon_{33} - 1) \tag{5.11b}$$

$$r_{42} = 2g_{44}\epsilon_0 P_s(\epsilon_{22} - 1) \tag{5.11c}$$

Since the lattice polarizability is much higher along the directions perpendicular to $\mathbf{P}_s$, either X or Y, than along $\mathbf{P}_s$ ($\epsilon_{11} = \epsilon_{22} \gg \epsilon_{33}$), one immediately infers from (5.11) that $r_{42} > r_{13}, r_{33}$. This justifies the marked anisotropy in the electrooptic response, depending on the direction of the applied field relative to $\mathbf{P}_s$.

The influence of the growth method on the electrooptic coefficients is a general effect that can be illustrated here for $BaTiO_3$. Most reported data have been obtained with crystals grown by the flux method. However, some recent values are available on solution-grown crystals. A comparison, taken from reference [6], is presented in Table 5.5. The cause for the differences is not clear. It may be due to the different chemical purity of the two types of samples.

The Pockels coefficient r_{42} is strongly temperature dependent near room temperature, in accordance with the resonance of ϵ_{11} at 9 °C. Figure 5.3 shows the reported experimental

Table 5.5 *Influence of growth method on r^T for $BaTiO_3$ [6].*

Growth method	r_{13}^T (pm/V)	r_{33}^T (pm/V)	r_{42}^T (pm/V)
KF-flux	24	80	1640
Top-seeded solution	19.5	97	—

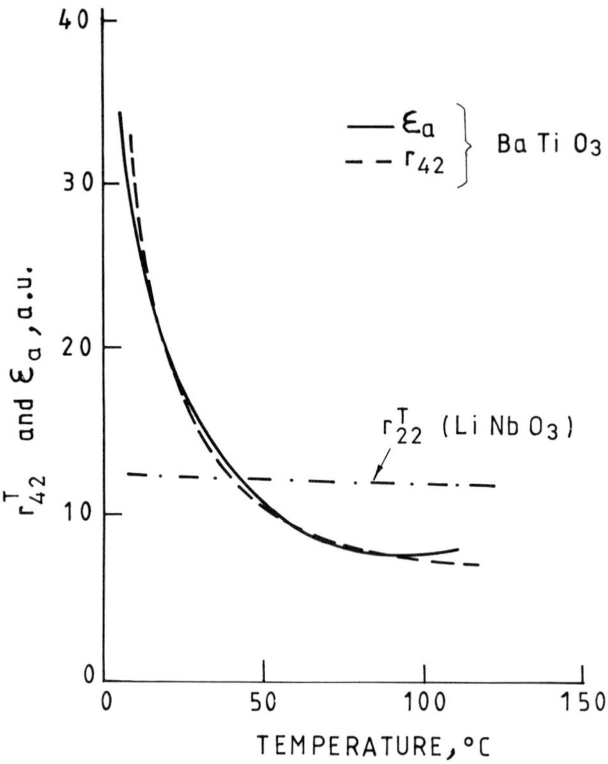

Figure 5.3 *Correlated resonance enhancement of r^T_{42} and ϵ_a near 9 °C for $BaTiO_3$, and slight nonresonant dependence of r^T_{22} on temperature for $LiNbO_3$. Data taken from references [6] and [7].*

behaviour. The large variation of the coefficient with temperature may impair the stability of electrooptic devices.

5.4.2. $LiNbO_3$

$LiNbO_3$ is a key electrooptic material for bulk and waveguide devices [7,23]. It exhibits a high figure of merit $n^3 r_{ij}$, as inferred from Table 5.1, and congruent single crystals of large size and good optical quality can be routinely grown. Unfortunately, it has a low damage threshold due to the photorefractive effect (see Chapter 10), and this limits its use for many nonlinear applications where high light intensities are required.

Structure

The room temperature ferroelectric structure of $LiNbO_3$, Figure 5.4, is related to that of perovskite. It consists of distorted oxygen octahedra sharing faces and forming a planar irregular hexagonal arrangement. The pile-up of the octahedra along the perpendicular direction follows the sequence, Li, Nb, and intrinsic vacancy, according to the nature of the octahedrally coordinated cation. The point-group symmetry is 3 (C_3), with the trigonal *c*-axis along the cation rows. It shows only a very slight deviation from the $3m$ (C_{3v}) group, due to a small differential rotation of the shared triangular faces around the *c*-axis. In fact, most analyses use the approximate $3m$ instead of the correct 3 symmetry.

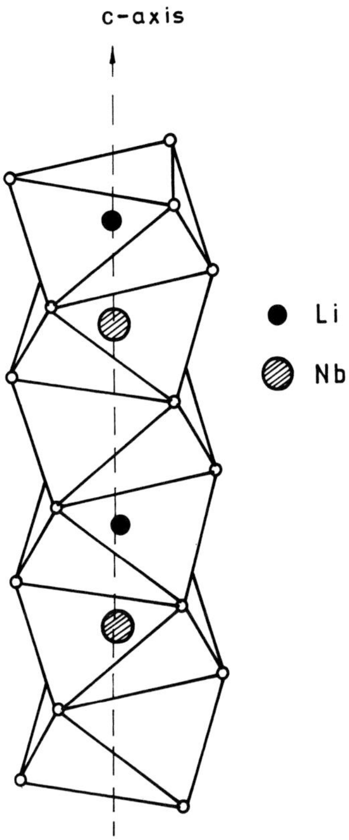

Figure 5.4 *Structure of $LiNbO_3$ showing the stacking of oxygen octahedra along the c-axis*

The high-temperature ($T > 1140\,^{\circ}$C) paraelectric phase becomes D_3 and is centrosymmetrical. Nb^{5+} ions occupy the centre of the corresponding octahedra whereas Li^+ ions lie at the centres of the triangular faces shared between the Li and vacant octahedra.

One important feature of $LiNbO_3$, as of many other oxides, is the strong departure of nominal stoichiometry. In accordance with the phase diagram, $LiNbO_3$ is normally grown with the congruent composition that assures homogeneous crystals. Other stoichiometries have been also grown, although the homogeneity and crystalline quality are doubtful. Therefore, more progress in the growth of stoichiometric crystals and the study of their properties is still needed. Anyhow, it appears that the Li deficiency in congruent crystals is accommodated by means of Nb_{Li} antisites and Nb vacancies [24].

Physical properties

Many physical properties of $LiNbO_3$ markedly depend on stoichiometry, e.g. lattice parameter, Curie temperature, absorption spectra, and luminescence yield [7,25]. As an example, the absorption edge shifts to lower wavelengths for decreasing Li deficiencies and this behaviour has been often used to measure the crystal stoichiometry [26].

Recently, an extensive effort, using a variety of techniques, has been carried out to clarify the nature of the intrinsic defects, to identify the location of impurities in the lattice and to ascertain their role on physical properties. One relevant impurity is Mg, since it greatly inhibits the photorefractive damage for concentrations above a threshold $c = 6\%$ [27]. It also shifts the absorption edge to lower wavelengths, as in stoichiometric crystals. It has been, then, proposed that Mg occupies Li sites and reduces the concentration of Nb_{Li} antisites.

Electrooptic properties

The electrooptic coefficients r_{ij}^{T}, r_{ij}^{S} and r_{ij}^{e}, for congruent $LiNbO_3$ have been given in Table 5.2. The bond anharmonic polarizability model of Section 5.2.2. can be now applied, as an example, to $LiNbO_3$. Relations (5.1) lead to [2],

$$\chi_{31}^{2)} = \{1.10\beta_1 - 1.11\beta_2\}/\epsilon_0 V \tag{5.12a}$$

$$\chi_{33}^{2)} = \{0.64\beta_1 - 1.80\beta_2\}/\epsilon_0 V \tag{5.12b}$$

$$\chi_{22}^{2)} = \{0.40\beta_1 - 1.20\beta_2\}/\epsilon_0 V \tag{5.12c}$$

V being the volume of the unit cell. β_1 and β_2 stand for the longitudinal polarizabilities along the 1.89 Å and 2.11 Å Nb–O bonds, respectively. Transverse polarizabilities have been neglected, and Li–O bonds (mostly ionic) have been ignored. From (5.12) and (4.70) one obtains the following relation among electronic electrooptic coefficients:

$$r_{22}^{e} = -0.18 r_{33}^{e} + 0.47 r_{13}^{e} \tag{5.13}$$

which is in reasonable accordance with the experimental data (absolute values) quoted in Table 5.3.

There has been some controversy in the literature with regard to the effect of stoichiometry on the electrooptic coefficients. Apparently no conclusive evidence of any correlation has been obtained.

The wavelength dispersion behaviour of the electrooptic coefficients is summarized in reference [7]. As an illustrative example r_{22}^{T} ranges from $-6.3\,\text{pm/V}$ at $0.633\,\mu\text{m}$ to $-9.1\,\text{pm/V}$ at $0.436\,\mu\text{m}$.

The temperature dependence of r_{ij}^{T} can be fitted by a law

$$-\log r_{ij}^{T} = aT + b \tag{5.14}$$

a and b being constants tabulated in reference [7]. For r_{22}^{T}, the dependence is included in Figure 5.3, together with that corresponding to r_{42}^{T} for $BaTiO_3$. At variance with this material, no resonant behaviour is observed for $LiNbO_3$, due to the absence of any close phase transition. The relative variation of the coefficient is

$$\frac{(\mathrm{d}r_{22}^{T}/\mathrm{d}T)}{r_{22}^{T}} \approx 4 \times 10^{-4} \tag{5.15}$$

There are very few data [7] on the effect of impurities on the Pockels coefficients. Within experimental error no effect has been found for Mg-doping. Contradictory results have been reported for Ti-doping. A substantial reduction of r_{33} has been measured in waveguides produced by most common methods (Ti in fusion, proton exchange and He implant), although it almost completely recovers after appropriate post-fabrication treatments.

5.4.3. KTP

The crystal structure at room temperature is orthorhombic and belongs to the accentric point group mm^2. It consists of chains of TiO_6 octahedra, linked at two corners, and

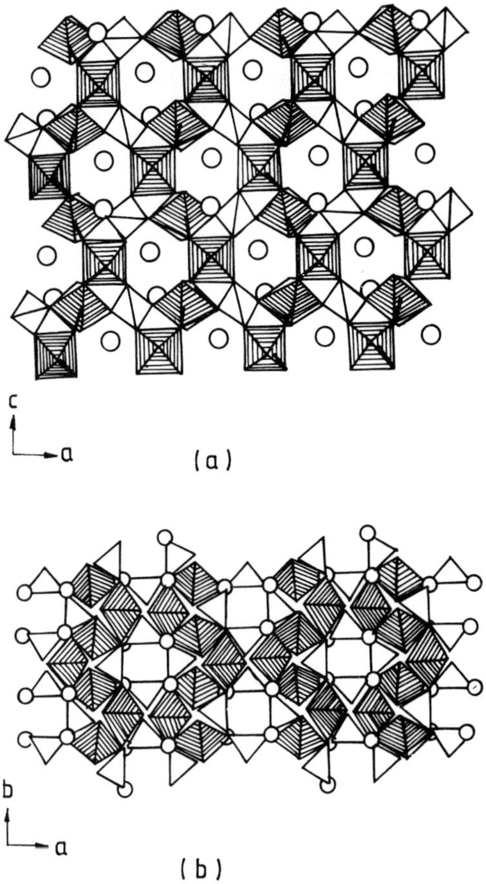

Figure 5.5 *Structure of KTP [28]. (a)* a–c *projection; (b)* a–b *projection. Shaded and open areas stand for Ti and P octahedra, respectively. Open circles are K atoms. Bold lines indicate Ti–O bonds.*

separated by PO_4 octahedra [28] (Figure 5.5). The solid melts and decomposes at 1150 °C. It has been grown by the flux and hydrothermal methods. The Curie temperature is 936 °C. The high electrooptic coefficients and the low dielectric constants make KTP a very promising material for electrooptic waveguide applications. It has a further advantage of a smaller optical damage sensitivity in comparison with $MgO : LiNbO_3$. In fact, demonstration devices have been already fabricated and tested.

5.5. Nonferroelectric oxides

5.5.1. Sillenites

$Bi_{12}SiO_{20}$ (BSO), $Bi_{12}GeO_{20}$ (BGO) and $Bi_{12}TiO_{20}$ (BTO) have the same body-centred structure [11], known as sillenite from the γ-phase of Bi_2O_3. Point-group symmetry is $m3$. Large and homogeneous crystals of BGO and BSO have been grown from the melt by the Czochralski method. Melting temperatures are 930 °C (BGO) and 895 °C (BSO). For BTO the above technique is not adequate and flux and hydrothermal methods have been successfully used.

BSO grows in a nonstoichiometric composition with a Bi excess of about 9%. It gives rise to a shoulder close to the main edge in the absorption spectra, which is responsible of the pale yellow coloration of pure crystals. The absorption shoulder can be eliminated by doping with several impurities, such as Al, Ga and P.

A relevant optical property of BSO and BGO, and at lesser extent BTO, is a very high rotatory power (see Table 2.3). Crystals with both clockwise and counterclockwise rotation senses have been reported.

5.5.2. Borate-family crystals

As a result of a successful crystal engineering effort [29] to search for nonlinear UV materials, two useful borate crystals, β-BaB_2O_4 (BBO) and LiB_3O_5 (LBO), have been grown and are commercially available.

BBO, the β-phase of barium metaborate, is a negative uniaxial crystal [30] with point group 3m (C_{3v}), that can be grown either by a high-temperature flux or a Czochralsky method [31]. It has a wide transparency range extending from 0.190 μm to 2.6 μm. The electronic, as well as the linear and nonlinear optical properties, derive [32] from the basic anionic group $(B_3O_6)^{3-}$, where boron is trigonally coordinated by oxygen (Figure 5.6a). Because of the conjugated π-orbital system, the group has a highly polarizable structure very adequate for nonlinear effects.

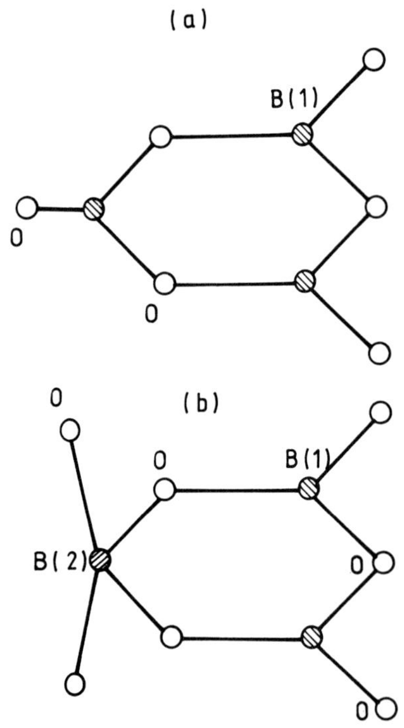

Figure 5.6 *Basic anionic groups in borate crystals: (a) BBO: $(B_3O_6)^{3-}$ with trigonally coordinated boron B(1); (b) LBO: $(B_3O_7)^{5-}$ with boron coordinated trigonally B(1) and tetrahedrally B(2). (From [33]).*

LBO is an optically biaxial crystal [34] with point group mm2 (C_{2v}), grown by high temperature self-flux, flux pulling and top-seeded solution growth. Like BBO it is chemically stable, mechanically robust and not hygroscopic. Its transparency window is even larger than for BBO, extending down to 0.16 μm in the UV range. The basic structural unit is the anionic group $(B_3O_7)^{5-}$, with boron either trigonally or tetrahedrally coordinated to oxygen (Figure 5.6(b)). Theoretical calculations [33] have shown that this structure accounts for the higher gap in comparison with BBO.

Recently, a new member of the borate family, $KBe_2BO_3F_2$ (KBBF), has been grown [33]. It crystallizes in the space group R32, the basic structure consisting of planar BO_3 groups (nonlinear active) and nearly tetrahedral BeO_3F groups. Optically, the crystal is negative uniaxial with a birefringence $\Delta n > 0.07$. The optical absorption edge occurs at 155 nm and first measurements have yielded $d_{11} \approx 2.5 d_{36}$ (KDP). It is a promising UV nonlinear material, although the physical characterization is still at a very early stage.

5.6. Electrooptic ceramics

An electrooptic ceramic [36–38] is a transparent aggregate of ferroelectric single crystal grains, presenting either Pockels or Kerr activity. Grain size can be varied over a wide range of values (typically from approximately 0.5 μm to 40 μm or more) depending on chemical composition and method of fabrication. In the paraelectric phase the structure is centrosymmetrical and has no spontaneous polarization. Below the Curie temperature the grains become ferroelectric through a suitable lattice deformation that generates a polarization vector **P**. However, although individual grains are anisotropic, the material is isotropic at a macroscopic scale, since crystallographic axes are oriented at random and anisotropy is averaged out.

Under application of an electric field, all domains that are favourably oriented with regard to the field grown at the expense of the others. Also domain reorientation occurs. With increasing field the process continues until a final saturation state is reached with a saturation polarization $\mathbf{P}_s$. Then the ceramic becomes anisotropic or poled. The withdrawal of the field does not essentially modify the structure which keeps the $\mathbf{P}_s$ value.

Electrooptic ceramics have a number of advantages over single crystals. They are:

(1) less expensive fabrication procedures;
(2) possibility of preparing a wider range of sizes and shapes;
(3) controlled orientation of the polarization (optic axis) in poled ceramics;
(4) custom design of optical and electrooptical properties by properly adjusting composition and/or fabrication parameters.

On the other hand, ceramics are less transparent than single crystals and homogeneity and reproducibility are often more difficult to guarantee.

As a general rule, the preparation of transparent ceramics is a difficult task, requiring very careful synthesis and fabrication procedures. A number of nonelectrooptic systems have been successfully prepared and investigated (e.g. for light window applications), including Al_2O_3-MgO, Y_2O_3-ThO_2, MgO, CaF_2, CaO, $MgAl_2O_4$ and aluminium oxynitride (ALON). However, most efforts have been devoted so far to solid solutions of the system lead–zirconate–titanate (PZT) and particularly to those with La doping (PLZT ceramics) because of their high transparency and very interesting electrooptic

Table 5.6 *Transparent electrooptic ceramics [37,38].*

Composition	Notation
$(Pb,La)(Zr,Ti)O_3$	PLZT
$(Pb,La)(Hf,Ti)O_3$	PLHT
$(Pb,Ba,Sr)(Zr,Ti)O_3$	PBSZT
$(Pb,Sn)(In,Zr,Ti)O_3$	PSIZT
$(Pb,Ba,La)Nb_2O_6$	PBLN
$(Pb,La)(Zn,Nb,Zr,Ti)O_3$	PBLNZT
$K(Ta,Nb)O_3$	KTN
$Pb(Sc,Nb)O_3$	PSN
$(Pb,La)(Mg,Nb,Zr,Ti)O_3$	PLMNZT
$(Ba,La)(Ti,Nb)O_3$	BLTN
$(Pb,La,Li)(Zr,Ti)O_3$	PLLZT
$(Sr,Ba)Nb_2O_6$	SBN

properties. Some other related electrooptic systems have been also prepared and are listed in Table 5.6. Up-do-date accounts of the state of the art can be found in references [39] and [40].

5.7. PLZT ceramics

5.7.1. Introduction

PLZT are transparent ceramics [36–40], first produced in 1969, resulting from the addition of La, as a chemical modifier, to PZT ceramics made up of $Pb(Zr_yTi_{1-y})O_3$ compositions. They have already found extensive commercial applications as optical shutters, modulators, displays, colour filters, image storage devices and arrays for data processing. Other dopants, e.g. Ba or Sn, are not as successful in achieving optical transparency. PLZT ceramics are preferentially fabricated by hot pressing in a vacuum/oxygen atmosphere that leads to a fully dense material. A number of reasons have been advanced to account for the superior performance of PLZT, such as the high La solubility in the PZT structure (leading to a broad range of homogeneous compositions) and the reduction in the anisotropy of the oxygen octahedra which decreases light scattering losses.

5.7.2. Compositions and microstructure

The general formula for PLZT ceramics is

$$Pb_{1-x}La_x(Zr_yTi_{1-y})O_3 \tag{5.16}$$

The basic structure is that of the perovskite ABO_3, where Pb^{2+} and La^{3+} occupy *A* sites and Zr^{4+} and Ti^{4+} the *B* sites octahedrally coordinated to O. Because La^{3+} has an excess of charge +1 over the Pb^{2+}, charge compensation is needed and supposedly achieved by the introduction of *B*-site vacancies. The fraction of those vacancies required to guarantee electrical neutrality in formula (5.16) is $x/4$. Consequently the complete formula should write

$$Pb_{1-x}La_x(Zr_yTi_{1-y})_{1-(x/4)}\square_{x/4}O_3 \tag{5.17}$$

where $\square$ stands for *B*-site vacancies.

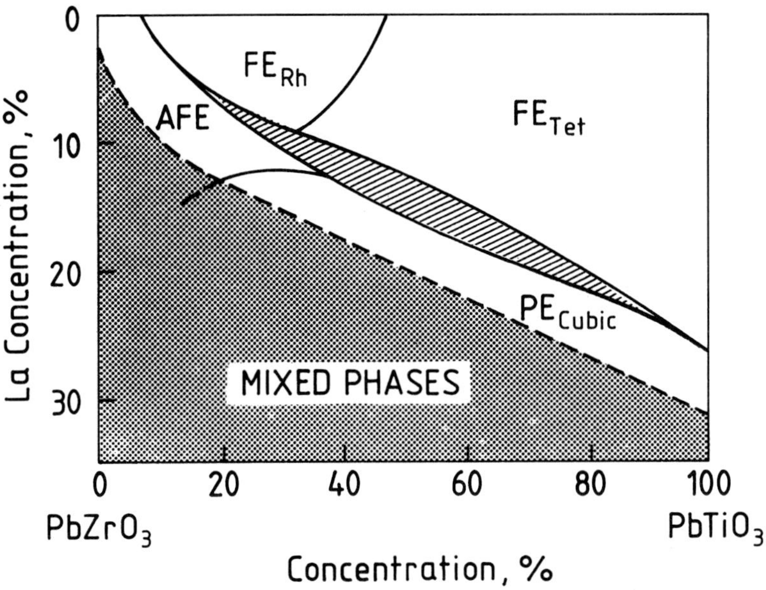

Figure 5.7 ***Room-temperature phase diagram for PLZT [36].***

The room-temperature phase diagram is shown in Figure 5.7. The PZT system changes from an antiferroelectric orthorhombic structure for pure $PbZrO_3$ to a rhombohedral ferroelectic for 8% $PbTiO_3$ and to a tetragonal ferroelectric for 47% $PbTiO_3$. The addition of La gives rise to a region of homogeneous solid solutions for low La concentrations bound by the dashed line in the diagram. Higher La concentrations produce heterogeneous mixtures. The shadowed area separating the paraelectric and antiferroelectric from the ferroelectric phases is a region of coexisting metastable phases and present some very interesting and useful electrooptic properties. The location of a PLZT system in the diagram (i.e. its composition) is specified by three successive numbers indicating respectively, from left to right, the La, Zr and Ti atomic concentrations as a percentage, e.g. 8/65/35.

PLZT ceramics exhibit a highly uniform microstructure consisting of randomly oriented grains with average sizes of a few microns. Densities reach up to 99.9% of the theoretical value.

5.7.3. Dielectric properties

The dielectric properties of PLZT ceramics markedly depend on composition and are intimately connected with the electrooptic behaviour to be described in Section 5.7.5. They are:

(1) high dielectric constants (5000–8000);
(2) low or moderate dielectric losses (a few percent);
(3) high electrical resistivities ($10^{11}\ \Omega \times$ cm);
(4) moderate dielectric breakdown fields (100 kV/cm).

The curves relating the induced polarization **P** to the applied field **E** show a variety of shapes depending on composition. The three basic types are illustrated in Figure 5.8. They include (a) narrow hysteresis loops (low coercive fields) for the rhombohedral ferroelectric

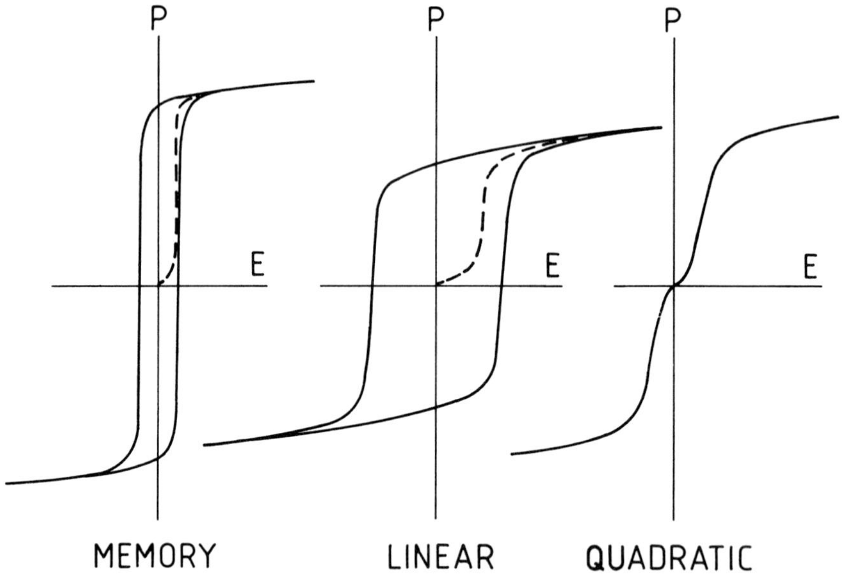

Figure 5.8 *Three basic types of hysteresis loops for PLZT [36].*

compositions (*memory type*), (b) squared hysteresis loops, having a large coercive field, for the tetragonal ferroelectric compositions (*linear type*) and (c) slim loop shapes for those compositions in the shadowed area of the phase diagram between the ferroelectric or paraelectric phases (*quadratic type*). Particularly relevant is the latter behaviour, frequently designated as *relaxor*, which is brought about by an electrically induced change from the paraelectric to the ferroelectric phases. On the other hand, the occurrence of a ferroelectric loop, either (a) or (b), defines a bistable system with two memory states associated with the $+P_s$ and $-P_s$ values. In case (b), the relationship between the saturating polarization P_s and the applied field is often linear (*Pockels behaviour*).

For a given material the coercive field E_c provides an indication of grain size, so that a low E_c implies large grains and *vice versa*.

The cyclic variation of the polarization along a $P(E)$ loop is accompanied by a corresponding strain cycle, caused by the shift of ions in the lattice. For a memory loop,

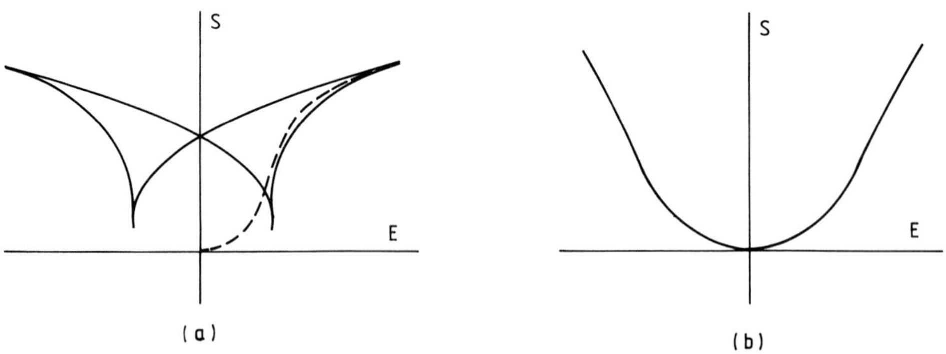

Figure 5.9 *Strain vs electric field cycles corresponding to: (a) memory hysteresis loop; (b) quadratic (slim) hysteresis loop.*

type (a), the strain cycle has a typical butterfly shape shown in Figure 5.9(a). On the other hand, for slim loops, type (c), the strain *vs* field curve is shown in Figure 5.9(b). This behaviour is that expected for an electrostrictive response.

Switching between memory states is a complex phenomenon. Switching times depend on sample size, microstructure and applied field, but are typically of the order of microseconds.

5.7.4. Optical properties

The main property of PLZT is optical transparency, which is satisfactorily achieved for compositions having the La concentration x near or above the boundary between the paraelectric and ferroelectric phases. Transmissions up to 70% in the range 0.5–0.7 μm are obtained (composition 9/65/35). For wavelengths below 0.5 μm strong absorption sets in due to the absorption edge of the material that lies at 370 nm (3.35 eV). However, the optical transmission depends significantly on the defects present in the sample, i.e. porosity and pore size.

The refractive index is very high, $n = 2.5$, accounting for a high reflectivity contribution to the transmission losses.

5.7.5. Electrooptic properties

Electrooptical properties are closely related to the dielectric behaviour described by the $P(E)$ curves. In fact, loop types (a), (b) and (c) give rise to three basic electrooptic responses, respectively termed, *memory*, *linear* and *quadratic*, as indicated in Figure 5.8.

For materials having a wide (high coercive field) hysteresis loop, type (b), one may work at each of the memory states and induce a linear change in P, and so a birefringence Δn, linearly dependent on E (without causing polarization reversal).

In memory ceramics, presenting a high remanent polarization and a low coercive field (type (a)), one operation mode involves switching from one memory state to the opposite by traversing the hysteresis loop. This is called the memory birefringence mode. Another useful configuration, named memory scattering, uses the change in field-induced scattering associated to configurational changes in domain size and orientation.

Particularly interesting is the case of compositions having a slim loop shape $P(E)$ curve. In that case the induced birefringence is a quadratic function of the applied field, i.e. Kerr-type behaviour is observed. In the absence of the field the material relaxes back to the nonpolarized situation. This effect is largely associated to the electrostrictive strains developed on field application. The electrooptic device could operate in transverse or

Table 5.7 *Pockels and Kerr coefficients for electrooptic ceramics [37].*

Ceramic composition	s (m^2/V^2)	r_c (m/V)
PLZT (8.5/65/35)	38.60×10^{-16}	
PLZT (9/65/35)	3.80×10^{-16}	
PLZT (9.5/65/35)	1.50×10^{-16}	
PLZT (10/65/35)	0.80×10^{-16}	
PLZT (12/65/35)	0.16×10^{-16}	
PLZT (8/40/60)		1.02×10^{-10}
PLZT (12/40/60)		1.20×10^{-10}
KTN (65/35)	0.17×10^{-16}	

longitudinal configuration, depending on the relative orientation of the light propagation and applied field directions.

Table 5.7 lists values of the linear (Pockels) and quadratic (Kerr) electrooptic coefficients measured for a number of PLZT compositions. One should note that the quadratic coefficients for PLZT ceramics are exceptionally high and exceed most other crystalline materials. It is also to be remarked that s coefficients exhibit a strong dependence on temperature (e.g. a factor 3.5 from $-40\,°C$ to $+80\,°C$) demanding for temperature compensation in some critical devices.

5.8. Semiconductors

The main difference of semiconductor materials with regard to the above discussed inorganic insulators is their lower energy gap ($<3\,eV$). This feature gives rise to appreciable carrier population in the conduction band and so to a higher electrical conductivity. On the other hand, the small gap makes semiconductors attractive for infrared applications, particularly in the optical communication technologies.

A very intense research effort has been devoted to GaAs. It has the zinc-blende structure with a direct gap at 800 nm, allowing for laser emission at the first window of the optical fibre communications technology. On the other hand, it presents higher carrier mobilities in comparison with silicon. Other relevant semiconductors are InP, InAs and InSb. Much attention has been also paid to II–VI compounds, particularly CdTe and ZnTe. A list of useful physical parameters for these and other semiconductors can be found in reference [41]. From the point of view of nonlinear applications in the infrared range chalcopirite crystals, such as $AgGaS_2$, $ZnGeP_2$ and $CdGeAs_2$, show remarkable performances.

5.8.1. Optical properties [41,42]

Here, we will restrict to photon energies above the vibrational region (medium infrared), which is determined by the electronic structure. There are various types of electronic transitions contributing to the optical response: interband, intraband, impurity–band and impurity–impurity. The relative importance of each kind depends on the particular material, physicochemical state (temperature, doping etc.) and light wavelength.

Interband transitions

Direct interband transitions connect states of equal momentum $\mathbf{k}$ in different bands. For a pure semiconductor, the lowest-energy allowed transition corresponds to the transfer of one electron from the top of the valence band to the minimum of the conduction band. It determines the so-called main absorption edge (M_0 Van-Hove singularity). For parabolic bands the absorption coefficient α presents a square-root dependence on ω

$$\alpha(\omega) \approx (\omega - \omega_g)^{1/2} \tag{5.18}$$

$\mathscr{E}_g = \hbar\omega_g$ being the gap energy. This law has been, in fact, observed for a number of materials.

Symmetry selection rules may forbid such direct transitions (at $\mathbf{k} = 0$). In this case, the absorption coefficient follows the spectral dependence:

$$\alpha(\omega) \approx (\omega - \omega_g)^{3/2} \tag{5.19}$$

In addition to these direct transitions, indirect transitions involving a change in **k** can be induced via the participation of a crystal phonon accounting for momentum conservation. They are markedly sensitive to temperature.

In many materials, the absorption near the edge does not follow any of the above rules. An exponential tail for the absorption *vs* frequency dependence (*Urbach law*) is often observed, e.g. in GaAs.

Other perturbations in the shape of the absorption edge are caused by heavy doping, due to the merging of the donor and/or acceptor levels into the nearby crystal bands.

Intraband transitions

A substantial concentration of carriers in the conduction and valence bands can be induced by high temperature or/and heavy doping. Therefore, phonon-assisted transitions between states of the same band can be induced by photon absorption. This free carrier absorption can be described by a classical Drude model that predicts a λ^2 spectral dependence. A more rigorous theory distinguishes scattering of electrons by acoustic modes, optical modes, or impurities, and leads to different potential laws λ^n for the wavelength dependence.

Impurity transitions

An optical transition between a filled trap and the conduction band manifests itself as a shoulder on the main absorption edge, with a threshold or minimum energy $\mathscr{E}_\mathrm{D}$ representing the optical ionization energy of the trap. A similar effect is associated to valence band–acceptor transitions, showing a low-energy edge at $\mathscr{E}_\mathrm{A}$ (separation of the acceptor level from the top of the valence band).

Donor–acceptor transitions are also possible, the transition energy being given by

$$\hbar\omega = \mathscr{E}_\mathrm{g} - \mathscr{E}_\mathrm{D} - \mathscr{E}_\mathrm{A} + \frac{q^2}{4\pi\epsilon_0\epsilon R} \tag{5.20}$$

where R is the donor–acceptor distance. Therefore, different frequencies will appear in the absorption (and luminescence) spectra corresponding to various pair separations.

Excitons

So far, the Coulomb interaction between the electron promoted to the higher excited level and the hole left in the initial state has been ignored. In fact, the electron–hole attractive potential causes the formation of a bound pair or exciton. For a direct gap semiconductor, the energy of the exciton ground state with regard to the top of the valence band is

$$\mathscr{E}_1 = \mathscr{E}_\mathrm{g} - \mathscr{E}_\mathrm{B} \tag{5.21}$$

$\mathscr{E}_\mathrm{B}$ being the binding energy.

The exciton may also have higher excited states well separated from the conduction band. For loose or Wannier excitons these states constitute a series labelled by an integer number with energies linearly dependent on $1/n^2$ in accordance with a simple hydrogenic model.

Excitons are often easily recognized in the low temperature absorption spectra as strong and sharp lines just below the edge. On increasing temperature, broadening of the excitonic levels takes place because of thermal ionization and the peaks in the spectra become broader

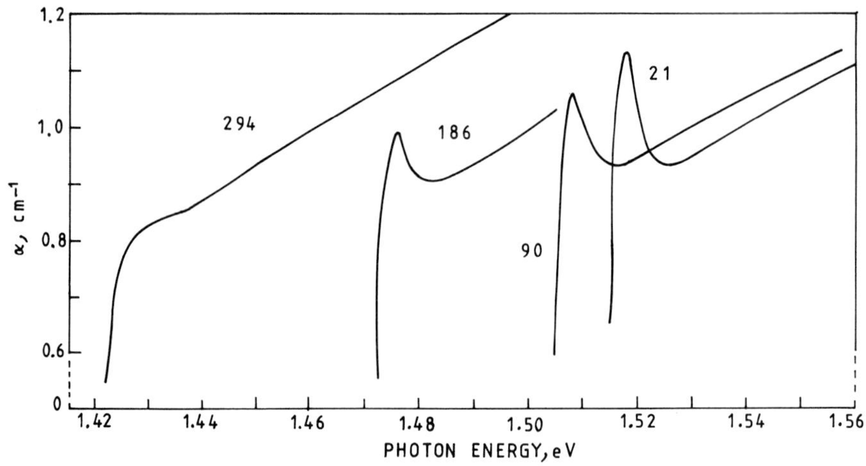

Figure 5.10 *Absorption spectra of GaAs, showing the broadening of the exciton peak with rising temperature in degrees Kelvin [43].*

until they finally merge into the interband background. This behaviour is illustrated in Figure 5.10 for GaAs.

Excitons can be trapped at impurities as well as form bound pairs or exciton molecules (*biexcitons*) or even condense into *electron–hole drops*.

5.9. Electrooptic behaviour of semiconductors

The effect of an electric field on the optical parameters of a crystalline semiconductor arises from several different contributions, associated with the various light excitations, namely (a) interband transitions, (b) exciton effects, (c) free carriers (intraband transitions) and (d) ionic or lattice relaxations. Their relative importance depends on the particular material and excitation wavelength. In fact, those mechanisms are in principle, not different from those operating in any inorganic dielectric (e.g. oxides), except for the role played by the smaller energy gap. This introduces a relatively high thermal free carrier concentration and gives rise to an appreciable intraband contribution to the third-order nonlinear response, particularly in low-gap (InSb, InAs, CdHgTe, etc.) or heavily doped semiconductors.

The dominant electronic contribution to the Pockels coefficients is associated with the transitions from the valence to the conduction band, and it is often described by a bond anharmonic model. Moreover, one should take into account the additional component due to electron coupling to vibrational modes to account for r_{ij}^{T} and r_{ij}^{S} (see Table 5.3).

Table 5.8 gives the nonresonant r_{ij}^{T} coefficients (i.e. for photon energies below the bandgap) for a number of useful semiconductor crystals. Other relevant physical parameters, such as energy gap, refractive index and static dielectric constant are also included.

An interesting even-order electrooptic effect also related to the interband transitions appears when the light frequency approaches the fundamental absorption edge. It is caused by light-assisted tunnelling between the valence and conduction bands and is known as the *Franz–Keldysh* effect [46,47]. In this frequency range one should also consider the effect of the electric field on exciton levels and exciton bandwidths.

Table 5.8 *Pockels coefficients of semiconductor crystals (pm/V).*

Crystal	Symmetry	$\mathscr{E}_g$ (eV)	n	r^T (pm/V)	λ (μ)	$\epsilon(0)$
GaAs [44]	$\bar{4}3m$	1.42	3.5	$r_{41} = -1.17$	1.06	13.2
InP [44]	$\bar{4}3m$	1.35	3.29	$r_{41} = -1.32$	1.06	12.6
GaP [44]	$\bar{4}3m$	2.26	3.11	$r_{41} = -1.10$	1.15	12
CdTe [4]	$\bar{4}3m$	1.56	2.77	$r_{41} = -4.5$	1.0	9.4
ZnS [4]	$\bar{4}3m$	3.68	2.47	$r_{41} = -1.9$	0.633	16
ZnSe [4]	$\bar{4}3m$	2.68	2.66	$r_{41} = 2.0$	0.55	9.1
ZnTe [4]	$\bar{4}3m$	2.27	3.1	$r_{41} = 4.45$	0.59	10.1
CdS [45]	$6m$	2.47	2.48	$r_{33} = 3.2$	1.15	10.3
				$r_{13} = 3.1$	1.15	
				$r_{42} = 2.0$	1.15	
$AgGaS_2$ [1]	$\bar{4}2m$	—	$n_o = 2.55$	$r_{63} = 3.0$	0.633	$\epsilon_{11}^S = 10$
			$n_e = 2.50$	$r_{41} = 4.0$	0.633	$\epsilon_{33}^S = 14$
$ZnGeP_2$ [2]	$\bar{4}2m$		$n_o = 3.11$	$r_{63}^e = -2.3$		
			$n_e = 3.15$	$r_{41}^e = -2.3$		

References are indicated against each crystal.

Therefore, in the most general case, the overall nonlinear susceptibility may be written

$$\chi^{NL} = \chi_f^{NL} + \chi_b^{NL} + \chi_{FK}^{NL} + \chi_{ex}^{NL} + \chi_{vm}^{NL} \tag{5.22}$$

where χ_f^{NL}, χ_b^{NL}, χ_{FK}^{NL}, χ_{ex}^{NL} and χ_{vm}^{NL} refer respectively to the free carrier, bound electron, Franz–Keldysh, exciton and vibrational mode mechanisms. They will be now briefly discussed (except for the latter contribution, already mentioned when discussing dielectric materials).

5.10. Physical mechanisms

5.10.1. Bound electrons: bond anharmonic models

The bond anharmonic models, already mentioned for the case of ferroelectric crystals, have been developed in much more detail for semiconductor crystals of the zinc-blende and wurtzite structures illustrated in Figure 5.11. These lattices are the low-temperature (zinc-blende) and high-temperature (wurtzite) crystalline phases of ZnS, made up of tetrahedral covalent bonds. Flytzanis and Ducuing [48] have used a LCAO description of the bond ground state in terms of hybrid sp^3 orbitals and calculated the α and β polarizabilities within the Unsold approximation. The signs obtained from the calculation are correct and the absolute values in reasonable agreement with experiment. One interesting outcome of the model is that α is nearly isotropic, whereas $|\beta_{\parallel}| \gg |\beta_{\perp}|$, i.e. second-order polarizabilities are mostly longitudinal. A simple application of the model is, as for dielectric crystals, to establish relations among nonlinear susceptibility coefficients. As an example, using equation (5.1), together with $\beta_{\perp} = 0$ and assuming that the zinc-blende and

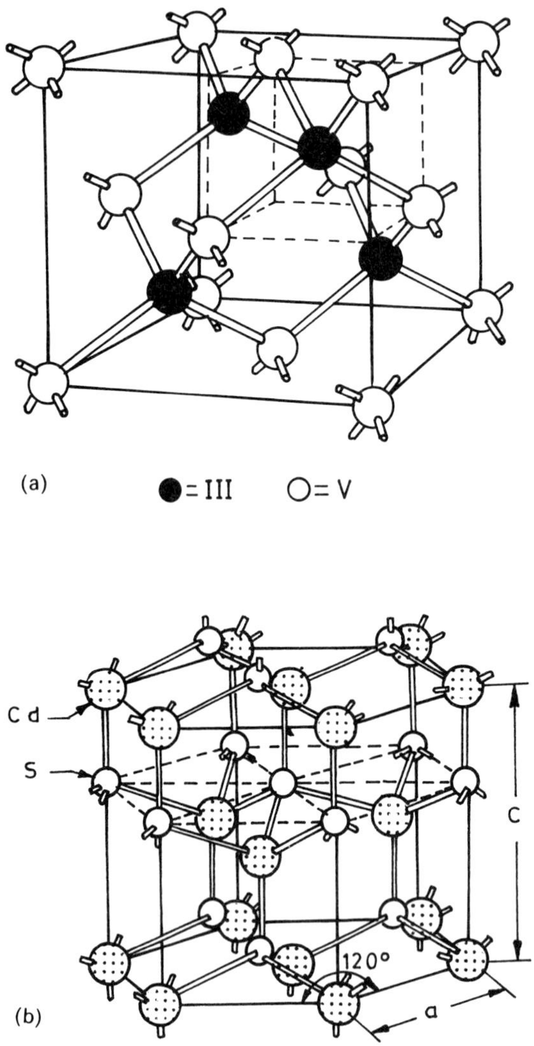

Figure 5.11 *Semiconductor structures: (a) zinc-blende; (b) wurtzite.*

wurtzite structures contain identical bonds, one obtains the following relations between the second-order nonlinear coefficients:

$$d_{33}\,(\text{wurtzite}) = \frac{2d_{14}\,(\text{zinc-blende})}{3} \tag{5.25a}$$

$$d_{31}\,(\text{wurtzite}) = \frac{-d_{14}\,(\text{zinc-blende})}{3} \tag{5.25b}$$

which immediately lead to

$$d_{33} = -2d_{31} \tag{5.26a}$$

or

$$r^{\mathrm{e}}_{33} = -2r^{\mathrm{e}}_{13} \tag{5.26b}$$

for the wurtzite structure. Relation (5.26b) is, in fact, well obeyed for ZnS and CdSe (see Table 5.3). More distorted wurtzite crystals, such as BeO and ZnO, show large deviations from it.

More information can be gained from the model when the bond anharmonicity is explicitly taken into account, i.e. specific expressions for the dependence of the linear polarizability on the applied field are given. One of the approaches that combine theoretical and empirical knowledge is briefly reviewed in Section 5.10.2. An alternative approach, known as bond orbital model, can be consulted in references [49,50].

5.10.2. Phillips spectroscopic models

These models use the Phillips model [51,52] for the linear susceptibility and the finite-field approach to obtain the bond polarizabilities as successive derivatives of the bond dipole moment with regard to the applied field. As a practical example, let us consider a zinc-blende semiconductor crystal. One starts from the expression for the linear susceptibility of a crystal containing N equivalent axially symmetrical bonds, with linear polarizabilities α_{zz} and α_{xz}, parallel and perpendicular to the bond axis, respectively. Away from resonance and taking into account the crystal bond structure one obtains,

$$\chi^{1)}_{ZZ} = \frac{4N(1+2\rho)\alpha_{zz}}{3} = \frac{\Omega_{\mathrm{p}}^2}{(\bar{\omega}_{ng}^2 - \omega^2)} \tag{5.27a}$$

i.e.

$$\alpha_{zz} = \frac{3}{4N(1+2\rho)} \frac{\Omega_{\mathrm{p}}^2}{(\bar{\omega}_{ng}^2 - \omega^2)} \tag{5.27b}$$

where $\rho = \alpha_{xz}/\alpha_{zz}$, and $\Omega_{\mathrm{p}} = \{Ne^2/m\epsilon_0\}^{1/2}$ is the plasma frequency. $\bar{\omega}_{ng}$ is an average frequency for the optical transitions between filled and unfilled levels. Equations (5.27) can be obtained from the perturbative formalism and the Unsold approximation described in Chapter 4. The parametric dependence on the field E is exclusively included in the average resonance frequency $\bar{\omega}_{ng}$ or effective gap $\bar{\mathscr{E}}_{\mathrm{g}} = \hbar\bar{\omega}_{ng}$, which are Stark-shifted by the field. It has been shown from molecular orbital theory that $\mathscr{E}_{\mathrm{g}}$ can be decomposed into a homopolar $\mathscr{E}_{\mathrm{h}}$ and a heteropolar C gaps, so that

$$\mathscr{E}_{\mathrm{g}} = (\mathscr{E}_{\mathrm{h}}^2 + C^2)^{1/2} \tag{5.28}$$

with

$$\mathscr{E}_{\mathrm{h}} \approx ad^5 \tag{5.29a}$$

$$C \approx b\left(\frac{Z_A}{R_A} - \frac{Z_B}{R_B}\right) \mathrm{e}^{-kd/2} \tag{5.29b}$$

d being the bond length, and R_A, Z_A, R_B, Z_B, the ionic radius and charge number of ions A and B, respectively. a and b are proportionality constants. $\mathrm{e}^{-k_{TF}d/2}$ is the Thomas–Fermi screening factor with

$$k_{TF} = \left(\frac{4k_{\mathrm{F}}}{\pi R_0}\right)^{1/2} \tag{5.30}$$

$k_{\mathrm{F}} = 3\pi^2 N$ being the Fermi wavenumber for the valence electrons and R_0 the Bohr radius. For identical atoms, i.e. homopolar bonds, $C = 0$.

In order to describe the effect of the electric field on the electronic structure of the bond, i.e. on the bond gap, two models have been proposed. They will be succinctly discussed to illustrate the strategies used in the theoretical analysis.

Bond charge model

The key point in this model [53] is to assign a certain electronic point charge q_b

$$q_b = -\frac{2q\mathcal{E}_g^2}{\mathcal{E}_g^2 + \hbar^2\Omega_p^2} \tag{5.31}$$

to the bond (q is the electron charge). The charge is located at distances R_A and R_B from ions A and B, respectively. Then, the effect of the field on the bond is simulated by a shift ΔR in the position of the bond charge, without altering the bond length. In other words, the field modifies the heteropolar, but not the homopolar, gap. The change in C is obtained by writing the bond dipole moment induced by an incremental field ΔE_z in the bond direction

$$p_z = q_b\Delta R = \alpha_{zz}\Delta E_z \tag{5.32}$$

Consequently,

$$\frac{\partial R_A}{\partial E_z} = -\frac{\partial R_B}{\partial E_z} = \frac{q_b}{\alpha_{zz}} \tag{5.33}$$

From this equation one may calculate $\partial C/\partial E_z$, and so $\partial \mathcal{E}_g/\partial E_z$ by derivating expression (5.29b) with regard to E_z. Then, β_{zzz} is finally obtained from (5.27b) through $\beta_{zzz} = \partial\alpha_{zz}/\partial E_z$ (see reference [53]).

Charge transfer model

It derives from an alternative physical picture of the electronic charge distribution in the bond [54,55]. For heteropolar bonds, the amount of transferred charge between the two atoms is

$$Q = \frac{-qC}{\mathcal{E}_g} \tag{5.34}$$

The application of an electric field modifies Q and consequently C, $\mathcal{E}_g$ and α_{zz}. Then, one follows the same reasoning as in the previous model to calculate β_{zzz}.

5.10.3. Franz–Keldysh effect

The effect of the field on direct interband transitions for photon energies close to the band gap was initially and independently studied by Franz [46] and Keldysh [47]. The developed model is known as the Franz–Keldysh effect. From the experimental point of view, one observes an additional field-induced absorption tail below the zero-field edge which can be described as a shift and broadening of the main edge of the unperturbed crystal. The responsible physical mechanism is the photon-assisted tunnelling between the valence and conduction bands, which is schematically illustrated in Figure 5.12.

Theoretical treatments of such a mechanism are given in several references [54,55], which should be consulted for details. For nondegenerate bands and isotropic symmetry, the final

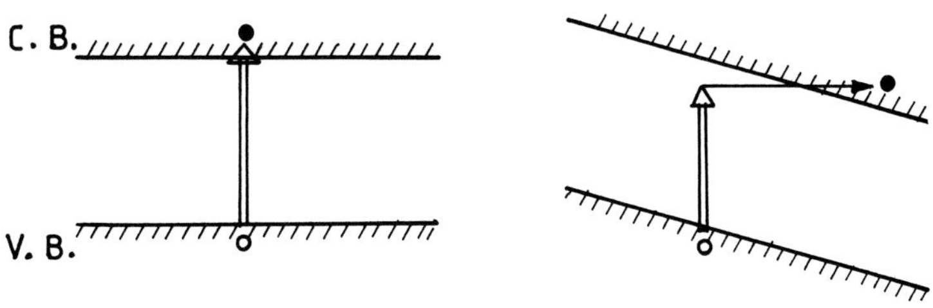

Figure 5.12 *Illustration of the Franz–Keldysh effect. Application of a field tilts the bands and permits tunnelling from an excited virtual band (VB) level to the conduction band (CB).*

formulae relating the field to the real and imaginary parts of the dielectric constant are

$$\Delta\epsilon_1(\omega, E) = \frac{B}{\omega^2}\xi^{1/2}G\left(\frac{\omega_g - \omega}{\xi}\right) \tag{5.35a}$$

$$\Delta\epsilon_2(\omega, E) = \frac{B}{\omega^2}\xi^{1/2}F\left(\frac{\omega_g - \omega}{\xi}\right) \tag{5.35b}$$

where $\xi = (q^2E^2/2\mu\hbar)^{1/3}$ and μ is the reduced mass of the electron–hole system. B is a constant involving material parameters and matrix elements between Bloch states of the valence and conduction bands. $F(\eta)$ and $G(\eta)$, with $\eta = (\omega_g - \omega)/\xi$, are the so-called *electrooptic functions of first* and *second class*, respectively. They essentially determine the field and frequency dependence of the effect and are plotted in Figure 5.13.

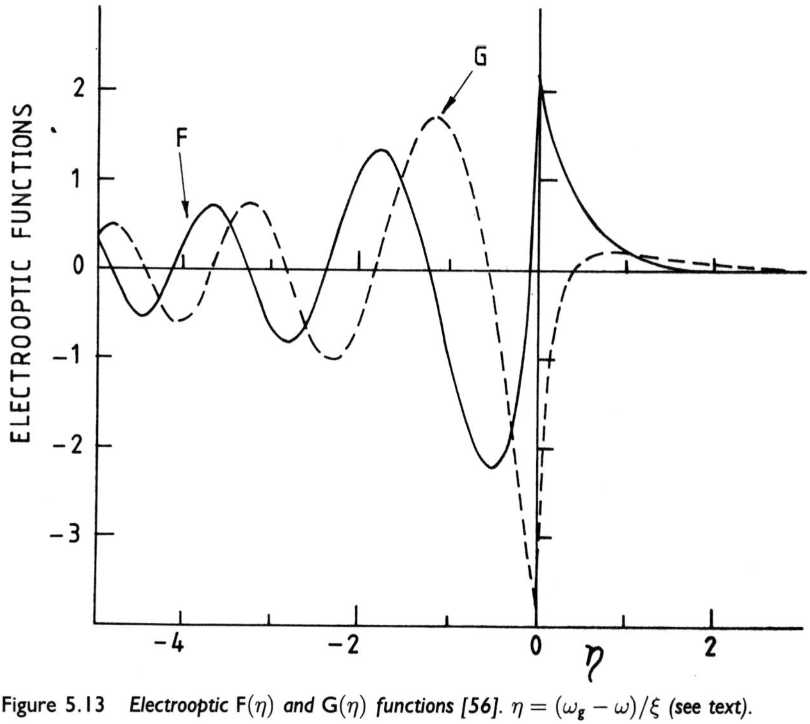

Figure 5.13 *Electrooptic* F(η) *and* G(η) *functions [56].* $\eta = (\omega_g - \omega)/\xi$ *(see text).*

Taking into account that for $n \gg \kappa$,

$$\Delta n = \frac{\Delta \epsilon_1}{2n} \tag{5.36a}$$

$$\Delta \alpha = \frac{4\pi \Delta \kappa}{\lambda} = \frac{2\pi \Delta \epsilon_2}{\lambda n} \tag{5.36b}$$

one obtains for the change in the refractive index n and the absorption coefficient α,

$$\Delta n(\omega, E) = \frac{B}{2n\omega^2} \xi^{1/2} G\left(\frac{\omega_g - \omega}{\xi}\right) \tag{5.37a}$$

$$\Delta \alpha(\omega, E) = \frac{2nB}{n\omega^2 \lambda} \xi^{1/2} F\left(\frac{\omega_g - \omega}{\xi}\right) \tag{5.37b}$$

Above the bandgap, $\omega > \omega_g$, the electrooptic functions cause an oscillatory contribution to the dielectric and optical parameters (*Stark oscillations*). However, in the zero-field limit it reduces to the well-known $(\omega - \omega_g)^{1/2}$ dependence, characteristic of a direct gap.

Below the gap, $\omega < \omega_g$, the $F(\eta)$ function decays exponentially. From the asymptotic behaviour as $\eta \longrightarrow \infty$ (wavelengths far from the edge and small fields) it can be shown that:

$$\Delta \alpha(\omega, E) \approx \left(\frac{E}{\omega_g - \omega}\right) e^{-\sigma(\omega_g - \omega)^{3/2}/E} \tag{5.38}$$

σ being a constant. Equation (5.38) accounts for the observed Franz–Keldysh tail in the absorption spectra. It is enhanced and extends to lower frequencies on increasing the field E.

Far below the edge, the refractive index change depends quadratically on the applied electric field, as one should expect for a mechanism which is invariant *vs* inversion of the field. For GaAs the corresponding quadratic electrooptic coefficient is $s = 4.5 \times 10^{-15}\ \mathrm{m^2/V^2}$ at $\lambda = 1.06\ \mu\mathrm{m}$.

Numerical calculations of Δn and $\Delta \alpha$ have been performed [57,58] for a number of semiconductors, that are useful for modulator design. Working near resonance conditions, both electrorefraction and electroabsorption effects show comparable magnitudes, their ratio depending on light frequency and electric field. In fact, one can distinguish separate regions where either losses (*electroabsorption regime*) or phase changes (*electrorefraction regime*) are dominant. Far enough from the edge, electrorefraction becomes dominant.

5.10.4. Exciton effects

The first effect of an applied field on the electronic structure of excitons is a Stark-shift of their energy levels. The problem can be formulated as for a single particle by using the effective mass approximation. The hamiltonian is that corresponding to a hydrogenic atom, together with the perturbation associated to the energy coupling with the electric field, $\mathcal{H}' = e\mathbf{r} \cdot \mathbf{E}$. Second-order perturbation theory shows that the Stark shift of energy level $\mathcal{E}$ is quadratic in the field E and negative

$$\frac{\Delta \mathcal{E}}{\mathcal{E}} \approx -(eR_0E)^2 \tag{5.39}$$

where R_0 is the Bohr radius of the exciton. As a consequence the exciton binding energy is decreased. Equation (5.39) holds as long as the field perturbation is sufficiently small, i.e.

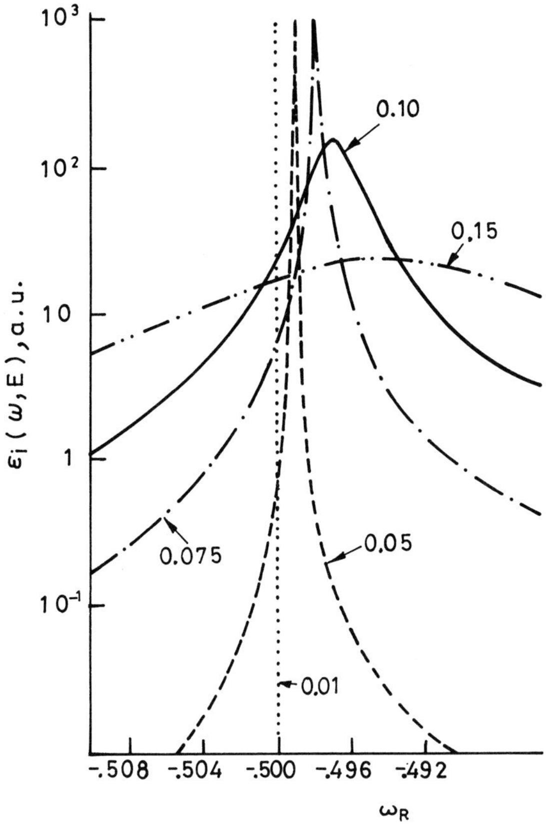

Figure 5.14 ***Stark effect on the first exciton level, showing energy shift and broadening (calculated) for different reduced fields marked on the corresponding curves [56].***

$eR_0E \ll 1$. In addition to the energy shift, splitting of excited levels also occurs. Results of the effective mass model for the electric-field-induced perturbation on the first exciton level are shown in Figure 5.14 for different values of the reduced electric field $E_R = qER_0/2\mathscr{E}_B$.

Another relevant feature associated to the presence of the field is the lifetime broadening of exciton levels. This can be easily visualized by plotting the electron–hole potential energy curve along the direction X of the field (Figure 5.15). Due to tunnelling of the exciton through the potential barrier, its lifetime becomes shorter than the value determined by thermal ionization. This effect leads to exciton ionization for high enough fields such as

$$eER_0 \approx \mathscr{E}_B \tag{5.40}$$

where $\mathscr{E}_B$ is the binding energy.

The region of the tail below the edge in the absorption spectrum and described by the Franz–Keldysh effect is also modified by the electron–hole interaction. The problem has been dealt with by several authors [59–61]. The absorption coefficient at the tail can then be written [42]

$$\alpha(\omega, E) = \alpha_{FK}M(\omega) \tag{5.41}$$

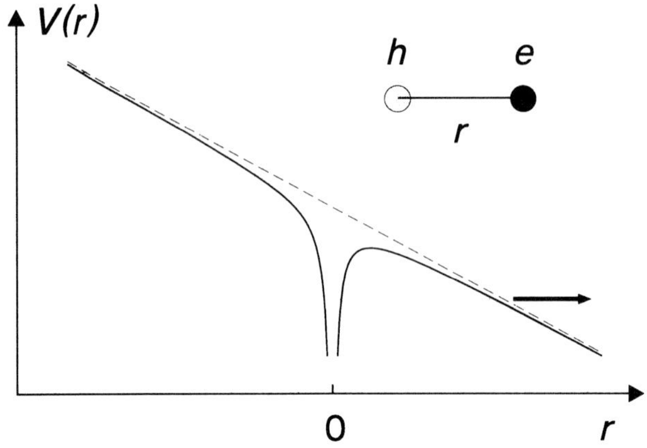

Figure 5.15 *Tunnelling from an exciton level into the conduction band induced by an applied field.*

$\alpha_{FK}(\omega)$ being the absorption curve predicted by the Franz–Keldysh model in the absence of exciton effects, and M an enhancement factor. Depending on the detuning $\epsilon = \omega - \omega_g$ and the field strength E, the incremental absorption coefficient given by (5.41) can reach values several orders of magnitude larger than predicted by the Franz–Keldysh formula. The dependence $M(\omega)$ is illustrated in Figure 5.16.

5.10.5. Free electron contribution (intraband effects)

As already mentioned, this mechanism [62,63] is a third-order nonlinear effect and so it only contributes to the Kerr electrooptic response. It may be important and even dominant in narrow-gap or heavily doped semiconductors.

The nonlinear response of free electrons is caused by: (a) nonparabolicity of the conduction band, (b) energy dependence of the electron scattering cross-sections and

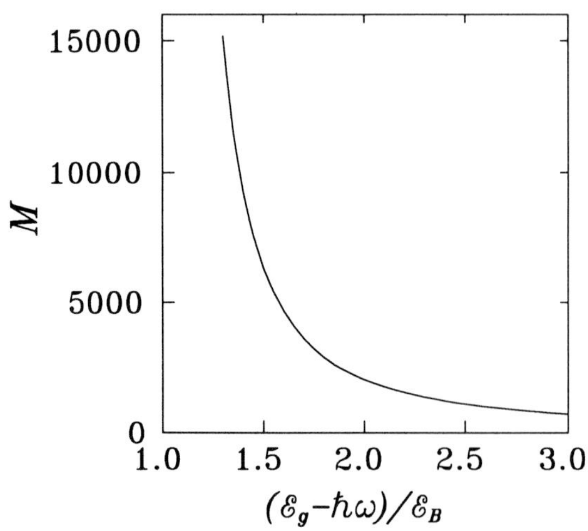

Figure 5.16 *Enhancement factor* M *for the Franz–Keldysh effect due to the exciton contribution. The reduced applied electric field is* $E_R = qR_0E/\mathcal{E}_B = 0.5$.

(c) intervalley electron redistribution. For moderate free electron concentrations ($\sim 10^{17}/cm^3$), the nonlinearity may become comparable to that associated with bound electrons. Unfortunately, we are not aware of any direct electrooptic experiments evaluating the quantitative importance of the free electron contribution, although abundant data are available for the optical regime.

5.11. Kerr materials: glasses

Although our focus has been centred so far on Pockels behaviour, all inorganic materials considered in this chapter also present Kerr activity. Representative s coefficients are given in Table 5.9. The large values for $BaTiO_3$ and KTN are noteworthy. Moreover, it has been discussed in Section 5.6.2 how some ferroelectric PLZT ceramic compositions show very high Kerr coefficients and so excellent device performance (Table 5.7). Some complementary information will now be given on glasses, as an example of useful Kerr materials. Other efficient quadratic materials are either organics (see Chapter 6) or quantum well semiconductor structures, to be discussed in Chapter 9.

5.11.1. Homogeneous glasses

Glasses constitute useful materials for optoelectronic applications since they combine good optical, thermal, mechanical and chemical qualities together with a great flexibility for conformation in arbitrary sizes and shapes. Moreover, silica glasses are particularly well suited to silicon technologies and much effort is being devoted to SiO_2 integrated optics devices [66,67].

The magnitudes of the s and B electrooptic coefficients depend significantly on glass composition [64]. The measured Kerr effect has been found to be larger for glasses having

Table 5.9 *Kerr electrooptic coefficients [64,65].*

Material	s_{ij} (m^2/V^2)	B (m/V^2)	λ (μm)
Water		5.1×10^{-14}	0.589
CS_2	$s_{44} = 0.5 \times 10^{-20}$	3.18×10^{-14}	0.546
CCl_4	$s_{44} = 1.5 \times 10^{-22}$	7.4×10^{-16}	0.633
Glass (SiO_2)		0.9×10^{-15}	0.633
Glass (SiO_2) heavy-metal doped ($1.9 \times 10^{22}/cm^3$)		1.4×10^{-14}	0.633
KDP	$s_{33} - s_{13} = 9.7 \times 10^{-18}$ $s_{31} - s_{11} = 3.9 \times 10^{-18}$ $s_{12} - s_{11} = 2.6 \times 10^{-18}$ $s_{66} = 0.9 \times 10^{-18}$		0.546
$BaTiO_3$	$s_{11} - s_{12} = 2.3 \times 10^{-15}$		0.633 ($T > T_c$)
$KTaO_3$	$s_{11} - s_{12} = 10^{-17}$		0.633 (−226)
$K(Nb_xTa_{1-x})O_3$	$s_{11} - s_{12} = 3 \times 10^{-15}$		0.633

high refractive indices, such as silica glasses containing heavy metals (Pb, Bi, Tl, Nb, Ta etc.) or telluride glasses. Some data are included in Table 5.9. Differences larger than one order of magnitude are observed.

5.11.2. Composite glasses

Particular attention should be given to the glasses containing nanocrystallites of metals (e.g. Au, Cu) or semiconductors (e.g. CdS, CdTe). An intensive research effort [68] has been made during recent years, stimulated by the high values ($\sim 10^{-8}$ esu) and very fast response times ($\sim 10^{-13}$ s) of the third-order susceptibility $\chi^{3)}_{ijk}(\omega : \omega, -\omega, \omega)$, i.e. optical Kerr effect. Consequently, they are attractive candidates for fabrication of all-optical logical gates in optical computers. Theoretical analysis distinguishes two factors contributing to the enhancement of the third nonlinear response. One, associated with local field effects, operates when the light frequency is close to the surface plasmon resonance of the particle given by $\epsilon'(\omega_{sp}) + 2\epsilon_o = 0$ (*dielectric confinement*). The other one is related to the influence of the nanometre size of the particle on quantum states and optical transitions (*quantum confinement*). It appears that for metal particles, dielectric confinement is dominant, whereas the opposite situation applies to semiconductor-doped glasses. In fact, this latter case provides an interesting example of three-dimensional confinement (*quantum dots*), which will be briefly discussed in Section 9.8.

Most work has so far been performed on the optical Kerr effect, i.e. $\chi^{3)}_{ijk}(\omega, -\omega, \omega)$. On the other hand, we are not aware of any determination of static Kerr susceptibilities, although electroabsorption experiments have indeed been performed. They have confirmed the main features of the quantum confinement model.

References

1. I. P. Kaminow and E. H. Turner, in *Handbook of Lasers* (ed R. J. Pressley), Chemical Rubber Co., Cleveland, Ohio (1971).
2. S. H. Wemple and M. diDomenico, in *Applied Solid State Science. Advances in Materials and Device Research*, vol. 3, (ed. R. Wolfe), Academic Press, New York (1972).
3. T. S. Narasimhamurty, *Photoelastic and Electrooptic Properties of Crystals*, Plenum Press, New York (1981).
4. Landolt-Börnstein, *Numerical Data and Functional Relationships in Science and Technology*, vol. 18 (eds K. H. Hellwege and A. M. Hellwege), Springer-Verlag, Berlin (1984).
5. D. Eimerl, *Ferroelectrics* **72**, 95 (1987).
6. M. B. Klein, in *Electrooptic and Photorefractive Materials* (ed P. Günter), Springer-Verlag, Berlin (1987).
7. C. J. G. Kirby, in *Properties of* $LiNbO_3$, EMIS Datareview Series No. 5, INSPEC, London (1989).
8. P. Günter, *Optics Commun.* **11**, 285 (1974); *Ferroelectrics* **75**, 5 (1987).
9. J. D. Bierlein and C. B. Arweiler, *Appl. Phys. Lett.* **49**, 917 (1986).
10. H. Y. Zhang, X. H. He, Y. H. Shih and S. H. Tang, *Optics Commun.* **86**, 509 (1991).
11. L. Arizmendi, J. M. Cabrera and F. Agullo-Lopez, *Int. J. Optoelectron.* **7**, 149 (1992).
12. H. Nakatani, W. Bosenberg, L. K. Cheng and C. L. Tang, *Appl. Phys. Lett.* **52**, 1288 (1988).
13. J. G. Bergman and G. R. Crane, *J. Solid State Chem.* **12**, 172 (1975).
14. N. Suzuki and K. Tada, *Jpn. J. Appl. Phys.* **23**, 1011 (1984).
15. R. J. Nelmes, Z. Tun and W. F. Kuhs, *Ferroelectrics* **71**, 125 (1987). (Also other papers in the same issue.)

16. J. C. Slater, *Ferroelectrics* **71**, 25 (1987).
17. S. Triebwasser, *Phys. Rev.* **101**, 993 (1956).
18. P. Günter, *Phys. Rep.* **93**, 199 (1982).
19. C. J. Bierley and F. W. Ainger, *Ferroelectrics* **91**, 181 (1989).
20. L. Hesselink, *Optics Photon. News*, April (1993).
21. R. Blinc and B. Zeks, *Soft Modes in Ferroelectrics and Antiferroelectrics*, North-Holland, Amsterdam (1974).
22. M. E. Lines and A. M. Glass, *Principles and Applications of Ferroelectrics and Related Materials*, Clarendon Press, Oxford (1977).
23. R. S. Weis and T. K. Gaylord, *Appl. Phys.* **A37**, 191 (1985).
24. S. C. Abraham and Marsh, *Acta Crystallogr.*, **B42**, 61 (1986).
25. A. Garcia-Cabañes, J. A. Sanz-Garcia, J. M. Cabrera, F. Agulló-López, C. Zaldo, R. Pareja, K. Polgar, K. Raksanyi and I. Foldvari, *Phys. Rev.* **B**37, 6085 (1988).
26. I. Foldvari, K. Polgar and A. Mecseki, *Acta Phys. Hung.* **55**, 321 (1984).
27. D. A. Bryan, R. Gerson and H. E. Tomaschke, *Appl. Phys. Lett.* **44**, 847 (1984).
28. J. D. Bierlein and H. Vanherzeele, *J. Opt. Soc. Am.* **B6**, 622 (1989).
29. C. T. Chen and G. Liu, *Ann. Rev. Mater. Sci.* **16**, 203 (1986).
30. C. T. Chen, B. C. Wu, A. D. Jiang and G. M. You, *Sci. Sin.* **13B**, 235 (1985).
31. L. K. Chen, W. R. Bosenberg and C. L. Tang, *J. Cryst. Growth* **89**, 553 (1988).
32. C. T. Chen and S. X. Chen, *Acta Phys. Sin.* **29**, 1000 (1980).
33. R. H. French, J. W. Ling, F. S. Okuchi and C. T. Chen, *Phys. Rev.* **B**44, 8496 (1991).
34. C. Chen, Y. Wu, A. Jiang, B. Wu, G. You, R. Li and S. Lin, *J. Opt. Soc. Am.* **B6**, 616 (1989).
35. C. Chen and B. Wu, Abstract MA7 presented to the meeting *Nonlinear Optics: Materials, Fundamentals and Applications*, Lahaina, Maui, Hawaii, 17–21 August (1992).
36. C. E. Land, P. D. Thacher and G. H. Haertling, in *Applied Solid State Science. Advances in Materials and Device Research*, Vol. 4, (ed R. Wolfe), Academic Press, New York (1974).
37. G. H. Haertling, in *Ceramic Materials for Electronics* (ed. R. C. Buchanan), Marcel Dekker, New York (1986).
38. A. Krumins and A. Sternberg, in *Electrooptic and Photorefractive Materials* (ed. P. Günter), Springer-Verlag, Berlin (1987).
39. G. H. Haertling, *Ferroelectrics* **131**, 1 (1992).
40. A. Sternberg, *Ferroelectrics* **131**, 13 (1992).
41. J. I. Pankove, *Optical Processes in Semiconductors*, Dover, New York (1975).
42. H. Haug and S. W. Koch, *Quantum Theory of the Optical and Electronic Properties of Semiconductors*, World Scientific, Singapore (1990).
43. M. D. Sturge, *Phys. Rev.* **127**, 768 (1962).
44. N. Suzuki and K. Tada, *J. Appl. Phys.* **23**, 1011 (1984).
45. M. Sugie and K. Tada, *J. Appl. Phys.* **15**, 421 (1976).
46. W. Franz, *Z. Naturfosch.* **13a**, 484 (1958).
47. L. W. Keldysh, *Sov. Phys. JEPT* **7**, 788 (1958).
48. C. Flytzanis and J. Ducuing, *Phys. Rev. Lett.* **26A**, 315 (1968); *Phys. Rev.* **178**, 1218 (1969).
49. W. A. Harrison and S. T. Pantelides, *Phys. Rev.* B**14**, 691 (1976).
50. W. A. Harrison, *Electronic Structure and the Properties of Solids*, Freeman, San Francisco (1980).
51. J. C. Phillips, *Rev. Mod. Phys.* **42**, 318 (1970).
52. J. C. Phillips and J. A. Van Vechten, *Phys. Rev.* **183**, 109 (1969).
53. B. F. Levine, *Phys. Rev. Lett.* **22**, 787 (1969).
54. C. L. Tang and C. Flytzanis, *Phys. Rev.* B**4**, 2520 (1971).
55. C. L. Tang, *IEEE J. Quantum Electron.* **QE-9**, 755 (1973).
56. M. Cardona, *Modulation Spectroscopy*, Academic Press, New York (1969).
57. B. R. Bennett and R. A. Soref, *IEEE J. Quantum Electron.* **QE-23**. 2159 (1987).
58. A. Alping and L. A. Coldren, *J. Appl. Phys.* **61**, 2430 (1987).
59. J. D. Dow and D. Redfield, *Phys. Rev.* B**1**, 3358 (1970).

60. A. Merkulov, *Sov. Phys. JEPT* **39**, 1140 (1974).
61. J. D. Dow, in *Optical Properties of Solids. New Developments* (ed. B. O. Seraphin), North-Holland, Amsterdam (1976).
62. D. S. Chemla, *Rep. Prog. Phys.* **43**, 76 (1980).
63. A. A. Borshch, M. Brodin and V. Volkov, *Refractive Nonlinearity of Wide-Band Semiconductors and Applications*, Harwood Academic, London (1987).
64. N. F. Borrelli, B. G. Aitken, M. A. Newhouse and D. W. Hall, *J. Appl. Phys.* **70**, 2774 (1991).
65. A. Yariv and P. Yeh, *Optical Waves in Crystals*, John Wiley, New York (1984).
66. S. Valette, P. Mottier, J. Lizet, P. Gidon, J. P. Jadot and D. Villani, *SPIE* **651**, 94 (1986).
67. S. Valette, *J. Mod. Opt.* **35**, 993 (1988).
68. C. Flytzanis, F. Hache, M. C. Klein, D. Ricard and Ph. Roussignol, in *Progress in Optics*, Vol. 29 (ed. E. Wolf), North-Holland, Amsterdam (1991).

Chapter Six

Organic Electrooptic Materials

6.1. Introduction

Organic materials constitute a vast world which has not yet been considered with sufficient attention by the solid state physics community. It includes the so-called liquid crystals [1–3], which have a relatively long tradition of research and development and already have a fair-sized niche in the electrooptics market. They are very efficient electrooptic materials, showing a variety of active processes, related to the complex cooperative dynamics of large organic molecules. The price paid for this high performance is the slow response times, typically in the millisecond range. More recently, ferroelectric liquid crystals have shown great promise because they can have responses as fast as a few microseconds. Liquid crystals and ferroelectric liquid crystals are considered today as a specific and well-defined field within organic materials, which will be discussed in the second part of this chapter (from Section 6.8 on). The discussion will provide the physical basis for a variety of devices described in Chapter 7.

The first part of the chapter is devoted to organic materials, whose dielectric and optical response is essentially (although not exclusively) associated with their electronic structure. They are now receiving increasing attention because they present a number of potential advantages with regard to inorganics for electrooptic and nonlinear optical applications. Some of them are:

(1) Higher values of the electrooptic and nonlinear coefficients for some well designed molecules and molecular arrangements. As an example, MNA crystals have an electrooptic coefficient much higher than those corresponding to $LiNbO_3$.
(2) Ultrafast (subpicosecond) nonresonant responses, making them particularly appropriate for optical signal processing and computing.
(3) Powerful chemical methods to engineer predesigned molecular structures, providing a very flexible tool for optimization of properties.
(4) Low relative values of the dielectric constant and refractive index, allowing for enhanced bandwidth in electrooptic devices.

In line with the rapid growth of the subject, some comprehensive books [4–6] and monographs [7–9] are now available. A very recent presentation of the state of the art can be found in reference [10].

6.2. Structural features of organic materials

It has been already mentioned in Chapter 4 that a major differential feature of organics is that molecules keep their identity to a much larger degree than in inorganics. This property, together with the mentioned outstanding capabilities of organic engineering, has promoted a new field of research known as *molecular electronics*. It has the long-term aim of designing and producing molecules (or molecular arrangements) having tailored (predetermined) optoelectronic properties.

Generally, the electronic structure is described in terms of bonds connecting pairs of neighbouring atoms. σ-bonds result from a combination of s and p atomic functions yielding a molecular orbital directed along the line joining the two atoms (Figure 6.1(a)). This line is a rotation symmetry axis for the orbital. On the other hand, π-orbitals combining p atomic functions along a perpendicular direction, form a weaker and more polarizable bond (Figure 6.1(b)). It is useful, then, to distinguish between saturated molecules having exclusively σ-bonds (e.g. saturated hydrocarbons) and π-bonded molecules with unsaturated carbons linked by double or triple bonds (one with σ and one or two with π character, respectively). Molecules having extended overlapped π-bonds are designated as *conjugated*. Their π electrons are not localized in a particular bond but move throughout the whole system like electrons in a box. A typical example is the benzene planar molecule (Figure 6.2).

A linear π-molecule constituted by N atoms has a set of N energy levels. In the ground configuration, half of them are fully occupied by pairs of electrons with antiparallel spins. The transitions from the occupied (bonding) to the unoccupied (antibonding) levels, cause strong optical absorption. The onset of this absorption occurs at a wavelength that increases with N, i.e. the conjugation length. This is a general behaviour, regardless of the type of molecule and the specific model to deal with the electronic structure.

In accordance with the weak intermolecular bonding, the polarization of an organic material can be adequately described by the sum of all individual molecular dipole

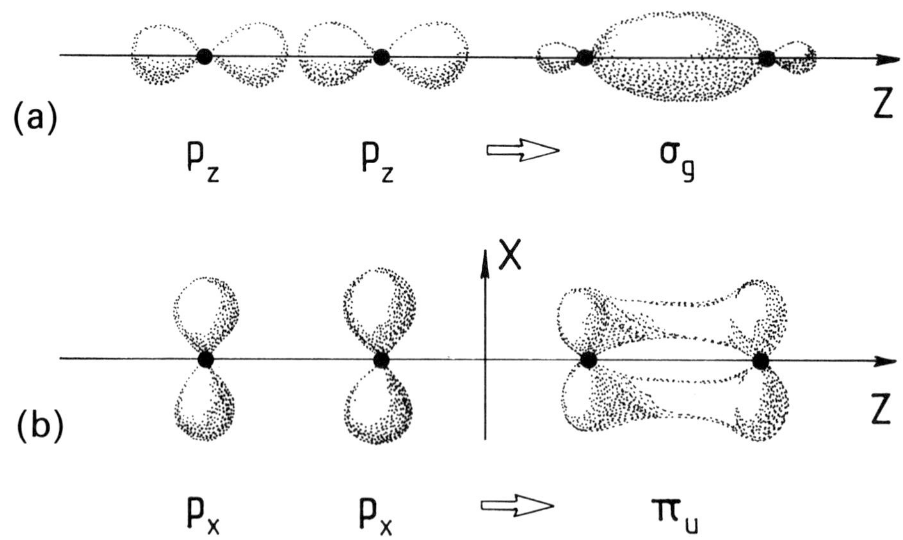

Figure 6.1 *Bond types in organic molecules: (a) σ; (b) π.*

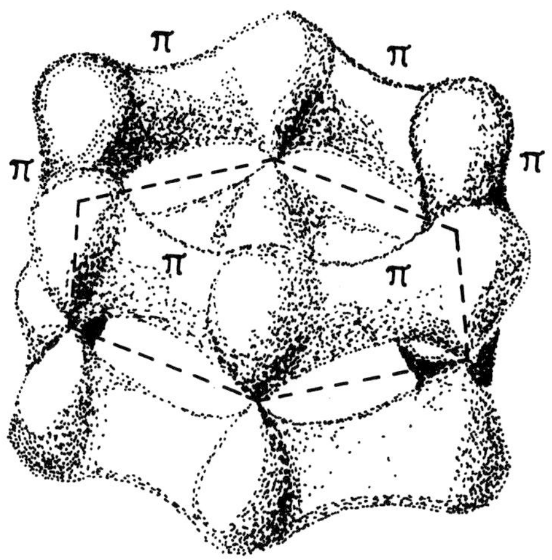

Figure 6.2 ***Conjugated π-orbital of benzene.***

contributions. Consequently, the overall electrooptic and nonlinear response is the result of two independent factors: the individual molecular parameters and the effect associated with the macroscopic molecular arrangement. The optimization of that response involves optimization of both factors, i.e. demands for an optimal molecular nonlinearity and a suitable molecular structure to maximize the addition of the individual contributions. These two aspects will be successively considered next.

6.3. Design criteria for electrooptic and nonlinear organic molecules

The design of optimized nonlinear organic molecules relies on an adequate theoretical modelling of their electronic structure and its relationship to molecular geometry.

In principle, the two main quantum-mechanical approaches described in Chapter 4, the finite-field and the sum-over-states (perturbative) methods, can be used for a detailed understanding of the linear and nonlinear optical behaviour of organic molecules. They involve either *ab-initio* calculations or those applying a variety of semiempirical schemes relying on different approximations, which are well-known to the quantum chemist. However, in order to make useful predictions and look for trends in the nonlinear optical properties, it is often preferable to use simple phenomenological models more directly connected with physical intuition. In fact, they can be derived from the general methods after suitable and reasonable simplifications of the mathematical formalism. In particular, the perturbative scheme designated the *sum of states* method, yields closed expressions for the hyperpolarizabilities containing parameters related to experimentally accessible magnitudes (equations (4.87)). Unfortunately, these general formulae involve all excited states and are of little practical use. However in many practical cases a low-lying excited state is dominant and the formalism is markedly simplified. This *two-level model* has found a fruitful application in discussing and predicting second-order nonlinearities (see Section 6.3.1).

We will now briefly describe some of these models, dealing successively with second- and third-order phenomena.

6.3.1. Second-order nonlinear response

For σ-bonded or saturated molecules, one may use a simple bond additivity model to compute Pockels and other second-order coefficients. Several schemes have been used, taking into account local field effects and the tensor character for the hyperpolarizabilities. However, β values for these molecules are small (10^{-31}–10^{-30} esu), and generally unsuitable for applications.

For π-electron molecules, the two-level model, developed by Oudar and Chemla [11] is particularly interesting. It applies to molecules having a low-lying charge transfer band and leads to the following Pockels (electronic) hyperpolarizability,

$$\beta_{zzz}(\omega : 0, \omega) = \frac{3q^2}{2\hbar m} \frac{\omega_{eg} f \Delta\mu}{\omega_{eg}^2(\omega_{eg}^2 - \omega^2)} \tag{6.1}$$

where ω_{eg} is the frequency of the optical transition between the ground (g) and excited (e) states, f the oscillator strength and $\Delta\mu$ the difference between the dipole moments of the two states. It has been assumed that the transition dipole moment μ_e, as well as $\Delta\mu$, are parallel to the z molecular axis. It is also considered that the operating frequency is far from resonance ($\omega \ll \omega_{eg}$), where damping parameters can be neglected. If one substitutes $\omega_{eg}^2 - 4\omega^2$ for ω_{eg}^2 in the second factor of the denominator the model yields $\beta_{zzz}(2\omega : \omega, \omega)$, which is responsible for second harmonic generation.

Equation (6.1) is very interesting, since it provides some relevant clues to attain a high second-order molecular polarizability. One should look for high strength transitions, involving a large change in dipole moment, i.e. an extensive electronic charge redistribution. Moreover, β_{zzz} can be theoretically estimated from (6.1), since ω_{eg}, f and $\Delta\mu$ are accessible to measurement. f is easily obtained from the area of the absorption band and $\Delta\mu$ can be inferred from solvatochromic experiments [12], where the shift in the position of the absorption band peak with the nature of a solvent is measured. The agreement with direct measurements is reasonable.

A prototype molecule to illustrate the two-level model is paranitroaniline (pNA) (Figure 6.3). The amino group NH_2 acts as a donor delivering electronic charge to the acceptor nitro group NO_2, through the delocalized π system of the benzene ring. In the ground state, a dipole moment already exists due to the transfer of ~0.23 electrons from the donor to the acceptor group. This transfer is markedly enhanced up to 0.7*e* in the excited state, leading to a change in dipole moment $\Delta\mu \sim 8$ debye.

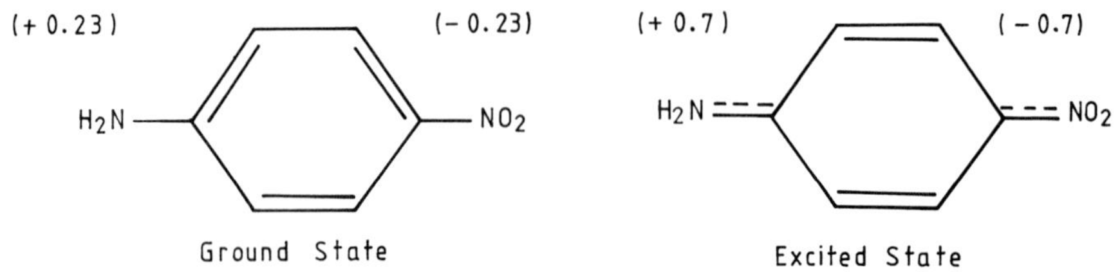

Figure 6.3 *Ground and excited states of paranitroaniline. The numbers in brackets stand for the transferred charge between the donor and acceptor groups (in units of the electron charge).*

Very often, nonsubstituted π-orbital molecules are centrosymmetrical and therefore the delocalized electron system presents a vanishing β. However, by substituting one or more molecular groups, one can break the centrosymmetry and develop a nonzero β. This has been substantiated by a large body of experimental work. A simple model, adequate to account for the effect of the substituent groups is the *equivalent internal field model* [13]. For the centrosymmetrical π molecule the field-induced dipole moment can be written

$$p_\pi = \alpha_\pi E + \gamma_\pi E^3 + \ldots \tag{6.2}$$

On the other hand, for the substituted molecule

$$p = \Delta\mu_x + \alpha E + \beta E^2 + \ldots \tag{6.3}$$

where $\Delta\mu_x$ is the *mesomeric* dipole moment associated with the presence of the substituted groups. This moment can be assumed to derive from an equivalent internal field E_x, acting on the centrosymmetrical molecule. According to (6.2) it should be given by

$$E_x = \frac{\Delta\mu_x}{\alpha_\pi} \tag{6.4}$$

Substituting $E + E_x$ into (6.2), and comparing with (6.3), one obtains the following expression for the second-order coefficient

$$\beta = \frac{3\gamma_\pi \Delta\mu_x}{\alpha_\pi} \tag{6.5}$$

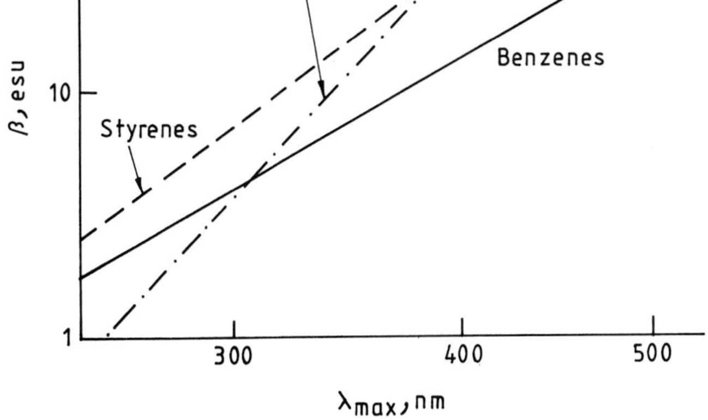

Figure 6.4 *Correlation between β and transparency range (wavelength of lowest energy transition) for some organic molecules. (From data supplied by Dr P. V. Vidakovic.)*

Table 6.1 *Measured second-order hyperpolarizabilities of organic molecules.*

Molecule	Structure	$\beta(2\omega : \omega, \omega)$, (esu)
Carbonyl diamide (Urea) [14]	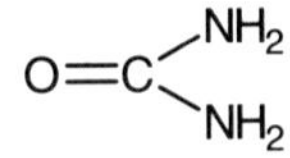	0.4×10^{-30}
Metanitroaniline (m-NA) [14]	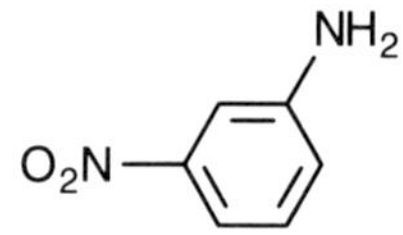	6×10^{-30}
Orthonitroaniline (o-NA) [14]	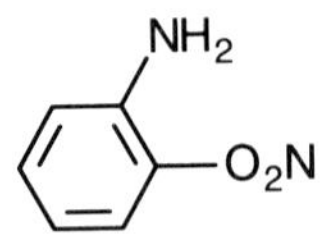	10×10^{-30}
Paranitroaniline (p-NA) [14]	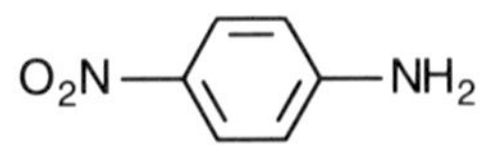	35×10^{-30}
2-Methyl-4-nitroaniline (MNA) [14]	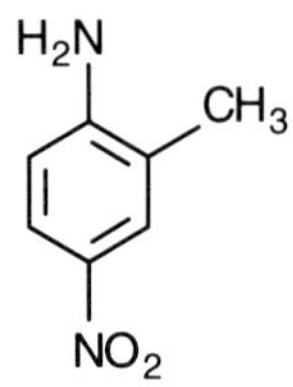	42×10^{-30}
3-Methyl-4-nitropyridine-1-oxide (POM) [14]	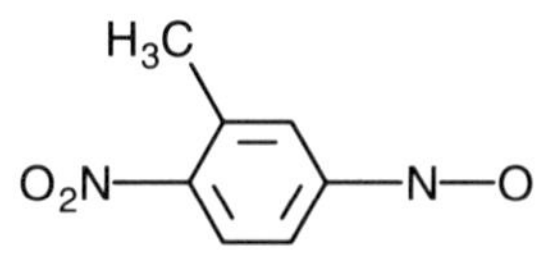	5×10^{-30}
Methyl-(2,4 dinitrophenyl)-aminopropanoate (MAP) [14]	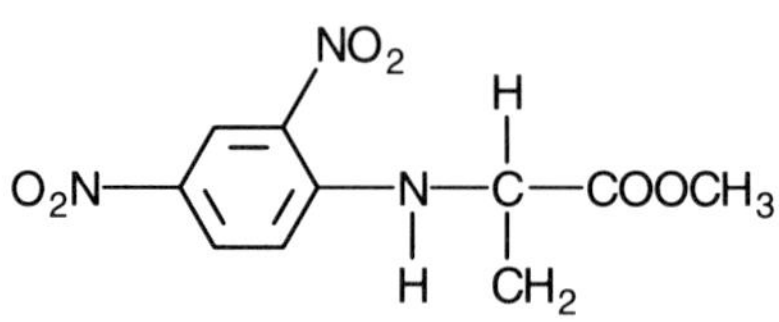	22×10^{-30}
4-(N,N-dimethylamino)-acetamidonitrobenzene (DAN) [15]	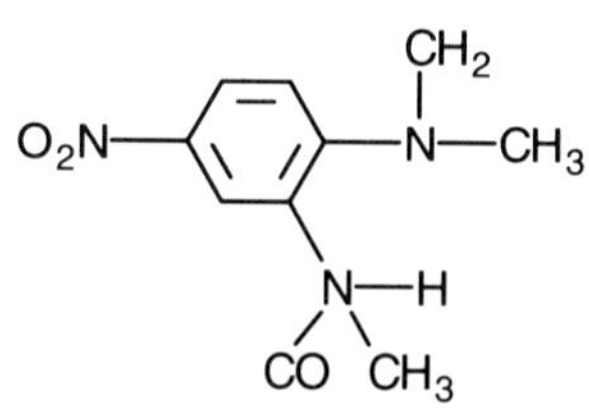	30×10^{-30}
N-(4-nitrophenyl)-L-prolinol (NPP) [14]	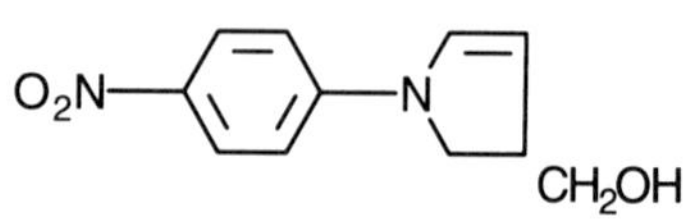	17×10^{-30}

Table 6.1 *Continued.*

Molecule	Structure	$\beta(2\omega : \omega, \omega)$, (esu)
Trans-4-dimethylamino-4′-nitrostilbene (DANS) [14]	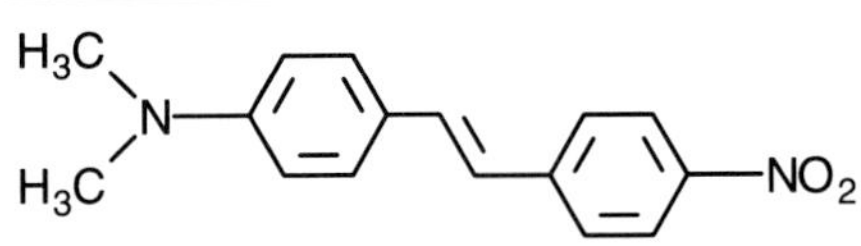	450×10^{-30}
3-(1,1-dicyanoethenyl)-1,5-dihydro-1H-pyrazole (DCNP) [16]	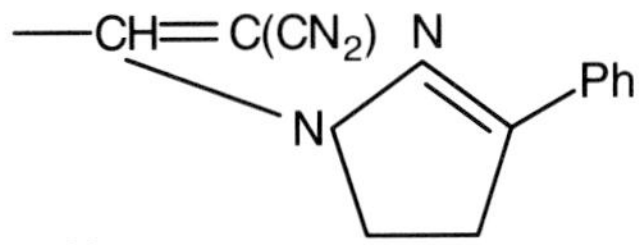	778×10^{-30}*
Merocyanine [5]	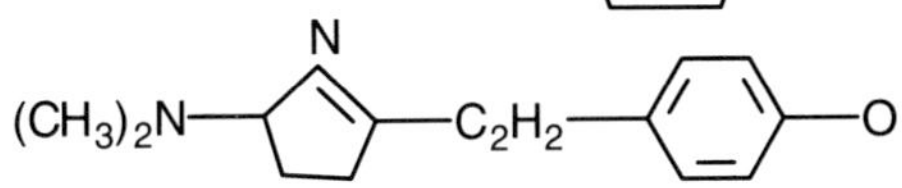	1000×10^{-30}

* Theoretical value.
Data taken at 1.06 μm.
References are indicated against each molecule.

establishing a simple linear relationship between β and the mesomeric dipole moment $\Delta\mu_x$. This useful correlation has been experimentally confirmed for monosubstituted benzenes and stilbenes. In fact, the sign of β is determined by that of $\Delta\mu$, being negative when the substitution involves an acceptor group and positive for a donor group. This internal field model is somewhat reminiscent of the polarization bias model used in Chapter 5 to discuss the Pockels coefficients in ferroelectric crystals.

In order to optimize the β-hyperpolarizability one may play with the chemical nature and number of substituents and their location in the molecule. In fact, many systematic studies have been carried out on the effect of these parameters [4,5].

Another alternative is to investigate the effect of the conjugation length, molecular planicity, bond alternation and aromaticity on β. As a general rule β is markedly enhanced with expanding conjugation, in qualitative accordance with a simple free electron model. This predicts $\alpha \approx N^3$ and $\gamma \approx N^5$ (Section 6.3.2), and then equation (6.5) immediately yields $\beta \approx N^3$. Unfortunately, the "gap" energy of the molecule decreases with increasing conjugation, leading to a smaller optical window. This behaviour is clearly illustrated in Figure 6.4 for several compounds. Therefore, some trade-off is necessary to achieve optimum overall performance for applications in the visible wavelength range.

Values of $\beta(2\omega : \omega, \omega)$ have been measured by the EFISH technique for a number of organic molecules and are listed in Table 6.1; esu units are used in accordance with the common practice in the literature. Unfortunately, values for the Pockels coefficients $\beta(\omega : \omega, 0)$ are not experimentally available. However, the dominant electronic contribution to these coefficients should differ exclusively from those listed in the table by the resonant factor associated to the harmonic frequency 2ω. Note the high values of β for several molecules, in particular DANS, DMA-PNB, and merocyanine.

New design routes

A different route to the design of nonlinear molecules has been proposed and implemented [10,17]. It relies on molecules with a null dipole moment but still having a nonzero β through the octupolar contribution to the tensor components.

6.3.2. Third-order nonlinear response

For third-order nonlinear phenomena, such as the Kerr effect, the situation concerning theoretical predictions is less developed than for second-order effects. Although the general theoretical schemes are equally applicable, the complexity of the mathematical procedures becomes much higher. In many cases, three- and four-level models are used to account for the values of γ and their dispersive behaviour [18,19].

As a first qualitative hint, one may use the Miller rule relating the hyperpolarizabilities to the linear polarizability. This criterion favours highly polarizable structures such as those corresponding to multiple bonds and particularly conjugated π-bonding. This is, indeed, in accordance with experiment since it has been found that the second- and third-order polarizabilities markedly increase with the size of the conjugated system. A simple model that provides quantitative support to this behaviour is the free electron model, as developed by Rustagi and Ducuing [20]. It simulates the π-electron system of a linear conjugated molecule by free electrons in an one-dimensional box or potential well whose length $2L$ is taken equal to the molecular dimension. The full expressions for the polarizabilities α and γ are

$$\sigma = \frac{16\pi\epsilon_0 L^4}{R_0} \sum_{n=1}^{N} \left\{ -\frac{2}{3\pi^2 n^2} + \frac{10}{\pi^4 n^4} \right\} \tag{6.6a}$$

$$\gamma = \frac{8.192\pi^3\epsilon_0^3 L^{10}}{R_0^3 e^2} \sum_{n=1}^{N} \left\{ -\frac{2}{9\pi^6 n^6} + \frac{140}{3\pi^8 n^8} - \frac{440}{\pi^{10} n^{10}} \right\} \tag{6.6b}$$

R_0 being the Bohr radius and $2N$ the number of π-electrons (N is the number of repeating bonds). For $N \gg 1$, one obtains to the lowest order in $1/N$,

$$\alpha = \frac{32\pi\epsilon_0 L^4}{3R_0\pi^2 N} \tag{6.7a}$$

$$\gamma = \frac{16.384\pi^3\epsilon_0^3 L^{10}}{45R_0^3 e^2\pi^6 N^5} \tag{6.7b}$$

indicating that both α and γ increase steeply with the molecular length $2L$. Since one may assume $N \approx L$, it is easily implied from (6.7) that

$$\alpha \approx L^3 \approx N^3 \tag{6.8a}$$

$$\gamma \approx L^5 \approx N^5 \tag{6.8b}$$

showing that the third-order polarizability is much more sensitive than the linear one to the increase in conjugation size. However, one should note again that the price paid for the large enhancement in γ is the reduction in the π–π^* gap or transparency window of the molecule.

Experimental values for γ are, generally, in the range 10^{-36}–10^{-35} esu, as inferred from EFISH, third-harmonic generation (THG) and four-wave mixing [5]. However, a strong increase with conjugation length, as predicted by the free electron model, has been found, and values as high as 10^{-32} esu have been reached [5].

New developments: organometallics

Organometallic molecules [21,22] (e.g. ferrocene derivatives) are being explored as potentially useful candidates for high second- and third-order nonlinearities. One possible advantage is their easy redox behaviour, associated with the metal ion, that can lead to large hyperpolarizabilities.

Recently, the synthesis of macroscopic amounts of carbon clusters C_{60} and C_{70} (fullerenes) has been achieved. Their unique structure and electronic configuration has promoted an intense activity to measure their nonlinear optical properties [22].

6.4. Macroscopic electrooptic (nonlinear) response

The optimization of the bulk or macroscopic nonlinear response, described by the r_{ijk} (or $\chi^{2)}_{ijk}$) and s_{ijkl} (or $\chi^{3)}_{ijk}$) coefficients requires, in addition to an optimal molecular response, an appropriate molecular arrangement that maximizes the sum of all elementary molecular contributions.

For Pockels or second-order nonlinear phenomena, the general relationship between $\chi^{2)}_{ijk}$ and β_{ijk}, derived in Section 4.14, writes

$$\chi^{2)}_{lmn}(\omega : \omega_1, \omega_2) = Nf(\omega)f(\omega_1)f(\omega_2)\langle b_{lmn}\rangle \tag{6.9}$$

where N is the molecular concentration, $f(\omega_1)$ appropriate local field factors and b_{ijk} stands for the β_{ijk} tensor referred to the laboratory axes (Figure 4.7). The relation between b and β is given by formula (4.90), which is

$$b_{lmn} = \sum_{lmn} a_{li}a_{mj}a_{nk}\beta_{ijk} \tag{6.10}$$

$a_{\mathrm{pq}} = \cos x_{\mathrm{q}}X_{\mathrm{p}}$ being the coordinate transformation matrix from the laboratory X_{p} to the molecular x_{q} axes. Starting from the molecular β, one should try to maximize $\chi^{2)}_{lmn}$ by playing with the structural parameters a_{pq}.

For an isotropic distribution of molecular orientations, having inversion symmetry, the overall second-order response is zero. It is, therefore, essential to establish an accentric arrangement of the molecules to assure Pockels activity of the material. There are several strategies to achieve this, including single crystal structures, oriented Langmuir–Blodgett films and polymer hosts incorporating suitably aligned active molecules. All these methods will be considered separately in the following sections.

For Kerr and third-order nonlinearities, the role of the molecular arrangement is not as critical as for second-order phenomena. Therefore, optimization of the individual molecular response is, very often, the dominant factor. Due to the high γ coefficients of long conjugated molecules, it appears that conjugated polymers may offer the best solution.

A detailed analysis of the connection between microscopic and macroscopic responses has been given by Oudar and Zyss [23].

6.5. Organic single crystals

Single crystal structures may provide organic substances with the lack of inversion symmetry required for Pockels activity and other second-order nonlinear phenomena. The objective of achieving such accentric structures has stimulated an intensive effort [4,9] in the field of *molecular* and *crystal engineering*, trying to make use of theoretical predictions and empirical knowledge. Several paths have been essayed with good success, including the introduction of side and chiral groups to force accentricity and the use of intermolecular hydrogen bonding to favour aligned (head to tail) arrangements. The effort has resulted in some optimized crystals, such as POM [24], NPP [24] and DCNP [25]. A summary of electrooptic (and nonlinear) coefficients is given in Table 6.2, together with some other interesting parameters: point symmetry, melting point, refractive index and transparency range. Dielectric constants are not given, but the reported values are $\epsilon < 5$, i.e. much lower than for most relevant inorganic crystals.

Electrooptic organic crystals are mostly grown from the melt, from solution and less frequently from the vapour phase. Solution growth is, in many cases, the best alternative. It yields large size crystals (several cm^3) with good quality (free from strain) over a period of a few weeks. However, one should remark that sustained efforts applied to improve crystal growth technology are essential, since the present state of the art has not yet reached the level of perfection and reproducibility required for industrial applications.

Anyhow, one should note the high r values for MNA, some of them superior to those of $LiNbO_3$. The list in Table 6.2 is headed by urea, which is generally taken as a reference material and has a wide transparency range extending into the UV. Crystals are commercially available for frequency doubling and parametric oscillation. Some disadvantages of organic crystals are: low melting points, mechanical softness and reduced transparency range. This latter feature may be less detrimental for electrooptical than for optical-mixing devices.

Table 6.2 *Pockels coefficients of organic crystals.*

Crystal	Point symmetry	Melting point (°C)	Refractive index	Transparency range (μm)	r (pm/V)
Urea [24]	$\bar{4}2m$	132.7	$n_o = 1.477$ $n_e = 1.583$	0.21–1.4	$r_{63}^T = 0.83$ $r_{41}^T = 1.9$
m-NA [24]	$mm2$	113	$n_x = 1.760$ $n_y = 1.720$ $n_z = 1.670$	0.48–2	$r_{33}^T = 16.7$ $r_{23}^T = 0.1$ $r_{13}^T = 7.4$
MNA [24]	m	131	$n_x = 2.0$ $n_y = 1.6$	0.5–2	$r_{11}^T = 67$
POM [24]	222	136	$n_x = 1.663$ $n_y = 1.829$ $n_z = 1.625$	0.5–2	$r_{41}^T = 3.6$ $r_{52}^T = 5.1$ $r_{63}^T = 2.6$
NPP [24]	2	116	$n = 1.68$	0.5–2	$r_{12} = 13$
DCNP [25,26]	m	194	$n_x = 1.9$ $n_z = 2.8$	0.6–2	$r_{33}^T = 87$
DMSM [27]	m		$n_y = 1.31$ $n_z = 1.55$	0.6–2	$r_{33}^T = 430$

Electrooptic data refer to $\lambda = 0.633\,\mu$.
References are indicated against each crystal.

6.6. Langmuir–Blodgett (LB) films

This technique [28–30] developed by Irving Langmuir [31] and Katherine Blodgett [32] allows oriented film preparation through controlled deposition of successive molecular monolayers on an appropriate substrate. The experimental procedure is schematically

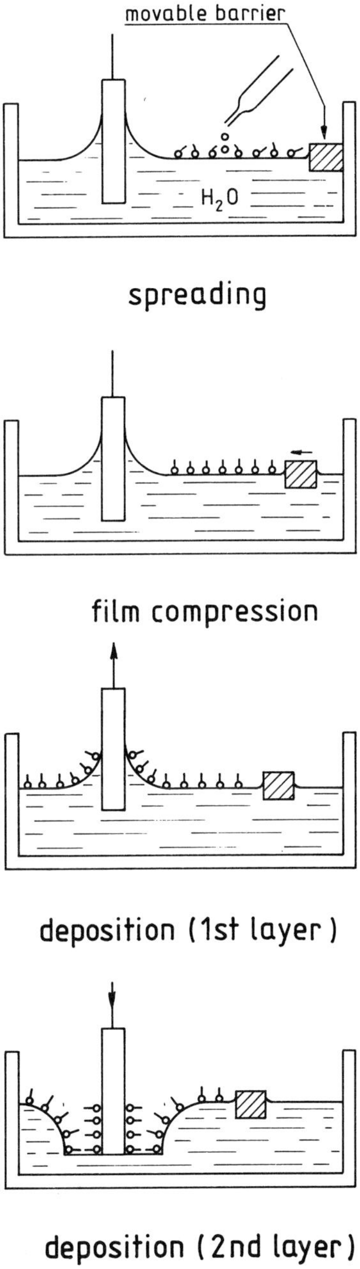

Figure 6.5 *Illustration of the various stages of the Langmuir–Blodgett process. (Adapted from [29].)*

illustrated in Figure 6.5. The first stage involves the formation of a monolayer of the organic compound, dissolved in an appropriate solvent, onto the surface of a water pool. Next, a certain surface pressure is applied to compress the monolayer laterally and induce an ordered arrangement of the molecules. The following stage involves dipping of a substrate plate into the liquid while the surface pressure is kept constant. During subsequent raising of the plate, molecules remain adhered to the plate surface, yielding a monolayer thick film. This latter stage can be repeated many times permitting the production of films up to $\sim 0.5\,\mu$m thickness. On the other hand, by using a trough with two separated pools, films consisting of alternate molecular layers can be prepared. In principle, the technique can be applied to molecules having hydrophobic and hydrophilic tails (designated as amphiphiles) and so a variety of compounds have been successfully deposited. The usual convention is to represent the hydrophobic and hydrophilic ends by a stick and a circle, respectively, as used in the diagram of Figure 6.5. Examples of hydrophobic molecules are long hydrocarbon chains. On the other hand, good hydrophilic behaviour is associated to carbonyl groups –COOH, as well as to –OH, and $-NH_2$. Three types of ordered molecular arrangements, designated as X, Y and Z, can be prepared by the LB technique and are schematically illustrated in Figure 6.6. Although glass is the most popular substrate, other materials can be also used such as quartz, silicon and polymers.

Recently, interest has arisen in the production of polymeric LB films. Two main strategies are being considered:

(1) Preparation of a monomer monolayer followed by polymerization either on the water surface or after deposition on the solid substrate. Polymerization can be accomplished by a suitable chemical agent or UV irradiation.
(2) Preparation of a polymer film monolayer directly on the air–water interface.

Most of the nonlinear measurements on LB films refer to harmonic generation (optical wave mixing). However, some electrooptic data have been recently obtained and are briefly summarized here. The Pockels susceptibility for a monolayer of hemicyanine deposited on a silvered substrate [33] amounts to

$$\chi^{2)}(\omega, 0) = 620 \times 10^{-9}\,\text{esu} = 270\,\text{pm/V}$$

The effect of successive alternate layers on the susceptibility has been investigated by Cresswell and coworkers [34]. Results for 1, 3 and 10 bilayers of 4-*n*-heptadecilamide-4′-nitrostilbene (4-HANS) are given in Table 6.3.

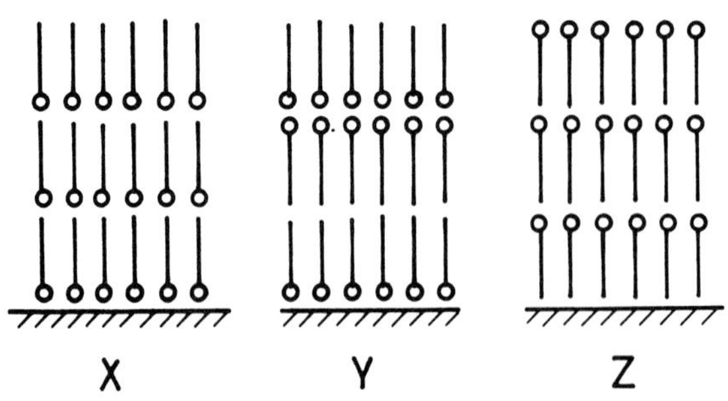

Figure 6.6 X, Y *and* Y *structures for Langmuir–Blodgett films [29].*

Table 6.3 ***Pockels susceptibilities for several 4-Hans/ JT11* bilayers [34].***

Number of bilayers	$\chi^{2)}(\omega:\omega,0)$ (pm/V)
1	5.2×10^{-12}
3	7.3×10^{-12}
10	6.4×10^{-12}

* 4-Hans: 4-*n*-heptadecilamido-4′-nitrostilbene; JT11: functionalized diarylalkyne.

Measurements have also been taken [35] for betaine and a solution of betaine in β-cyclodextrone. For a monolayer of pure betaine,

$$\chi^{2)}(\omega,0) = 163\,\text{pm/V}$$

The effect of successive layers of the betaine/β-cyclodextrone solution has also been studied.

It should here be remarked as a final comment that before LB films could be routinely used in optoelectronic devices, their structural quality and stability should be improved, particularly when a large number of layers are involved. Therefore, much research effort has still to be devoted to film preparation and characterization.

6.7. Polymeric materials

For both second- and third-order nonlinearities, polymeric materials may offer an excellent alternative, which has recently stimulated an intensive research effort [36,37]. The main advantage lies in the good processability of polymers to yield large volume samples and extended films. Unfortunately, some of the reports in this field do not reveal the nature of the investigated material due to industrial constraints requiring confidentiality.

(a) For *Pockels applications*, a rather simple route [38] to obtaining an active electrooptic material is by means of a solution of appropriate molecules into a glassy polymer host like PMMA (*guest–host* system). Molecules are partly aligned by applying a field during or after heating the material above the glass transition temperature T_g. The field can be applied through electrodes or by using a corona discharge [39]. The sample is next cooled down to room temperature while keeping the field applied. The mechanical and chemical properties of the system are determined by the polymer host whereas the nonlinear response is associated with the dissolved molecules. Thin films $\sim$1 μm thick are easily prepared on a variety of substrates from soluble polymers by the conventional spinning method. Values $r_{33} = 2.5$ pm/V at $\lambda = 0.633\,\mu$m have been measured [8] for an azodye (Disperse Red 1) in PMMA. The poling field was 0.62 MV/m.

The main drawbacks of this approach are the limited solubility of the active molecules and the loss of orientational order with ageing. In many cases, the electrooptic properties almost completely disappear after a few days or weeks. In order to solve these problems it is generally preferable to incorporate the active molecules chemically as side-chains to the polymer backbone, or even into the main chain itself. This latter approach raises some difficulties for efficient poling and so the side-chain method has been preferentially pursued.

Side-chain polymers permit us to introduce a high concentration of molecules without forming aggregates or precipitates. Electric field poling of the nonlinear molecules is easy and reorientation relaxation is markedly reduced because of the strong anchoring to the backbone. As an example, a poled side-chain polymer exhibiting a value $r_{33} = 15\,\mathrm{pm/V}$ (at $\lambda = 0.633\,\mu\mathrm{m}$), stable for more than one year, has been recently reported [40]. On the other hand, an electrooptic coefficient $r_{33} = 28\,\mathrm{pm/V}$, close to that of $LiNbO_3$, has been measured [37] for an Azko polymer at $\lambda = 1.3\,\mu\mathrm{m}$.

Another novel approach to improving orientational stability is to crosslink the polymer host after poling. This provides the whole structure with enhanced rigidity, inhibiting the reorientation of the side-chains [41].

(b) For *third-order nonlinear effects*, where molecular ordering is not critical, liquids, liquid crystals and solutions of high γ molecules are appropriate substances. However, conjugated polymers, providing extended π-bonding, are particularly useful for applications. Therefore, they have been the subject of extensive research [5]. Much work has been carried out on polydiacetylenes, in the form of either crystalline material, or films (LB, cast or vacuum deposited). Reported $\chi^{3)}$ values, measured by TGH, DFWM and EFISH, range from 10^{-11} esu to near 10^{-8} esu. Other conjugated polymers such as polyacetylene, or polytiophene, yield a similar span of $\chi^{3)}$ values.

There are other structures favouring high third-order nonlinearities. In particular, metallo-phthalocyanines [42] are macrocycles, providing a planar delocalized π-electron system tending to form discotic structures (Figure 6.7). Metal ions, occupying the central site in the phthalocyanine moiety, introduce charge transfer bands that may contribute to the nonlinear response. One advantage of these molecules is their high chemical and thermal stability. Polycrystalline films prepared by vacuum evaporation have yielded $\chi^{3)} \approx 10^{-11}$–$10^{-10}$ esu, for wavelengths far from resonance. Resonant values at the π–π^* transition reach 10^{-9} esu. Apparently, no Kerr electrooptic coefficients have been reported yet.

A different and promising approach consists of polymers containing semiconductor crystallites (PbS, CdS, etc.) and other aggregates [10]. Quantum size effects appear to lead

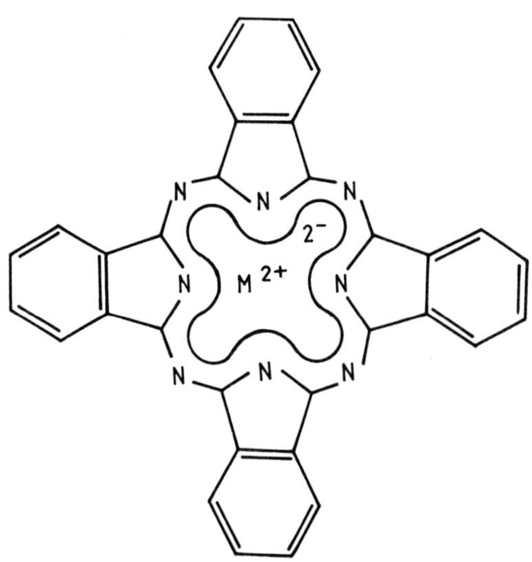

Figure 6.7 ***Planar structure of a metal-phthallocyanine.***

to enhanced $\chi^{3)}$ values, as already discussed in Chapter 5 for inorganic glasses doped with metal and semiconductor nanoparticles.

6.7.1. Inorganic hosts

Incorporation of organic molecules into inorganic matrices may offer some advantages in comparison with organic hosts, due to superior environmental, mechanical, thermal and optical performances. In particular, a successful effort has been devoted to incorporate organic dyes into silica (and other oxide) glasses, prepared by the sol–gel method [43–45]. The low temperatures involved in this process may become compatible with the stability of many organic molecules. Moreover, this technique permits easy deposition onto a solid substrate as well as adequate shaping of the sample [46]. Finally, by a slight modification of the synthetic route, large electric fields can be applied to achieve molecular orientation and electric-field induced second-harmonic generation has been obtained [47]. Not only glasses, but inorganic crystal structures, favouring ordering of guest organic molecules, have been designed and produced [48].

Another strategy, apparently not yet exploited for nonlinear optics applications, is the incorporation of active organic molecules into *intercalation compounds* (e.g. layer silicates). They have shown interesting engineering capabilities in the preparation of efficient materials for ionic transport [49].

6.8. Liquid crystals: an introduction

Liquid crystals (LC), known since 1888, constitute a peculiar state of matter with properties intermediate between those corresponding to classical solids and liquids [1–3,50]. Consequently, they are often designated as intermediate phases or *mesophases*. In fact they appear as stable phases during a short temperature range, when a crystalline solid is heated to reach the normal liquid state. This behaviour refers to the most common liquid crystals, called *thermotropic*, which are the ones to be discussed here at some length. They present flow behaviour, as ordinary liquids, but also possess a certain degree of long-range orientational order, which determines an average symmetry (lower than R_3) and leads to optical anisotropy. From the microscopic point of view most liquid crystals are made up of elongated molecules, containing a rigid structure and a flexible (aliphatic) section. We will refer primarily to these compounds in the text, although other molecular structures will be briefly mentioned at the end of the section.

According to the type of long-range order, three liquid crystal classes are usually defined: nematic, cholesteric and smectic, although cholesteric liquid crystals are often considered as a special type of nematics.

In nematic liquid crystals, molecules are located at random in space but they keep an average common orientation (Figure 6.8(a)). Even for polar molecules, it is assumed that the system is invariant under inversion, i.e. head to tail exchange in the molecules. This is a consequence of the weakness of the dipolar interactions. The symmetry of that molecular configuration is $D_{\infty h}$.

For cholesteric liquid crystals (Figure 6.8(b)), the average orientation of the molecular axis is not constant, but describes an helix. This structure is derived from the nematic one when the liquid is constituted by chiral molecules or contains them in solution. At a local level the symmetry is as for a nematic crystal.

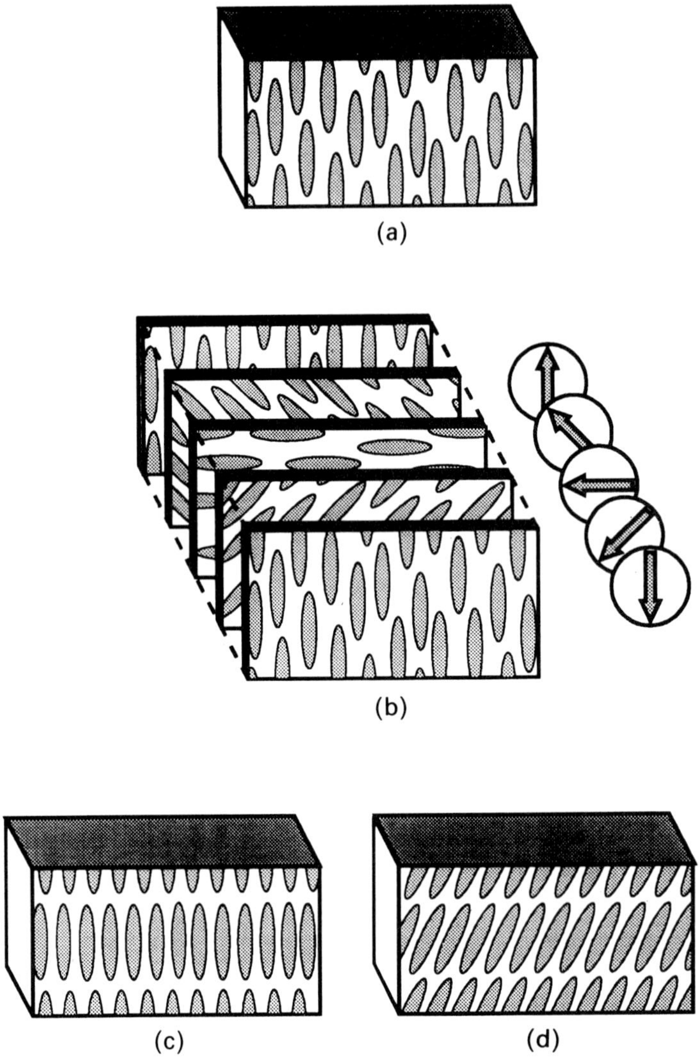

Figure 6.8 *Molecular arrangements in liquid crystals: (a) nematic; (b) cholesteric; (c) smectic A; (d) smectic C.*

Finally, in the smectic liquid crystals, average orientational order exists, as for nematics. However, molecules are also arranged in planes, either perpendicular (smectic A) or tilted (smectic C) with regard to the molecular axis (Figure 6.8(c) and (d), respectively). Evidence for the layered structure is obtained from the X-ray diffraction patterns showing the typical Bragg reflections. In each of the planes no positional order exists and so molecules behave as in a bidimensional liquid. The symmetry is $D_{\infty h}$ for smectic A. On the other hand, for smectic C symmetry is C_{2h} in the case of nonchiral molecules and C_2 for chiral ones, the binary axis being perpendicular to the molecular axis and contained in the layer planes. More complex smectic configurations exist presenting positional ordering in the layer planes. A total of 14 smectic phases have been

Table 6.4 *Illustrative liquid crystal compounds. Birefringence values Δn for some nematic phases are indicated between brackets.*

Molecules	Phases and temperature ranges (°C)
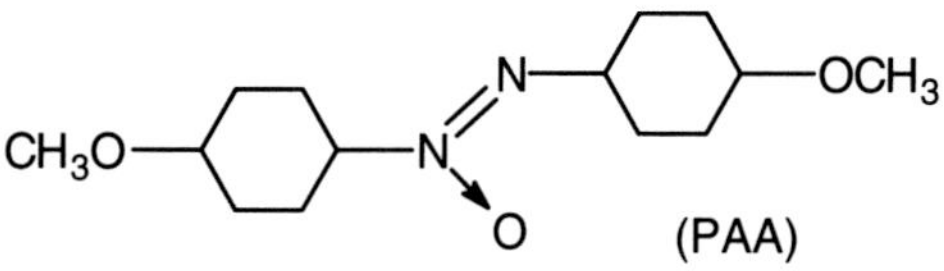 (PAA)	K–118–N–135.5–I ($\Delta n = 0.26$)
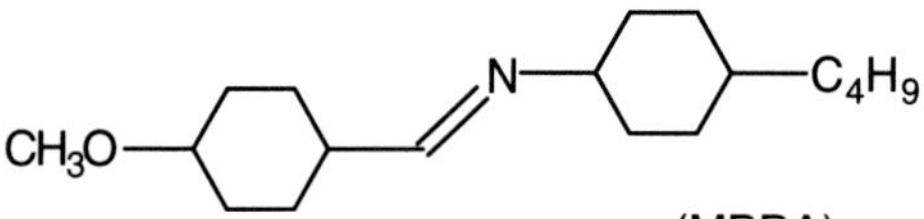 (MBBA)	K–22–N–47–I ($\Delta n = 0.19$)
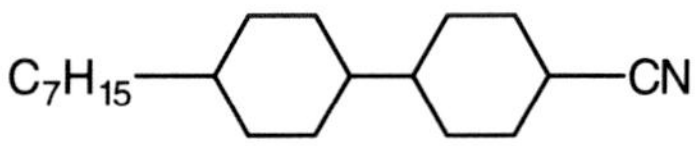	K–28.5–N–42–I ($\Delta n = 0.16$)
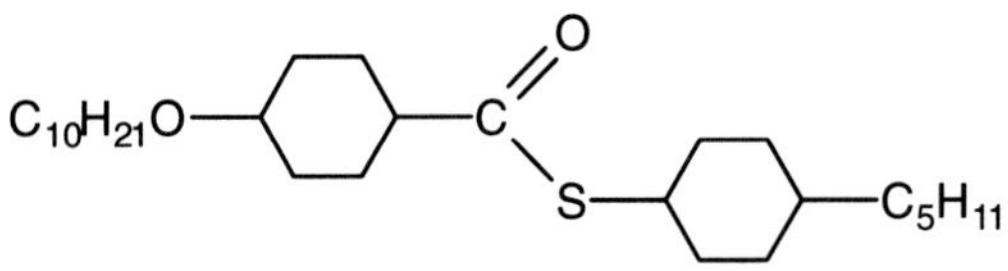	K–60–S_C–63–S_A–80–N–86–I

K: crystal, S: smectic phase, N: nematic phase, I: isotropic liquid.

identified which are designated by capital roman letters. The corresponding chiral phases are labelled by adding a superscript star, e.g. C*.

A few illustrative compounds forming LC phases, together with the corresponding temperature ranges are given in Table 6.4. A more detailed description of the properties associated with the nematic and cholesteric phases is presented in successive sections.

In addition to the most common rod-like (calamithic) LC, briefly described above, other mesophases have been identified. In particular, disc-like molecules also form LC phases called *discotic*. Nematic, cholesteric and smectic configurations have been reported. As an example, phtalocyanine molecules (see Figure 6.7), when appropriately substituted, give rise to columnar mesophases with hexagonal packing [51].

The search for new LC materials and the efforts to correlate molecular structure and physical behaviour are still actively pursued [52]. More than 20 000 rod-like compounds, with an immense variety of structural features, have been already reported.

Liquid crystal polymers

Polymeric materials may also present LC phases [53,54]. The structure of these polymers includes a rigid segment (as in ordinary rod-like LC molecules), responsible for the orientational order, and flexible chains with random complex arrangements. The rigid segments can be positioned in the main polymer chain or attached to it forming side-chains. Nematic, cholesteric and smectic phases have been found in LC polymers.

6.9. Nematic liquid crystals

6.9.1. Energy considerations: continuum elastic model

As mentioned in the introduction to this subject, it is convenient to think of a nematic liquid in terms of a model constituted by rigid rods of length a equal to the long molecular dimension. The orientation of each rod is then characterized by a unit vector $\mathbf{n}$ called a *director*. The equilibrium nematic configuration (see Figure 6.8(a)) corresponds to a parallel orientation of all $\mathbf{n}$ vectors. However, in many cases the effect of wall constraints and/or external perturbing fields distorts such a basic configuration and the molecules adopt a new equilibrium pattern. However, since the change in $\mathbf{n}$ is generally very small over distances $\sim a$, it is adequate to describe the distorted configuration by a continuous function $\mathbf{n}(\mathbf{r})$ and use a continuum elastic model to discuss the thermodynamics of the system.

It can be shown (see for example the book by de Gennes [1]) that any general distortion leading to a perturbed $\mathbf{n}(\mathbf{r})$ configuration can be decomposed into three basic modes called *splay*, *twist* and *bend*, illustrated in Figure 6.9. For such a general distortion, the free energy density F is

$$F = \tfrac{1}{2}K_1(\operatorname{div}\mathbf{n})^2 + \tfrac{1}{2}K_2(\mathbf{n}\cdot\operatorname{rot}\mathbf{n})^2 + \tfrac{1}{2}K_3(\mathbf{n}\times\operatorname{rot}\mathbf{n})^2 \qquad (6.11)$$

where the first, second and third terms, respectively, correspond to the energy densities associated with the splay, twist and bend modes. The coefficients K_1, K_2 and K_3, are appropriate elastic coefficients that are always positive to ensure compatibility with the

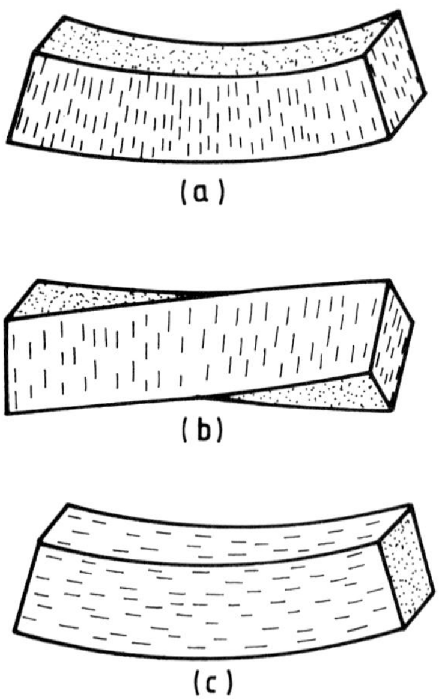

Figure 6.9 *Distortion modes in a liquid crystal: (a) splay; (b) twist; (c) bend [55].*

equilibrium nematic configuration. Although these coefficients are, in general, different, the typical order of magnitude is 10^{-11} N. It is clear from (6.11) that the nematic structure, $\mathbf{n}(\mathbf{r})$ = constant yields the absolute minimum $F = 0$ for the free energy, and so represents the equilibrium configuration. For a more general situation, involving certain constraints, the integral $\int_V F\,\mathrm{d}V$ over the volume of the sample has to be minimized in order to obtain the equilibrium *director* pattern.

6.9.2. Molecular anchoring at walls

Molecular anchoring at cell walls provides an useful means to define a preferential orientation in the nematic liquid crystal. In most practical cases, molecular alignment is either parallel (planar or homogeneous) or perpendicular (normal or homeotropic) to the wall surface. There are a number of methods to obtain either type of molecular orientation. For homogeneous alignment one may use directional rubbing of the wall surface. On the other hand, addition of suitable detergents to the liquid or wall treatment with polyamides or certain lipids (e.g. lecithin) imposes a normal orientation of the molecules.

In dealing with the orientation behaviour of the molecules close to the walls (e.g. at distances of the order of molecular dimensions), one should distinguish between strong and weak anchoring, depending on the comparative value of the molecule–wall and molecule–molecule interactions. When the former predominates the anchoring is strong, and it is weak in the opposite case.

6.9.3. Twisted nematics (TN)

A particularly interesting and useful configuration is that arising from two orthogonal planar alignments imposed by constraints at the upper and lower walls of the sample cell. Distortion of the liquid to adapt itself to these constraints produces a twist mode, defined by the twist angle θ as a function of the distance z to the lower wall. The components of the director $\mathbf{n}$ are

$$n_x(z) = \cos\theta(z), \qquad n_y(z) = \sin\theta(z), \qquad n_z(z) = 0 \tag{6.12}$$

and the associated energy density

$$F = \frac{1}{2}K_2\left(\frac{\mathrm{d}\theta}{\mathrm{d}z}\right)^2 \tag{6.13}$$

One has to minimize

$$\int F\,\mathrm{d}z = \frac{1}{2}K_2\int\left(\frac{\mathrm{d}\theta}{\mathrm{d}z}\right)^2\mathrm{d}z \tag{6.14}$$

leading to the equilibrium condition

$$\frac{\mathrm{d}\theta}{\mathrm{d}z} = \text{constant} \tag{6.15}$$

i.e. the equilibrium distortion corresponds to an uniform twist, and so the structure is designated as *twisted nematic* (see Figure 6.12(a) below). In most respects this structure resembles that of a cholesteric liquid crystal and for the study of its physical properties one should refer to Section 6.11.

Nematic structures with twist angle higher than $\pi/2$ (*supertwisted nematics*) have also been prepared and present remarkable electrooptic properties [56].

6.9.4. Dielectric and optical properties

For a nematic liquid crystal the dielectric permittivity and so the refractive index are anisotropic, showing axial symmetry around the director **n**. For the dielectric permittivity one can define two principal values $\epsilon_{\|}$ and $\epsilon_{\perp}$ corresponding to applied electric fields parallel or perpendicular to **n**, respectively. Consequently the crystal is optically uniaxial. The magnitude of the birefringence $\Delta n = n_o - n_e$ is usually very large in comparison with inorganic materials such as quartz or $LiNbO_3$ and typically lies in the range 0.1–0.3 (see Table 6.4).

On the other hand Δn is a function of light wavelength (dispersive behaviour) and shows a strong dependence on temperature. An empirical dependence

$$\Delta n = \Delta n_o (1 - T_r)^m \tag{6.16}$$

has been used, where T_r is the reduced temperature $T_r = T/T_{NI}$, i.e. normalized to the transition temperature to the isotropic phase. Δn and m are fitting parameters. Note that for $T \geq T_{NI}$ the birefringence is cancelled.

6.10. Electrooptics of nematic liquid crystals

Electric fields induce a variety of effects in nematic liquid crystals, some of them being quite complex. Their occurrence depends on the material, type of doping, etc. They will now be briefly discussed.

6.10.1. Variable birefringence. Fredericksz transition

Let us consider the simple case of a perfectly insulating nematic liquid, free of ionic charges. The effect of the field can be described within the continuum elastic model by taking into account an additional dielectric energy term in the expression (6.11) for the energy density. The additional term can, with regard to the random molecular orientation, be written

$$F(\mathbf{E}) = -\tfrac{1}{2}\epsilon_0 \Delta\epsilon (\mathbf{n} \cdot \mathbf{E})^2 \tag{6.17}$$

where $\Delta\epsilon = \epsilon_{\|} - \epsilon_{\perp}$ stands for the dielectric anisotropy or difference between the dielectric components parallel and perpendicular to the long molecular axis. Due to this term, the field tends to align the molecules parallel or perpendicular to it, depending on whether the anisotropy is positive ($\Delta\epsilon > 0$) or negative ($\Delta\epsilon < 0$), respectively. However, the effect of the field competes with the alignment imposed by the wall constraints. A detailed analysis shows that for weak fields the wall alignment predominates and the field cannot modify the initial molecular orientation. However, above a certain critical potential V_c, molecular reorientation develops in the bulk of the sample, causing a change in the dielectric and optical parameters of the medium. This is the so-called *Fredericksz transition*, which is second order. For a planar molecular alignment at walls, the critical potential V_c can be easily obtained from minimization of the free energy, including the term (6.17). It is found that

$$V_c = \left\{ \frac{\pi^2 K_{11}}{\epsilon_0(\epsilon_{\|} - \epsilon_{\perp})} \right\}^{1/2} \tag{6.18}$$

Above this value, the molecular director experiences a rotation θ towards the field direction, which is a function of the applied potential (Figure 6.10). For high enough potentials, the whole sample is aligned with the field and the effect is saturated.

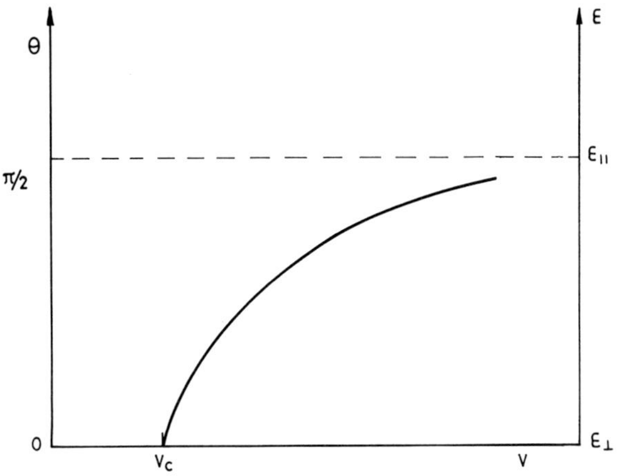

Figure 6.10 *Rotation θ of the molecular director* **n** *as a function of applied voltage for a nematic crystal. The Fredericksz transition occurs at* V_c. *Note the change in dielectric constant from $\epsilon_\perp$ to $\epsilon_\|$, associated with the molecular rotation.*

The field-induced molecular rotation makes the dielectric constant evolve from $\epsilon_\perp$ to $\epsilon_\|$ on increasing the field. Consequently, the index tensor is also modified. In the absence of a field and planar alignment along X, the principal axes for light propagation normal to the cell faces (Z-axis) are along the molecular director (Y-axis) and along the X-axis. The refractive indices are n_e (extraordinary) and n_o (ordinary), respectively. When the field is applied, the ordinary index is not altered, whereas n_e becomes $n_e(\theta)$, given by

$$\frac{1}{n_e^2(\theta)} = \frac{\cos^2\theta}{n_e^2} + \frac{\sin^2\theta}{n_o^2} \tag{6.19}$$

θ being the field-induced rotation of the molecular director. Therefore, the cell can be considered as a *voltage-controlled phase modulator*. When the field is removed, molecules return to their initial orientation along Y, showing a *memory* behaviour.

6.10.2. Dichroic absorption by dissolved dyes

Let us consider organic dyes whose absorption depends on the molecular orientation with regard to the light electric field (dichroic dye). When incorporated into the liquid crystal, the dye molecules tend to align with the LC director. Therefore, an applied electric field causing molecular reorientation, induces a strong change in the absorption spectra and provides a useful electrooptic (electroabsorption) effect [56].

6.10.3. Convective instabilities: dynamic light scattering

The application of an electric field to nematic liquid crystals containing mobile charge carriers gives rise to complex hydrodynamic phenomena that strongly alter the equilibrium molecular configuration. These effects have interesting optical consequences and have been widely used in display systems. Schematically, the observed phenomena can be described as follows [1]. When the applied potential reaches a threshold value, independent of sample thickness, a periodic distortion of the nematic alignment is produced. A cellular flow pattern develops, indicating the surge of a convective instability. At higher voltages the flow becomes turbulent and the long-range nematic order is completely destroyed. An optical consequence is the occurrence of a strong and diffuse scattering of light (*dynamic scattering*).

The model generally invoked [1] to account for this behaviour was put forward by Helfrich, and Tsvetkov and Carr. The essential idea is that mobile charges accumulate at the nodes of the flow pattern, indicating an additional space charge field. This field enhances the distortion and consequently instability results. The threshold voltage is related to the elastic coefficients providing the restoring molecular torque.

6.11. Cholesteric liquid crystals

6.11.1. Energy considerations: continuum elastic model

The continuum elastic model can, indeed, be applied to cholesteric liquid crystals. However, the absence of any mirror symmetry (chiral molecules) implies the presence of an additional term proportional to $\mathbf{n} \cdot \mathrm{rot}\,\mathbf{n}$ in the expression (6.11) for the energy density [1]. This can be now conveniently written

$$F = \tfrac{1}{2}K_1(\mathrm{div}\,\mathbf{n})^2 + \tfrac{1}{2}K_2(\mathbf{n} \cdot \mathrm{rot}\,\mathbf{n} + p_0)^2 + \tfrac{1}{2}K_3(\mathbf{n} \times \mathrm{rot}\,\mathbf{n})^2 \tag{6.20}$$

The new specific contribution arising from chirality results from the cross product in the second (*twist*) term, which also yields a nonrelevant constant value $K_2 p_0^2/2$.

Considering a pure twist distortion with Z-axis, i.e. $n_x = \cos\theta(z)$, $n_y = \sin\theta(z)$, $n_z = 0$, the energy density takes the form

$$F = \tfrac{1}{2}K_2\left(\frac{\mathrm{d}\theta}{\mathrm{d}z} - p_0\right)^2, \tag{6.21}$$

that after minimization, $F = 0$, leads to the following solution for the equilibrium twist pattern

$$\theta = p_0 z + z_0 \tag{6.22}$$

i.e. the configuration is a helix with Z-axis and geometrical period

$$\Lambda = \frac{2\pi}{p_0} = \frac{\Lambda}{2} \tag{6.23a}$$

However, the invariance of physical properties *vs* inversion of the molecular director determines the pitch or physical period

$$\mathcal{P} = \frac{\pi}{p_0} \tag{6.23b}$$

In particular, the refractive index shows a periodic pattern along Z with period or pitch $\mathcal{P}$. It can be varied by using external agents as temperature, electric fields and magnetic fields.

6.11.2. Dielectric and optical properties

From the dielectric and optical point of view, a cholesteric liquid behaves locally as a nematic, i.e. is uniaxial with the local director $\mathbf{n}$ being the optic axis. So the dielectric tensor can be written

$$\epsilon_{ij}(\mathbf{r}) = \epsilon_\perp \delta_{ij} + \Delta\epsilon n_i(\mathbf{r}) n_j(\mathbf{r}) \tag{6.24}$$

where $\epsilon_\parallel$ and $\epsilon_\perp$ are the principal values along and perpendicular to $\mathbf{n}$, respectively.

However, the occurrence of a helical configuration for $\mathbf{n}$ introduces some interesting and practically relevant features. One has to solve the equation for a monochromatic wave with the dielectric tensor (10.24). For a plane wave propagating along Z, it is convenient to take

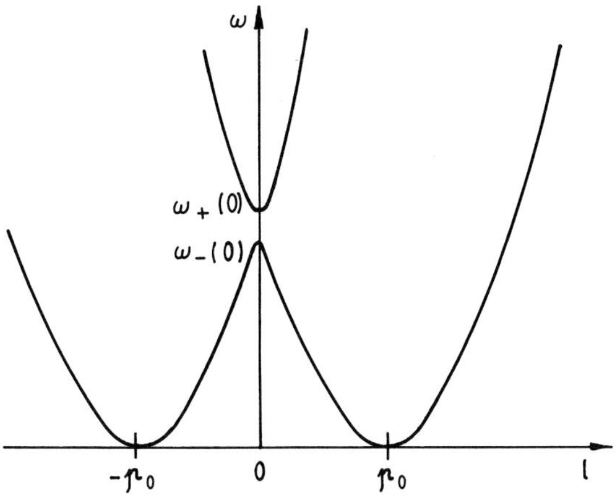

Figure 6.11 *Branches of the dispersion law in a cholesteric liquid crystal [1].*

a basis for the polarization states constituted by two circularly polarized modes $E^{\pm}$, $E^{\pm} = E_x + iE_y \sim e^{i(l \pm \not{k}_0)z}$, where l is an arbitrary real number (linearly related to the propagating wavevector). The dispersion relation $l(\omega)$ is obtained from the equation

$$(-\bar{k}^2 + l^2 + \not{k}_0^2)^2 - 4\not{k}_0^2 l^2 - \Delta_k^2 = 0 \tag{6.25}$$

where

$$\bar{k}^2 = \left(\frac{\omega}{c}\right)^2 \bar{\epsilon} \tag{6.26a}$$

$$\Delta_k^2 = \left(\frac{\omega}{c}\right)^2 \Delta_\epsilon \tag{6.26b}$$

with $\bar{\epsilon} = (\epsilon_{||} + \epsilon_{\perp})/2$ and $\Delta_\epsilon = (\epsilon_{||} - \epsilon_{\perp})/2$. It consists of two branches schematically illustrated in Figure 6.11. One can see a forbidden frequency region between $\omega_+(0)$ and $\omega_-(0)$, given by

$$\omega_+(0) = \frac{c\not{k}_0}{n_o} \tag{6.27a}$$

$$\omega_-(0) = \frac{c\not{k}_0}{n_e} \tag{6.27b}$$

The peculiarities of light propagation in a cholesteric LC essentially depend on the relative values of the twist rate coefficient $\not{k}_0$ of the structure and on the light dephasage rate associated to the birefringence Δn (i.e. twist rotation *vs* phase retardation). The details of the analysis can be consulted in the appropriate literature [1]. Here, only the main results, corresponding to useful operative regimes, will be quoted. The same discussion also applies to a TN configuration.

(a) $\not{k}_0 \ll k\Delta n$ *or* $\lambda \ll \Delta n \Lambda$ *(Mauguin limit)*

In this region the optical path retardation along a pitch of the molecular helix $(\Delta n \Lambda/2)$ is much larger than the light wavelength, i.e. the optical retardation rate is much faster than the molecular reorientation rate. The eigensolutions for a propagating wave along Z have the electric vector following the helicoidal director pattern. In other words, the twisted molecular pattern "guides" the wave polarization through the structure. An ordinary linearly polarized wave at the entrance face ($z = 0$) of the cell remains an ordinary wave all the way through its path to the exit face ($z = l$). The outgoing wave is linearly polarized along the director $\mathbf{n}(l)$. The dephasage introduced by traversing the liquid crystal cell with thickness l is

$$\delta\phi_1 = \frac{\omega l}{n_o} \tag{6.28}$$

The same situation is applicable to an incoming extraordinary wave, that keeps that character all the way through, attaining a dephasage

$$\delta\phi_1 = \frac{\omega l}{n_e} \tag{6.29}$$

For an incident wave, linearly polarized in an arbitrary direction, the outgoing wave will be, in general, elliptically polarized.

In many practical devices (see next chapter) the liquid crystal cell is placed between crossed polarizers.

(b) $\Delta n/n \ll 1$ ($p_0 > k\Delta n$ or $\lambda > \Delta n\mathscr{P}$)

From the general dispersion equation (6.25), it is found [1] that the propagating eigenmodes are two waves, very nearly circularly polarized, with wavevectors

$$k_1 = \bar{k} + \frac{(\Delta_k)^4}{8kp_0(k + p_0)} \tag{6.30a}$$

$$k_2 = \bar{k} + \frac{(\Delta_k)^4}{8kp_0(p_0 - k)} \tag{6.30b}$$

The situation is, in a way, similar to that described in Chapter 2 when discussing light propagation in optically active crystals. There, the pitch of the structure has atomic dimensions and condition $\lambda > \Delta n\mathscr{P}$ is, indeed, well obeyed.

In this regime, an incident linearly polarized wave will also be linearly polarized at the exit face, but with the polarization plane rotated an angle ψ given by

$$\psi = \tfrac{1}{2}(k_1 - k_2)l = \frac{p_0}{32}\,\frac{n_e^2 - n_o^2}{(n_e^2 + n_o^2)\lambda'(1 - \lambda'^2)} \tag{6.31}$$

with

$$\lambda' = \frac{\lambda}{\Lambda} = \frac{p_0}{k} \tag{6.32}$$

In comparison with optically active inorganic materials (see Table 2.3), equation (6.31) predicts extremely large specific rotations (about 10^4 times higher). Moreover, the rotatory power has a very strong dispersive behaviour, diverging at frequencies ω_+ and ω_- limiting the gap in the frequency spectrum, where propagating modes are not allowed (*Bragg regime*).

(c) Bragg regime

For the same situation considered in (b), the Bragg regime corresponds to frequencies $\omega \approx \omega_+ \approx \omega_- = cp_0/n$, i.e. $\lambda = 2\pi c/p_0 = 2\Lambda$. It ensures constructive interference for the waves reflected from all cholesteric layers and accounts for the intense reflected beam (first-order Bragg reflection). Higher-order reflections ($2\Lambda = m\lambda$) are not observed. On the other hand, it can be shown that the reflected wave is circularly polarized with the electric field reproducing the helix pattern of the molecular director.

The phenomenon of selective Bragg reflection is responsible for the coloration of cholesterics and constitutes the basis for a number of practical applications. As an example, by depositing a cholesteric film on a given substrate and measuring the Bragg reflected wavelength, the surface temperature of the substrate can be determined and monitored.

6.11.3. Electrooptics of cholesteric liquid crystals

The effect of an electric field on a cholesteric liquid crystal can be conveniently discussed with the continuum elastic model or by just directly considering the electrostatic torques induced by the field. Let us first assume a bulk cholesteric sample consisting of molecules with negative dielectric anisotropy, i.e. $\Delta\epsilon = \epsilon_{\parallel} - \epsilon_{\perp} < 0$. In order to minimize the free energy (6.17), molecules will tend to arrange perpendicularly to the field in order to maximize the induced electric dipole moment. Consequently, the axis of the director helix

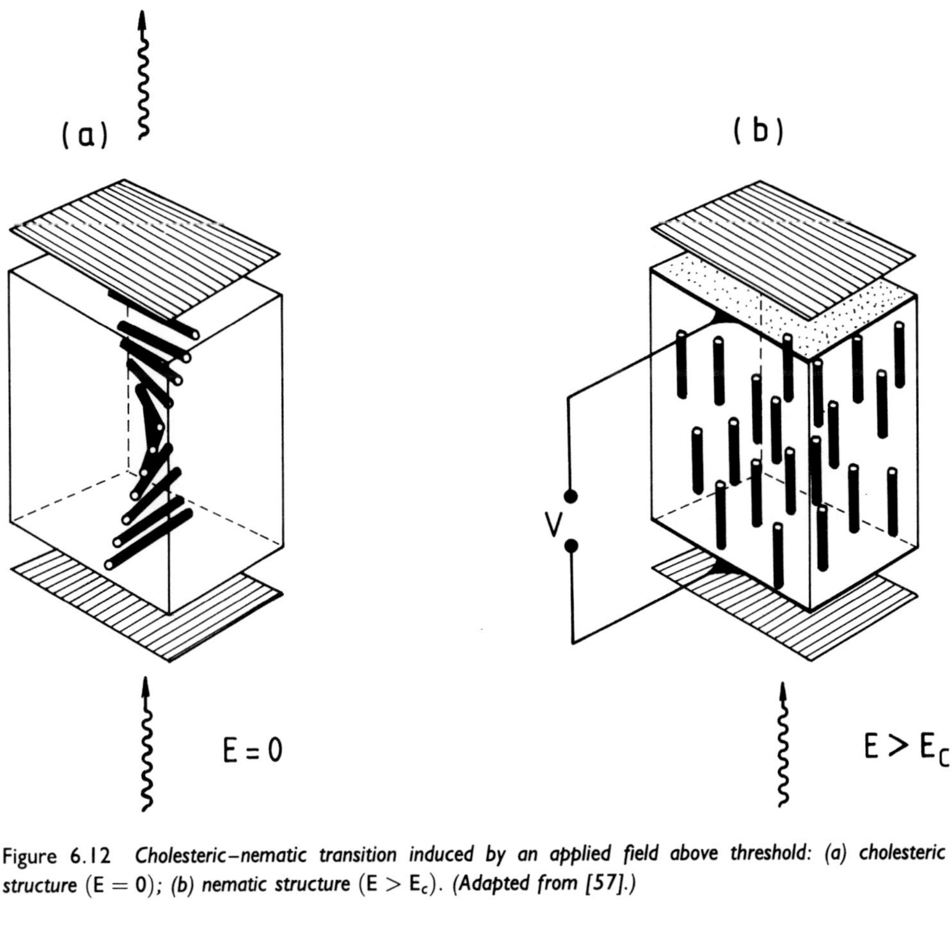

Figure 6.12 *Cholesteric–nematic transition induced by an applied field above threshold: (a) cholesteric structure* ($E = 0$)*; (b) nematic structure* ($E > E_c$). *(Adapted from [57].)*

becomes oriented along the field so that the associated electrostatic energy contribution is zero regardless of the pitch value.

The situation is different for molecules with positive dielectric anisotropy $\Delta\epsilon > 0$. In such a case, molecules tend to align parallel to the field and overcome the anchoring forces. The process is governed by the dielectric coupling energy density (6.17). In fact, at a given critical field the helix is completely unwound and a cholesteric-nematic transition has taken place, Figure 6.12. The critical field can be calculated with the continuum model and yields [56]

$$E_c = \frac{\pi^2}{\mathscr{P}} \left(\frac{K_2}{\epsilon_0 \Delta\epsilon} \right)^{1/2} \tag{6.33}$$

$\mathscr{P}$ being the unperturbed pitch. The change in both, orientation and pitch of the helix, induces an associated change in the propagation properties of light, in accordance with the discussion offered above.

For the supertwisted structures mentioned in Section 6.9.3, the effect of an electric field shows peculiar features and leads to bistable behaviour [56].

6.12. Ferroelectric liquid crystals

6.12.1. Structural considerations

The existence of ferroelectric liquid crystals (FLC) was demonstrated in 1975 by Meyer and coworkers [58]. This development has opened a new field with exciting possibilities for technological applications. Although the search for ferroelectricity in liquid crystals is a relatively old subject, all previous efforts had failed because of the weakness of the polar interactions (leading to the $\mathbf{n}/-\mathbf{n}$ invariance). However, a different approach

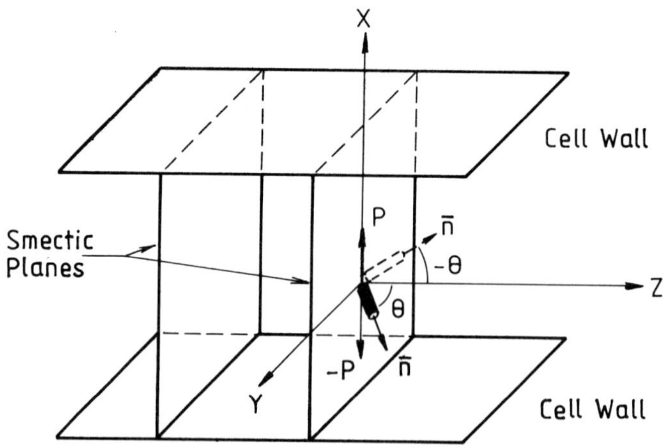

Figure 6.13 *Illustrating the two stable orientations of the molecular director* $\mathbf{n}$ *in a smectic C* liquid crystal, at angles* $+\theta$ *and* $-\theta$ *with regard to the Z-axis. The associated molecular polarizations are along* $+X$ *and* $-X$.

Table 6.5 ***Illustrative FLC molecules.***

Molecules†	Temperature range of S_C^* (°C)	$\lvert P_s \rvert$ (nC/cm^2)
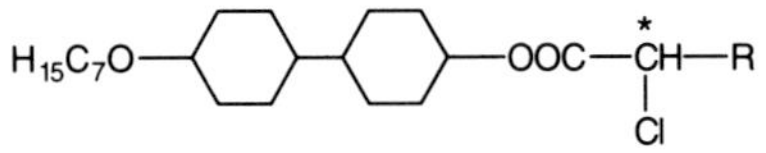	—	290
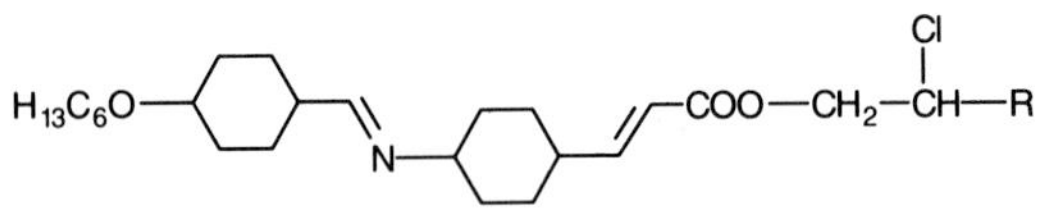		50
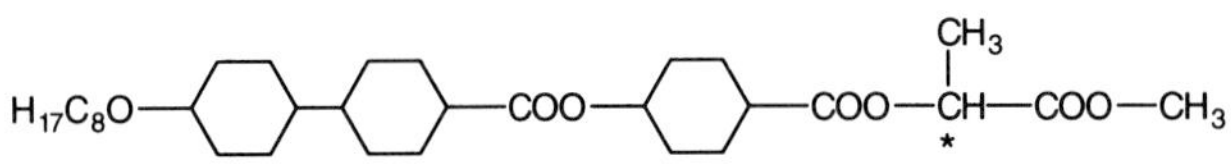	68–120	65
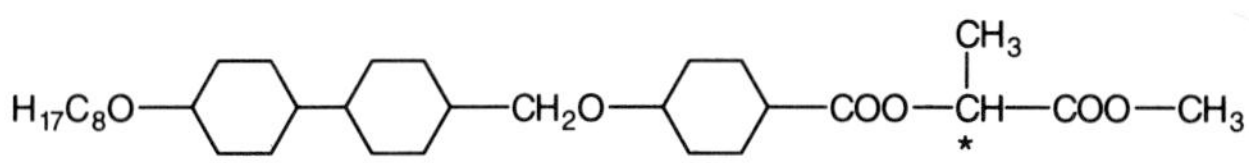	66–91	30
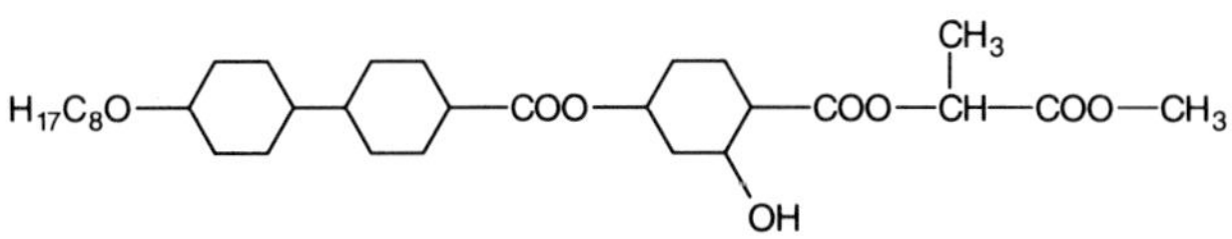	28–86	180

Data supplied by Dr J. Etxebarria (University of the Basque Country).
† R stands for $-CH(CH_3)-CH_2-CH_3$.

involving improper ferroelectric ordering in the tilted smectic C phase has been finally successful.

Let us suppose that the chiral molecules in a smectic C* crystal have an electric dipole moment perpendicular to the long molecular axis (or molecular director). The C_2 symmetry of the molecular arrangement imposes (due to a rotational bias) that the dipole moment should lie along the C_2 axis, i.e. perpendicular to the long molecular axis and contained in the smectic layer planes. Consequently, a net polarization appears (Figure 6.13). Inverting the tilt angle results in a reversal of the sense of polarization. A lucid comparison between solid and liquid crystalline ferroelectrics can be found in reference [59].

In accordance with the above mechanism for the ferroelectricity, the net polarization depends on the magnitude of the molecular dipole moment and the degree of hindered rotation. In practice, the polarization is quite small due to a low hindrance barrier, although some successful efforts have been made to improve this situation. Values higher than 200 nC/cm^2 have been obtained.

Since ferroelectric liquid crystals possess, in general, helical structure, the total polarization of the sample becomes zero. However, it is possible to cancel the helical structure and so obtain a net polarization. The most common method is to use surface anchoring forces to stabilize the unwound structure in the case of a thin sample. The resulting materials are designated [60] as *surface stabilized ferroelectric liquid crystals* (SSFLC).

Most FLC consist of rod-like molecules with a central mesogenic unit bound to flexible terminal chains. In one of these chains the chiral carbon atoms are included. Table 6.5 shows the structure of a few illustrative LCF molecules, together with the saturation polarization P_s and the temperature range of the S_C^* phase.

Other smectic structures (I* and F*) are also ferroelectric. On the other hand, antiferroelectric ordering has been reported for a smectic C* liquid crystals [61].

6.12.2. Electrooptics of surface stabilized ferroelectric liquid crystals (SSFLC) [62,63]

In an SSLC the active medium is confined between the two closely spaced ($\sim 2\,\mu$m) boundary planes of the cell. Leaving apart other complicating features, the director **n** is forced to be aligned parallel to both planes, which are, in turn, perpendicular to the smectic (XY) layers (Figure 6.13). Two stable polarization states are possible with the polarization **P** pointing up or down (along the X-axis) in the cell. The corresponding director **n** lies at angles $+\theta$ and $-\theta$, respectively, with regard to the normal to the smectic layers (Z-axis).

Ferroelectric switching between the two polarization states can be achieved by applying an electric field. The $P \cdot E$ torque causes the director **n** to describe a half-cone around the Z-axis from the $+\theta$ to the $-\theta$ position. This motion implies that the polarization $\mathbf{P}_s$ switches from the up to the down orientation. The net result of the field has been the rotation of **n** by an angle 2θ (double of the tilt angle θ) in the plane of the cell. Since **n** defines the optic axis, this rotation of the index ellipsoid immediately leads to electrooptic behaviour. Assuming that the cell is placed between crossed polarizers and is illuminated by a beam along the X-axis, the outcoming intensity I_s is related to the incoming intensity I_0 by

$$I_s = I_0 \left\{ (\sin 4\theta) \sin\left(\frac{\pi \Delta n l}{\lambda} \right) \right\}^2, \tag{6.34}$$

l being the cell thickness, λ the light wavelength and Δn the liquid crystal birefringence.

The switching process has a characteristic response time τ that can be approximately written as

$$\tau = \frac{\eta}{P_s E} \tag{6.35}$$

where η refers to the rotational viscosity accounting for the damping of the director conic motion. This formula indicates that switching times depend inversely on the applied field and directly on the viscosity η. The experimental results for the switching behaviour are rather complex and not completely well understood [64,65]. However, the main relevant result is that switching times in the microsecond and even submicrosecond region can be obtained, opening up a new range of liquid crystal electrooptic devices.

6.13. Recent developments: polymer dispersed liquid crystals (PDLC)

A recent development in the field of organic composite materials, offering very attractive

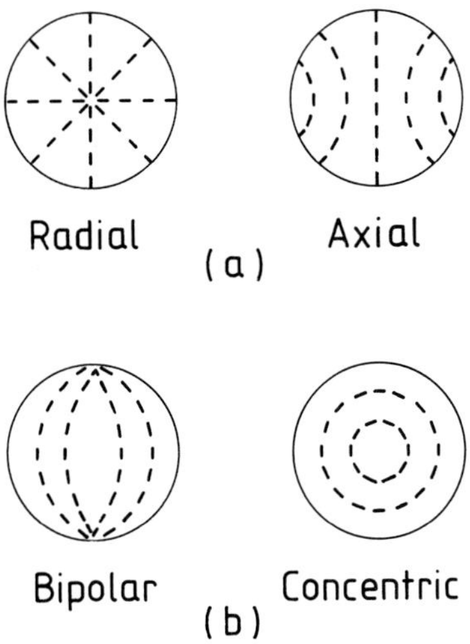

Figure 6.14 *Molecular arrangements inside LC drops dispersed in a polymer: (a) radial anchoring; (b) tangential anchoring. (Adapted from [69].)*

perspectives for electrooptical applications, is the dispersion of liquid crystals drops or cylinders into polymer hosts [66–69]. Particular interest has arisen in their use for switchable windows in building and transport industries. The research being performed is also challenging from a basic point of view in relation to the study of size and surface effects.

The dispersion of liquid crystals into the polymer can be achieved by several techniques such as phase separation, microencapsulation and use of microporous membranes. Phase separation is the simplest and more useful method. It permits the preparation of spherical droplets with a uniform and precisely controlled diameter in the range 0.01–20 μm.

The study of the stable nematic structures that can be accommodated inside the spherical or cylindrical droplets is a topic of basic research. The theoretical analysis starts from the expression for the free elastic energy (6.11), supplemented by the anchoring energy term,

$$\frac{1}{2}\int_V W_0 \sin^2\theta \, \mathrm{d}S \tag{6.36}$$

where W_0 is the strength and θ the molecular orientation at the cavity walls (referred to the wall normal). Depending on drop size and anchoring strength, a variety of nematic director configurations become stable, as observed through the polarizing optical microscope [69]. For example, for high strength and spherical particles, stable configurations are illustrated in Figure 6.14(a) for radial and in Figure 6.14(b) for tangential anchoring.

The application of an electric field [71,72] introduces the corresponding coupling term in the free energy and may lead to a change in the director configuration. It is, just, this effect which gives rise to interesting applications (see next chapter). In fact, one may draw stability diagrams using two coordinates, magnitude of electric field and drop size or anchoring strength.

References

1. P. G. de Gennes, *The Physics of Liquid Crystals*, Clarendon Press, Oxford (1974).
2. W. H. De Jeu, *Physical Properties of Liquid Crystalline Materials*, Gordon and Breach, New York (1980).

3. L. M. Blinov, *Electrooptical and Magnetooptical Properties of Liquid Crystals*, John Wiley, New York (1983).
4. D. S. Chemla and J. Zyss (eds), *Nonlinear Optical Properties of Organic Molecules and Crystals*, Vols 1 and 2, Academic Press (1987).
5. P. N. Prasad and D. J. Williams, *Introduction to Nonlinear Effects in Molecules and Polymers*, John Wiley, New York (1991).
6. H. Kuhn and J. Robillard (eds), *Nonlinear Optical Materials*, CRC Press (1992).
7. G. R. Meredith, in *Nonlinear Optics: Materials and Devices* (ed. C. Flytzanis and J. L. Oudar), Springer-Verlag, Berlin (1986).
8. B. K. Nayar and C. S. Winter, *Opt. Quantum Electron.* **22**, 297 (1990).
9. R. T. Bailey, F. R. Cruickshank, D. Pugh and J. N. Sherwood, *Int. J. Optoelectron.* **5**, 89 (1990); *Acta Cryst.* **A47**, 145 (1991).
10. S. Miyata (ed.), *Nonlinear Optics: Fundamentals, Materials and Devices*, North-Holland, Amsterdam (1992).
11. J. L. Oudar and D. S. Chemla, *J. Chem. Phys.* **66**, 2664 (1977).
12. R. N. De Martino, E. W. Choe, G. Khanarian, D. Haas, T. Ushe, G. Nelson, J. Stamatoff, D. Stutz, C. C. Teng and H. Yoon, in *Nonlinear Optical and Electroactive Polymers* (eds P. N. Prasad and D. R. Vlrich), Plenum, New York (1988).
13. J. L. Oudar and D. S. Chemla, *Opt. Commun.* **13**, 164 (1975).
14. J. F. Nicoud and R. J. Twieg, in *Nonlinear Optical Properties of Organic Molecules and Crystals*, Vol. 1 (eds D. S. Chemla and J. Zyss), Academic Press, Orlando (1987).
15. J. C. Baumert, R. J. Twieg, G. C. Bjorklund, R. A. Logan and C. W. Dirk, *Appl. Phys. Lett.* **51**, 1484 (1987).
16. J. O. Morley and D. Pugh, in *Organic Materials for Nonlinear Optics* (eds D. Bloor and R. A. Hann). London, Royal Society of Chemistry (1989), pp. 28–39.
17. J. Zyss, *Nonlinear Optics* **1**, 3 (1991); *J. Chem. Phys.* **98**, 6583 (1993).
18. B. J. Orr and J. F. Ward, *Mol. Phys.* **20**, 513 (1971).
19. M. G. Kuzyck and C. W. Dirk, *Phys. Rev.* **41A**, 5098 (1990).
20. K. C. Rustagi and J. Ducuing, *Opt. Commun.* **10**, 258 (1974).
21. H. S. Nalwa, *Appl. Organomet. Chem.* **5**, 349 (1991); *Adv. Mater.* **5**, 341 (1993).
22. J. R. Lindle, R. G. S. Pong, F. J. Bartoli and Z. H. Kafafi, *Phys. Rev.* **B13**, 9447 (1993).
23. J. L. Oudar and J. Zyss, *Phys. Rev.* **A26**, 2016 (1982).
24. K. D. Singer, S. L. Lalama, J. E. Sohn and R. D. Small, in *Nonlinear Optical Properties of Organic Molecules and Crystals*, Vol. 1 (ed. D. S. Chemla and J. Zyss), Academic Press, Orlando (1987).
25. S. Allen, T. D. McLean, P. F. Gordon, B. D. Bothwell, M. B. Hursthouse and S. A. Karulov, *J. Appl. Phys.* **64**, 2583 (1988).
26. S. Allen, in *Organic Materials for Nonlinear Optics* (eds R. A. Hann and D. Bloor), Royal Society of Chemistry, London (1989).
27. T. Yoshimura, *J. Appl. Phys.* **62**, 2028 (1987).
28. R. H. Tredgold, *Rep. Prog. Phys.* **50**, 1609 (1987).
29. B. Tieke, *Adv. Mater.* **2**, 222 (1990).
30. I. H. Peterson, *J. Phys. D: Appl. Phys.* **23**, 379 (1990).
31. I. Langmuir, *J. Am. Chem. Soc.* **39**, 1848 (1917).
32. K. B. Blodgett, *J. Am. Chem. Soc.* **57**, 1007 (1935).
33. G. H. Cross, I. R. Peterson. I. R. Girling, N. A. Cade, M. J. Goodwin, N. Carr, R. S. Sethi, R. Marsden, G. W. Gray, D. Lacey, A. M. McRoberts, R. M. Scrowston and K. J. Toyne, *Thin Solid Films* **156**, 39 (1988).
34. J. P. Cresswell, J. Tsibouklis, M. C. Petty, W. J. Feast, N. Carr, M. Goodwin and Y. M. Lvov, *SPIE* **1337**, 358 (1990).
35. V. Dentan, Ph.D. Thesis, Université de Paris-Sud, Orsay (1991, unpublished).
36. A. J. Heeger and D. R. Ulrich, *Nonlinear Optical Properties of Polymers*, North-Holland, New York (1988).

37. G. R. Möhlmann, *Europhys. News* **21**, 83 (1990).
38. M. G. Kuzyk, J. E. Sohn and C. W. Dirk, *J. Opt. Soc. Am.* **B7**, 842 (1990).
39. J. Comizzoli, *J. Electrochem. Soc.* **134**, 424 (1987).
40. J. R. Hill, P. Pantelis, F. Abbasi and P. Hodge, in *Organic Materials for Nonlinear Optics* (ed. R. A. Hann and D. Bloor), Royal Society of Chemistry (1989).
41. M. Eich, B. Reck, D. Y. Yoon, C. G. Wilson and G. C. Bjorklund, *J. Appl. Phys.* **66**, 3241 (1989).
42. A. B. Leves and C. C. Leznoff (eds), *Phthalocyanines. Properties and Applications*, VHC Publishers, Vol. 1, Weinheim (1989); Vol. 2, Berlin (1993).
43. D. Avnir, D. Levy and R. Reisfeld, *J. Phys. Chem.* **88**, 5956 (1984).
44. R. Reisfeld, R. Zusman, Y. Cohen and M. Eyal, *Chem. Phys. Lett.* **47**, 142 (1988).
45. R. Reisfeld, *SPIE Proc. Sol-Gel Optics* (eds J. D. MacKenzie and D. R. Ulrich), **1328**, 29 (1990).
46. C. Sanchez, J. Livage, M. Henry and F. Babinneau, *J. Non-cryst. Solids* **100**, 650 (1988).
47. E. Toussaire, J. Zyss, P. Griesmar and C. Sanchez, *Nonlinear Optics*, **1**, 349 (1991).
48. R. Masse, M. Bagieu-Beucher, J. Pecaut, J. P. Levy and J. Zyss, Abstract presented to the 2nd French–Israeli Symposium on Nonlinear Optics, Port-Bacarés, January (1992).
49. E. Ruiz-Hitzky and P. Aranda, *Adv. Mater.* **2**, 545 (1990).
50. P. J. Collings, *Liquid Crystals. Nature's Delicate Phase of Matter*, Princeton University Press, N. J. (1990).
51. D. Masurel, C. Sirlin and J. Srilin, *Nouv. J. Chem.* **11**, 455 (1987).
52. D. Demus, *Liquid Cryst.* **5**, 75 (1989).
53. H. Finkelmann, *Phil. Trans. R. Soc.* **A309**, 105 (1983).
54. H. Ringsdorf, B. Schlarb and J. Venzmer, *Angew. Chem.* **100**, 117 (1988).
55. I. Ch. Khoo, *Prog. Optics* **26**, 105 (1988).
56. E. P. Raynes, in *Electrooptic and Photorefractive Materials* (ed. P. Günter), Springer-Verlag, Berlin (1987).
57. J. Fünfschilling, *Condensed Matter News*, **1**, 12 (1991).
58. R. B. Meyer, L. Liebert, L. Strzelecki and P. Keller, *J. Phys. Lett*, **36L**, 69 (1975).
59. R. Blinc, *Condensed Matter News* **1**, 17 (1991).
60. N. A. Clark and S. T. Lagerwall, *Appl. Phys. Lett.* **36**, 899 (1980); Ferroelectrics **59**, 25 (1984).
61. A. D. L. Chandani, E. Gorecka, Y. Ouchi, H. Takezoe and A. Fukuda, *Jpn. J. Appl. Phys.* **28**, L1265 (1989).
62. J. S. Patel, *Opt. Engineering* **26**, 129 (1987).
63. L. A. Beresnev and L. M. Blinov, *Ferroelectrics* **92**, 335 (1989).
64. J. Wahl and S. C. Jain, *Ferroelectrics* **59**, 161 (1984).
65. X. Jiu-Zhi, M. A. Handschy and N. A. Clark, *Ferroelectrics* **73**, 305 (1987); *Liq. Cryst.* **2**, 707 (1987).
66. J. W. Doane, *MRS Bulletin* 16, 22 (1991).
67. G. P. Crawford, M. Vilfan, I. Vilfan and J. W. Doane, *Phys. Rev.* **A43**, 835 (1991).
68. G. Srajer, F. Lonberg and R. B. Meyer, *Phys. Rev. Lett.* **64**, 1102 (1991).
69. G. P. Crawford and J. W. Doane, *Condensed Matter News* **1**, 5 (1992).
70. R. Ondis-Crawford, E. P. Boyko, B. G. Wagner, J. H. Erdman and J. W. Doane, *J. Appl. Phys.* **69**, 6380 (1991).
71. J. H. Erdmann, S. Zumer and J. W. Doane, *Phys. Rev. Lett.* **64**, 1907 (1990).
72. S. Kralj and S. Zumer, *Phys. Rev.* **A45**, 2461 (1992).

Chapter Seven

Bulk Electrooptic Applications

7.1. Introduction

By modifying the absorption coefficient and refractive index of a material, electrooptics provides the means to alter the amplitude, phase and direction of a light beam. Consequently, a large variety of optoelectronic devices for light control and manipulation can be realized. Devices considered in this chapter are:

- Deflectors and scanners
- Phase, polarization and intensity modulators
- Frequency shifters
- Optical shutters
- Spatial light modulators
- Displays

Each of these applications has appropriate *figures of merit* and so may demand different design criteria and rely on different electrooptic materials.

A great stimulus for the development of electrooptic devices resulted from the invention of lasers in 1959 and the fast subsequent growth of optoelectronic technologies, mostly in connection with the fields of optical communications and optical processing and computing.

In this chapter bulk electrooptic devices will be dealt with. Although many of them are more efficiently operated in waveguide technologies (because of the advantages of miniaturization), they are still needed for bulk optics applications. Waveguide and integrated optics devices will be discussed separately in Chapter 8. On the other hand, applications specifically based on the photorefractive effect will be the subject of Chapter 11. Finally, devices using the electrooptic behaviour of semiconductor quantum wells and superlattices (nanostructures) will be described in Chapter 9.

7.2. Deflectors and scanners

These devices perform a continuous (*analogic*) or point to point (*digital*) deflection of a light beam, governed by an electric field [1–4]. This is accomplished through a field-induced change in refractive index. Main applications are in the area of digital laser displays. In laser technology, deflectors can be also used as intracavity or external Q-switches and mode-lockers for pulse generation and shaping. Both Pockels (e.g. KDP, ADP, $LiNbO_3$ and $LiTaO_3$) and Kerr (e.g. nitrobenzene) materials have been used. However, the high fields required for a nitrobenzene Kerr cell and problems associated to charge transport and electrode processes make the use of Pockels single crystals preferable.

7.2.1. Analogue deflectors

The basic set-up is constituted by a prism illuminated by a collimated beam, as schematically illustrated in Figure 7.1. θ stands for the beam deviation caused by the prism, whereas $\delta\theta$ is the field-induced deflection. It can be shown that

$$\delta\theta = \frac{\Delta n_{\mathrm{p}}(l_1 - l_2)}{n_0 D} \tag{7.1}$$

Δn_{p} being the field-induced change of the prism refractive index and n_0 the index of the external medium. D is the width of the beam and l_1 and l_2 the path lengths of the limiting rays inside the prism. The highest $\delta\theta$ occurs for the light beam incident along the direction for maximum deviation. Then, using the full prism aperture ($l_1 = l$, $l_2 = 0$), (7.1) becomes

$$\delta\theta = \frac{\Delta n_{\mathrm{p}} l}{n_0 D} \tag{7.2}$$

Typical deflections are quite small, $\delta\theta \approx 10^{-3}$–$10^{-5}$ rads for fields $E \approx 10^3$–10^4 V/cm.

It is often more convenient to express the deflection in terms of the ***number of resolvable spots*** N_{R}. This refers to the number of exit beams that can be resolved within the total deflection $\delta\theta_{\mathrm{m}}$ caused by the applied field. Assuming diffraction-limited divergence θ_{D} for the beam, i.e.

$$\theta_{\mathrm{D}} = \frac{\alpha\lambda}{n_0 D} \tag{7.3}$$

where $\alpha \sim 1$ depends on the beam amplitude profile, then

$$N_{\mathrm{R}} = \frac{\delta\theta}{\theta_{\mathrm{D}}} = \frac{\Delta n_{\mathrm{p}} l}{\alpha\lambda} \tag{7.4}$$

For typical conditions $N_{\mathrm{R}} \approx 1$. Therefore, multiple-prisms deflectors [4] have to be used in practical devices. In these cascade deflectors, N_{R} is enhanced by a factor $2L/l$, L being the full length of the prism train. N_{R} values over 150 have been achieved, and up to ~1000 estimated from proposed designs.

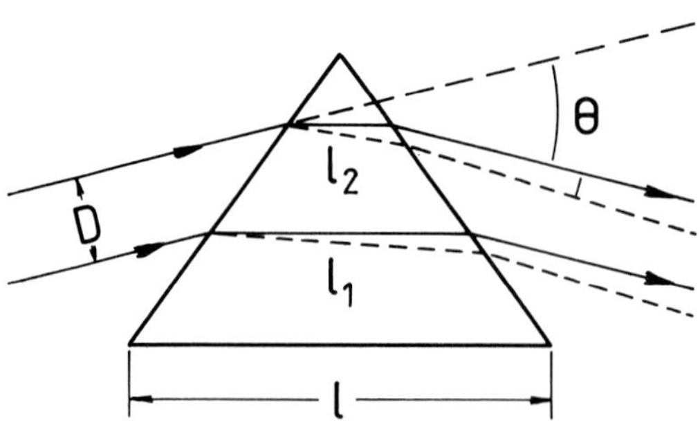

Figure 7.1 *Deflection of a light beam by a prism.*

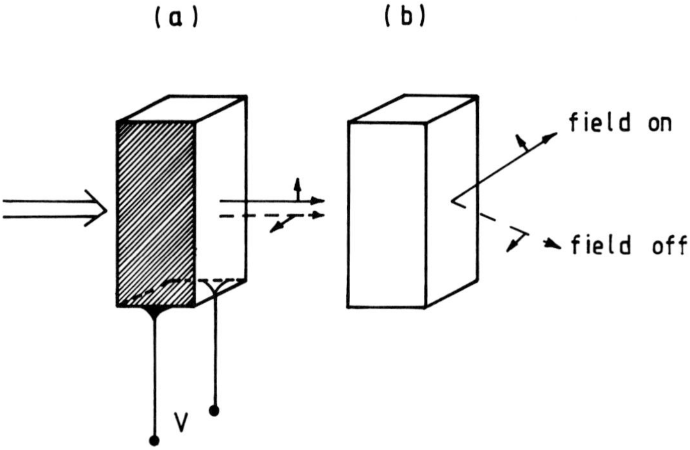

Figure 7.2 *Scheme of a digital light deflector: (a) electrooptic polarization switch; (b) polarization discriminator.*

7.2.2. Digital deflectors

These operate on discrete deflection positions. The basic set-up [2] consists of an electrooptic *polarization switch* followed by a birefringent *polarization discriminator*, as illustrated by Figure 7.2.

The polarization switch is a device producing either one of two orthogonal polarization states depending on whether the field is applied or not. The operating principle is described in Section 7.5. It consists of a linear polarizer followed by a slab of electrooptic crystal operating under half-wave voltage (V_{π}) conditions. The generated polarization state acts as input to the polarization discriminator. This yields an outgoing beam whose direction is different according to the input polarization state. Several alternatives exist for this device: (a) Split-angle birefringence plate, (b) total internal reflection polarizer, (c) Wollaston prism and (d) birefringent wedge. Devices (a) and (d) are schematically illustrated in Figure 7.3.

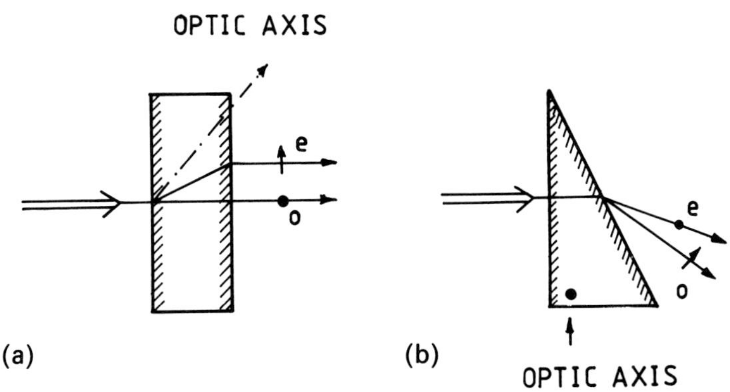

Figure 7.3 *Two types of polarization discriminator: (a) split-angle birefringence plate; (b) birefringence wedge [2].*

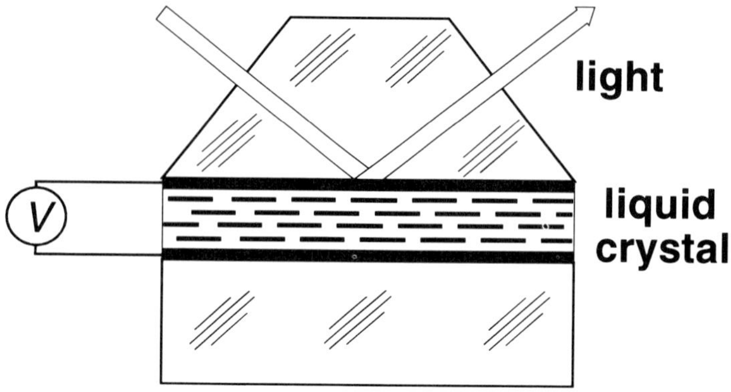

Figure 7.4 *Light deflector using a liquid crystal film. (Adapted from [7].)*

By using a cascade of n elements (stages), it is possible to operate with 2^n discrete beam positions. Moreover, one can cover a two-dimensional pattern of $2^n \times 2^m$ output spots by using two sets of n and m deflector units, respectively, producing beam displacements in perpendicular directions. As an example, a system of 20 stages leading to a raster (pattern) of 1024 × 1024 spot positions was successfully developed [5] at Philips (Hamburg).

A problem with digital deflectors is that of background light going into undesired positions (crosstalk). It appears as a consequence of defective collimation (convergent beams) or voltage fluctuations. For most applications, deflectors are designed with a maximum cross-talk level of about or less than 1%.

Some other strategies have been also pursued for discrete beam deflection. As an example [6], the electric field-induced (or -frustrated) total reflection of a light beam by a *liquid crystal* layer between two glass prisms has been proposed and tested (Figure 7.4). An analogue version, providing continuously varying deflection has been also reported [8].

7.2.3. Index gradient deflectors

Instead of using the deflection produced at the interfaces between electrooptic media, one may use a continuous refractive index gradient to deflect the light beam. Let us consider a linear index gradient $\mathrm{d}n/\mathrm{d}x$ along X and a light beam propagating along Z (Figure 7.5).

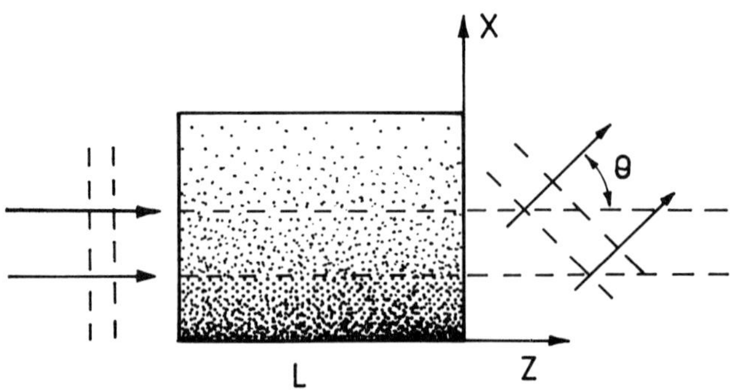

Figure 7.5 *Operation of an index gradient deflector.*

For small gradients and beam deviations, a simple geometrical analysis leads to a ray deflection θ after traversing a length L of the medium

$$\theta = \frac{dn}{dx} L \tag{7.5}$$

The practical generation of a constant index gradient can be accomplished by using a quadrupolar electrode arrangement. The electrodes should have a hyperbolic (or near) profile. With this kind of device, an $LiNbO_3$ streak camera with 20 ps resolution has been built [9]. On the other hand, quadrupole field deflectors are commercially available with deflection efficiencies of 1.5–3 μrad/V. The maximum number of resolvable spots at $\lambda = 500$ nm lies in the range from 50 to 100.

7.3. Light modulators

Light modulation [10–14] imposes some information on the phase, polarization state or amplitude (intensity) of an optical beam. The operating principles of a light modulator can be conveniently discussed with reference to the results presented in Chapter 3. Pockels and Kerr modulators have been built, although Pockels devices based on $LiNbO_3$, $LiTaO_3$, KDP, and KD*P single crystals have found more widespread use. Other materials, such as ceramics, for high-power shutters, and semiconductors, for infrared applications, have also been considered. Liquid crystals have the premium of a very high electrooptic response. However, their very low response time is a severe handicap that has prevented their use for high-frequency applications.

7.4. Phase modulators

Their function is to modulate the phase of a light wave, without affecting its polarization state. They constitute the most direct application of the electrooptic effect. Longitudinal as well as transversal geometries can be used. For the analysis, let us consider a KDP crystal whose perturbed index ellipsoid has axes X', Y' and Z for a field applied along Z. A light wave $\mathbf{E}_L = \mathbf{u}_{x'} A \cos(\omega t - kz)$, with linear polarization along X' and wavevector along Z, does not change its polarization state during propagation. On the other hand, the outgoing and input phases differ by the amount

$$\Delta\phi(E) = \omega(n_0 + \Delta n)\frac{l}{c} \tag{7.6}$$

l being the crystal length along Z, n_0 the unperturbed index and Δn the field-induced index change for the considered geometry. When the applied field is sinusoidally modulated, i.e. $E = E_m \sin \Omega t$, the outgoing phase becomes also modulated. Using expression (3.62) for the index change, i.e.

$$\Delta n = -\tfrac{1}{2} n_0^3 r_{63} E \tag{7.7}$$

the dephasage $\Delta\phi$ can be written as

$$\Delta\phi = \Delta\phi_0 - \Phi \sin \Omega t \tag{7.8}$$

with

$$\Delta\phi_0 = \Delta\phi(0) = \omega \frac{n_0 l}{C} \tag{7.9a}$$

$$\Phi = \tfrac{1}{2}\omega l n_0^3 r_{63} E_m \tag{7.9b}$$

Therefore, the phase of the outgoing wave is also modulated at frequency Ω. Unfortunately, the outgoing phase-modulated wave is no longer monochromatic and side-bands appear in the frequency spectrum at $\omega + m\Omega$ (m integer). This is a consequence of the nonlinear mixing of the two frequencies ω and Ω. After a suitable shift of the time origin $\Delta t = \Delta\phi_0/\omega$, the emerging wave can be written (ignoring the vector character)

$$\begin{aligned} E(l,t) &= A\cos(\omega t + \Delta\phi) = A\cos(\omega t + \Phi\sin\Omega t) \\ &= A\{\cos(\Phi\sin\Omega t)\cos\omega t - \sin(\Phi\sin\Omega t)\sin\omega t\} \end{aligned} \tag{7.10}$$

and using the Bessel function expansion for $\cos(M\sin\Omega t)$ and $\sin(M\sin\Omega t)$ (mentioned in Section 3.14), one finally reaches

$$\begin{aligned} E(l,t) = A\{&J_0(\Phi)\cos\omega t + J_1(\Phi)\cos(\omega+\Omega)t - J_1(\Phi)\cos(\omega-\Omega)t \\ &+ J_2(\Phi)\cos(\omega+2\Omega)t + J_2(\Phi)\cos(\omega-2\Omega)t + J_3(\Phi)\cos(\omega+3\Omega)t \\ &- J_3(\Phi)\cos(\omega-3\Omega)t + \ldots\} \end{aligned} \tag{7.11}$$

explicitly showing the occurrence of combined frequencies $\omega \pm m\Omega$. These side-bands are symmetrical with regard to the central frequency ω.

7.4.1. Frequency shifters

If an electric field which varies linearly with time is applied to a Pockels material, a *frequency shifter* device is obtained. Let us consider the same situation as in the previous section and apply the field

$$E(t) = \frac{\mathrm{d}E}{\mathrm{d}t}\, t \tag{7.12}$$

$\mathrm{d}E/\mathrm{d}t$ being a constant. It causes a dephasage

$$\Delta\phi = -\frac{\pi}{\lambda} n_0^3 r_{63} l \frac{\mathrm{d}E}{\mathrm{d}t}\, t \tag{7.13}$$

λ being the wavelength in vacuum. This dephasage, imposed on an incident plane wave, $A\,\mathrm{e}^{i(kz-\omega t)}$, leads to an outgoing wave

$$E(z,t) = A\,\mathrm{e}^{i(kz-\omega t+\Delta\phi_0+\Delta\phi)} \tag{7.14}$$

where $\Delta\phi_0$ and $\Delta\phi$ are respectively given by (7.9a) and (7.13). The mathematical structure of (7.14) results in a different plane wave, having a frequency shift $\Delta\omega$ given by

$$\Delta\omega = -\frac{\mathrm{d}(\Delta\phi)}{\mathrm{d}t} = \frac{\pi}{\lambda} n_0^3 r l \frac{\mathrm{d}E}{\mathrm{d}t} \tag{7.15}$$

In order to estimate a typical value for the shift $\Delta\omega$, let us take $r = 10^{-11}$ m/V, $l = 10^{-2}$ m, $\lambda = 10^{-6}$ m and $\mathrm{d}E/\mathrm{d}t = 10^{-12}$ V/(ms). One obtains $\Delta\omega \approx 10^5\ \mathrm{s}^{-1}$.

7.4.2. Pulse chirping

On the other hand, one may impose on the modulator a field that varies quadratically with time, i.e.

$$E(t) = E_0 - \frac{1}{2}\frac{d^2E}{dt^2}t^2 \tag{7.16}$$

d^2E/dt^2 being a constant. Then, following the same argument given above, it is found that the frequency of the light is modified by the amount

$$\Delta\omega = -\frac{\pi}{\lambda}n_0^3 rl \frac{d^2E}{dt}t \tag{7.17}$$

i.e. the change in frequency varies linearly with time. For a light pulse the front side of the pulse has a different frequency from that of the rear side. This effect is known as *frequency chirping*, well known in microwave technology. It is used as a first stage in pulse compression devices [15].

7.5. Polarization modulators

By using an input linear polarizer in conjunction with an electrically controlled retardation plate, one may build a polarization modulator, i.e. a device providing a modulated polarization state. In fact, the light wave oscillates between two limiting elliptical polarization states at the frequency of the applied field.

A particularly relevant case is the *polarization switch* mentioned in Section 7.2.2 on digital light deflectors. The retardation plate is modulated with a field amplitude V_π. For $V = 0$, the transmitted beam retains the input linear polarization, whereas for $V = V_\pi$ the polarization direction is rotated by $\pi/2$.

7.5.1. Optical isolator

This is a device that transmits light in one direction but blocks the backwards reflected beam. Therefore, it operates like a one-way valve. This effect can be realized by combining a linear polarizer and a quarter-wave plate. The transmitted light is circularly polarized.

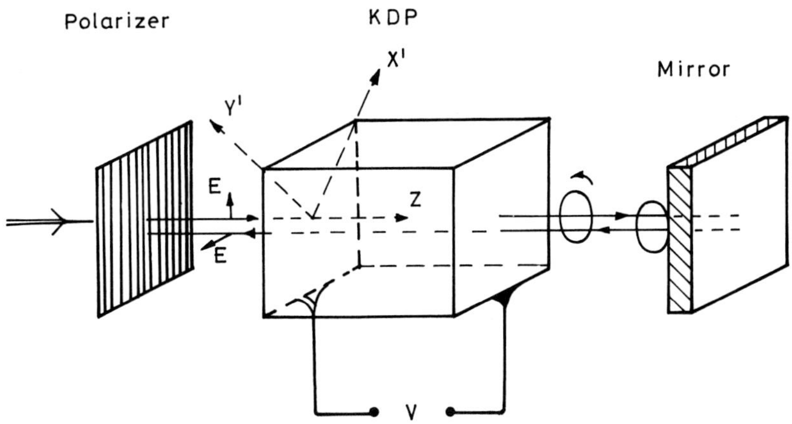

Figure 7.6 *Schematic layout for an optical isolator.*

After being reflected by a mirror, the $\pi/2$ retardation between the two orthogonal linearly polarized components is maintained and so the circular polarization. Then, the quarter-wave plate introduces an additional $\pi/2$ dephasage in the reflected beam leading to a total π retardation. This corresponds to a linear polarization rotated by $\pi/2$ with respect to the initial one and so the beam is blocked by the input polarizer.

One may substitute an electrooptic crystal slab for the quarter-wave plate, as illustrated in Figure 7.6. Then, by switching on and off a half-wave voltage V_π, the electrooptic phase retardation is shifted from $\delta = 0$ (field off) to $\delta = \pi/2$ (field on). In the first case, the backwards reflected beam keeps the input polarization and is transmitted. On the other hand, when the field is on ($\delta = \pi/2$), the discussion of the previous paragraph applies and the reflected light is blocked. Therefore, the device operates as a *field-controlled isolator*. It constitutes one method to *Q-switch* a laser and produce nanosecond pulses [3].

7.6. Amplitude (intensity) modulators

The operation principles of an amplitude modulator can be conveniently discussed with reference to Section 3.11. It consists of a polarization modulator, causing a modulated retardation between two orthogonal polarization states, followed by a polarization analyser or discriminator. This latter element is often a linear polarizer crossed with respect to the linear input polarizer. It converts the polarization modulation to amplitude (intensity) modulation. If $\delta(t)$ is the retardation imposed by the field, the intensity transmitted by the device is

$$I = I_0 \sin^2\left(\frac{\delta}{2}\right) \tag{7.18}$$

which has general validity when losses are ignored. For a modulated δ, the curve oscillates between maximum I_{max} and minimum I_{min} intensities as illustrated in Figure 3.7.

Longitudinal as well as *transverse* modulators can be used and are commercially available. Most often they are built from KD*P, $LiTaO_3$ and $LiNbO_3$ single crystals.

Optical shutters or *switches* are particular modulators where the minimum transmitted intensity $I_{min} = 0$.

7.6.1. Longitudinal Pockels modulators

A suitably oriented plate of the chosen electrooptic material (KDP in the example of Section 3.11a) is placed between crossed linear polarizers. The geometry is as schematically illustrated in Figure 3.5(a). The field is applied along the light propagation direction (Z-axis) through either transparent (e.g. ITO) or ring-shaped electrodes attached to the entrance and exit faces of the device. The input polarizer is aligned along the X-axis of the index ellipsoid (i.e. at 45° from the field-perturbed X'- and Y'-axes). For a length l of crystal, the retardation or phase difference δ between the two component waves polarized along X' and Y' is

$$\delta = \omega n_o^3 r_{63} \frac{V}{c} \tag{7.19}$$

which in terms of the half-wave voltage $V_\pi = \lambda/2n_o^2 r_{63}$, becomes

$$\delta = \frac{\pi V}{V_\pi} \tag{7.20}$$

Table 7.1 *Half-wave voltages for some common materials.*

Material	Configuration	V_π (kV)	λ (μm)
Pockels effect			
KDP [16]	Longitudinal	8.5	0.633
KD*P [16]	Longitudinal	4.4	0.633
ADP [16]	Longitudinal	10.5	0.633
$LiNbO_3$ [17]	Longitudinal	5.4	0.633
$LiTaO_3$ [17]	Longitudinal	2.7	0.546
$Ba_2NaNb_5O_{15}$ [18]	Transverse ($d = l$)	1.7	0.633
GaAs [19]	Transverse ($d = l$)	10.8	1.06
InP [19]	Transverse ($d = l$)	11.4	1.06
CdTe [19]	Transverse ($d = l$)	5.0	1.06
MNA [20]	Transverse ($d = l$)	1.3	0.633
DCNP [20]	Transverse ($d = l$)	0.37	0.633
Kerr effect			
Nitrobenzene† [17]	Transverse	48.9	0.633
PLZT‡ [21]	Transverse	0.3	0.633

† Cell size $d = 3.2$ cm, $l = 5$ cm.
‡ Cell size $d = 0.5$ mm, $l = 0.375$ mm; composition 9/65/35.

V_π is exclusively dependent on material and wavelength but not on sample dimensions (see Table 7.1).

7.6.2. Transverse Pockels modulators

For these modulators one should refer to Section 3.11b and Figure 3.5(b), where a KDP crystal was considered. The modulating electric field is now applied along the Z-axis, perpendicularly to the direction of light propagation (along Y'). The phase retardation achieved after a length l of material is given by (3.67)

$$\delta = 2\pi(n_e - n_o)\frac{l}{\lambda} - \pi(n_e^3 r_{63} - n_o^3 r_{63})\frac{lV}{\lambda d} \tag{7.21}$$

V being the potential applied through a crystal thickness d. Expression (7.21) leads to a half-wave voltage,

$$V_\pi = \frac{d}{l}\frac{\lambda}{n_e^3 r_{63} - n_o^3 r_{63}} \tag{7.22}$$

that can be made smaller than the value corresponding to longitudinal configuration by an amount d/l. For a modulating voltage of small amplitude $V = V_m \sin\Omega t$ ($V_m \ll V_\pi$)

$$\delta(E) = \frac{l}{d}\frac{\pi V_m}{V_\pi} \tag{7.23}$$

which is enhanced by the factor l/d with regard to (7.20). Unfortunately, the ratio l/d has limitations imposed by the divergence of the light beam. Most often one is operating with a gaussian mode from a laser. Then, it is shown that the optimum configuration to maximize l/d is obtained by focusing the beam waist at the centre of the crystal length with a diameter $2w_0$ given by $w_0 = (\lambda l/2\pi n_0)^{1/2}$, i.e. l corresponds to the mirror separation in a confocal resonator. Under these conditions the required thickness d corresponds to the beam diameter size at the end faces, i.e.

$$d^2 = \frac{4\lambda l}{\pi n_0} \tag{7.24}$$

which establishes the relation between d and l for optimum operation.

7.6.3. Kerr modulators

For isotropic Kerr materials a transversal geometry must be used to generate a phase retardation δ. The expression for δ as a function of E is, according to (3.78),

$$\delta\lambda = \frac{2\pi\Delta n l}{\lambda} = 2\pi B l E^2 \tag{7.25}$$

l being the propagation length and B the Kerr constant. The half-wave voltage to induce $\delta = \pi$ is

$$V_\pi = \frac{d}{(2Bl)^{1/2}} \tag{7.26}$$

for a thickness d of the medium. Then, equation (7.25) can be written

$$\delta(E) = \frac{\pi V^2}{V_\pi^2} \tag{7.27}$$

Most common Kerr materials are nitrobenzene and some PLZT compositions. V_π values are included in Table 7.1. Very promising are the semiconductor nanostructures to be discussed in Chapter 9.

7.6.4. Some practical considerations

A few practical considerations for the design of crystal modulators are now briefly summarized.

- In the frequency range from d.c. to about 1 MHz, piezoelectric resonances may appear in the modulator operation.
- Crystals of high optical quality (free of defects and strain) are required for good performance. The scattered light raises the light level transmitted in the extinction configuration and degrades the extinction ratio (see Section 7.7).
- In many commercial systems several crystals are used to reduce the operating voltage and improve the temperature stability by cancelling the thermally induced birefringence.
- Antireflection coatings are often applied to reduce reflection losses. Also, an index matching fluid is frequently used inside the crystal housing.

7.7. Other noncrystalline devices

In addition to the above-described Pockels and Kerr modulators, other modulating devices using liquid crystals and PLZT have been built for specific applications. A few of them will now be discussed briefly.

7.7.1. PLZT shutters

These consist of thin plates of polished quadratic PLZT ceramic with deposited interdigital electrodes (to reduce operating voltage), sandwiched between crossed polarizers. Their main features are: operating voltages greater than 100 V, fast response, wide-viewing angle, large temperature range and high extinction ratios [21,22]. A number of products have been developed, including nuclear flashblindness goggles, thermal/flash protective goggles and electronic welding helmets.

7.7.2. Variable liquid crystal wavelength filters and phase retarders

Taking into account the relation between half-wave voltage and light wavelength, it is possible to build a variable spectral filter governed by an adjustable electric field. Some commercial devices using liquid crystals are already available in the market. It is also possible to use the field-controlled birefringence of a nematic liquid crystal to fabricate [23] a variable phase retarder without any moving parts, at variance with the Soleil–Babinet compensator.

7.7.3. Smart windows

Polymer-dispersed liquid crystals, discussed in Section 6.13, are being investigated for the development of switchable (*smart*) windows, whose transmittance is controlled by an electric field [24]. The device consists of a polymer sheet sandwiched between transparent electrodes. In the absence of a field the index mismatch between the LC droplets and the polymer host makes the window translucent. When the field is applied the molecules are oriented inside the droplets and match the refractive index of the polymer. Then, the device becomes transparent.

7.8. High-frequency modulators

The speed of response, and so the bandwidth, of a *lumped* modulator is limited by the time constant of the electrical circuitry and the finite transit time of the light through the electrooptic crystal. These limitations may become very serious for frequencies near or above 100 MHz. On the other hand, the intrinsic electrooptic response of the material is much faster when dealing with electronic and optical phonon processes in a crystal like KDP or $LiNbO_3$.

The limitations associated to the $\tau = RC$ time constant of the electrical circuit are largely determined by the capacitance of the crystal slab. In fact, when $\omega > (RC)^{-1}$, where R_0 is the internal resistance of the voltage source, only a fraction of the voltage is applied to the crystal. This limitation can be markedly reduced by decreasing the size and so the capacity of the device. Therefore, one should turn to integrated optics modulators to achieve superior performance (see Chapter 8).

On the other hand, when the applied voltage varies during the transit of the light wavefront, the phase retardation is no longer given by expressions (7.20) or (7.23) and the operation of the modulator is impaired.

In view of those limitations, several bulk (not integrated) approaches have been developed for the design of high-frequency modulators to operate at around or above 100 MHz. Accordingly, the *resonant* and the *travelling-wave* modulators will be now described.

7.8.1. Resonant modulators

The input voltage requirements of electrooptic modulators can be markedly reduced by using *resonant-frequency modulation*. In resonant modulators, the crystal plate constitutes a capacitor which is part of a resonant circuit. The low radio-frequency losses in $LiNbO_3$ and $LiTaO_3$ enable the design of high-Q resonant circuits [11,25] that operate up to more than 10 GHz, although with a narrow bandwidth. There is a variety of applications for these modulators including phase-sensitive detection, laser stabilization, FM mode-locking and linewidth-broadening experiments.

7.8.2. Travelling-wave modulators

Transit time limitations can be eliminated by adopting a travelling-wave configuration [3,11,26], i.e. the modulating field is applied as a microwave propagating through a waveguide (Figure 7.7). The idea is to have the optical and modulating (microwave) beams propagating at the same speed. In these conditions the light wavefront will "see" all the time the same modulating field as in a conventional low-frequency modulation experiment. This argument can be further supported by a more quantitative analysis.

Let us consider the complex expression for a travelling microwave field of frequency Ω

$$E(z,t) = A\,e^{i\{k_M z - \Omega t\}} = A\,e^{ik_M(z - v_M t)} \tag{7.28}$$

$k_M = \Omega/v_M$ being the wavevector and v_M the wave velocity. The transit of a light wavefront through the length l of the waveguide lasts from an initial instant t to $t + t_L$. $t_L = l/v_L$ is the light transit time, corresponding to the light velocity $v_L = c/n$ in the electrooptic medium. Then, the phase delay or retardation introduced during the transit is

$$\delta(t_L) = \int_t^{t+t_L} \delta(t')\,dt' \tag{7.29}$$

where the retardation $\delta(t')\,dt'$ associated with the infinitesimal path $dl = v_L\,dt'$ is the real part of $\tilde{\delta}$ given by

$$\tilde{\delta}(t')\,dt' = CE(z = v_L(t' - t), t')v_L\,dt' \tag{7.30}$$

C being a constant dependent on geometry and optical and electrooptical parameters. By carrying out the integration, it is found that

$$\tilde{\delta}(t_L) = Cv_L t_L A\,e^{-i\Omega t}\left(\frac{e^{ik_M(v_L - v_M)t_L} - 1}{ik_M(v_L - v_M)}\right) \tag{7.31}$$

The factor within brackets is strongly resonant at $v_L = v_M$ and abruptly decreases on increasing the velocity mismatch $v_L - v_M$. At perfect matching, $v_L = v_M$, the factor becomes 1. Then, $\hat{\delta}$ (and δ) takes the value corresponding to an in-phase field $E(0,t) = A\,e^{-i\Omega t}$ equal to that provided by the microwave at the entrance of the waveguide. This confirms the qualitative argument advanced above.

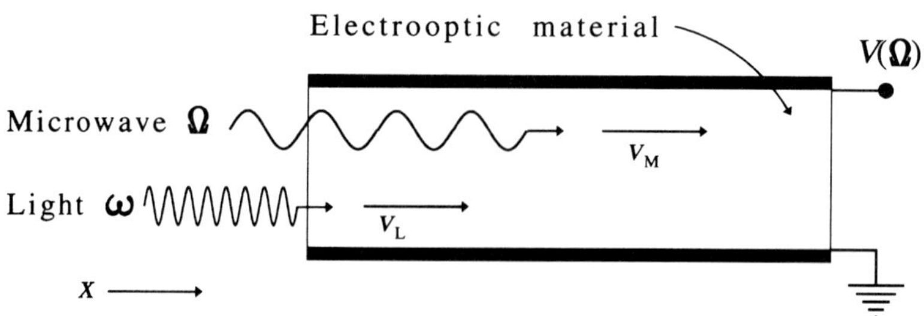

Figure 7.7 *Operation of a travelling-wave modulator.*

To achieve velocity matching, partial filling of the microwave cavity or zig-zag light propagation can be used.

7.9. Fabry–Perot modulators

The modulation capabilities of a given material can be greatly enhanced by using a Fabry–Perot interferometer configuration, where the light is forced to experience multiple passages between the entrance and exit mirrors. This improved performance will be substantiated in the following discussion.

7.9.1. Amplitude modulators

A Fabry–Perot amplitude modulator operates as the multiple-beam etalon shown schematically in Figure 7.8. The light beam of intensity I_0 is incident normally on the material plate of thickness l, which is bound by mirrors of reflectivities R_1 and R_2. For the simpler case $R_1 = R_2 = R$, the transmittance of the interferometer (ignoring losses) is given by

$$T = \frac{I_T}{I_0} = \frac{(1-R)^2}{(1-R)^2 + 4R\sin^2\Delta\phi} \tag{7.32}$$

$\Delta\phi$ being the phase-shift introduced by each passage through the length l

$$\Delta\phi = \frac{2\pi n l}{\lambda} \tag{7.33}$$

Equation (7.32) is often written

$$T = \frac{1}{1 + F\sin^2\Delta\phi} \tag{7.34}$$

where $F = 4R/(1-R)^2$ is the *fringe contrast* of the interferometer. The dependence $T(\phi)$ given by (7.34) is the so-called *Airy function*, illustrated in Figure 7.9. Maximum and

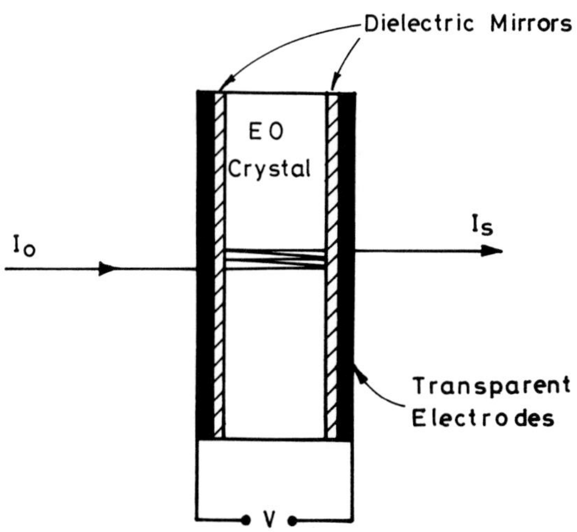

Figure 7.8 *Fabry–Perot modulator.*

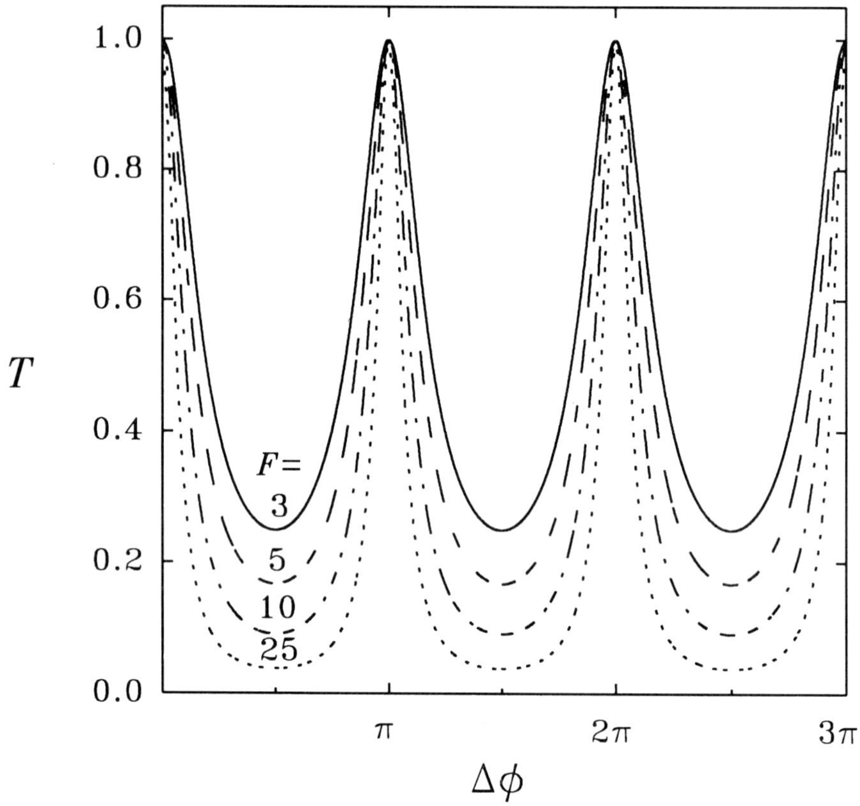

Figure 7.9 *Airy function corresponding to the response of a Fabry–Perot modulator. The corresponding values of the parameter* F *are marked on each curve.*

minimum transmittances

$$T_{\max} = 1 \tag{7.35a}$$

$$T_{\min} = \frac{1}{1+F} \tag{7.35b}$$

are respectively obtained for $\Delta\phi = m\pi$ and $\Delta\phi = \frac{1}{2}(2m+1)\pi$, m being an integer number.

Let us now specifically consider that the interferometer material is a KDP plate with the quaternary Z-axis normal to the mirrors. The situation corresponds to the longitudinal configuration discussed in Section 3.11.2. In the presence of an applied field E between the transparent electrodes, the refractive index for light polarized along X' is

$$n = n_0 - n_0^3 r_{63} \frac{E}{2} \tag{7.36}$$

leading to

$$\Delta\phi = 2\pi n_0 \frac{l}{\lambda} - \pi n_0^3 r_{63} \frac{V}{\lambda} = \Delta\phi_0 - \frac{\pi V}{2V_\pi} \tag{7.37}$$

The second term is the field-induced dephasage during each passage. n_0 and $\Delta\phi_0$ stand for the values in the absence of the field.

The effect of the field on the interferometer transmittance can be now easily illustrated with reference to Figure 7.9. As an example, the voltage $V_{1/2}$ required to decrease the transmittance from $T = 1$ to $T = \frac{1}{2}$ is

$$V_{1/2} = 2\Delta\phi_{1/2}\frac{V_\pi}{\pi} = \frac{2V_\pi}{\pi}\arcsin\left(\frac{1}{F}\right)^{1/2} \tag{7.38}$$

where $\Delta\phi_{1/2} = \arcsin(1/F)^{1/2}$ is the phase shift separating the two transmittance values. A practical device operates at one inflexion point of the $T(V)$ curve and applies a modulated voltage $V = V_m \sin\Omega t$. After derivation of (7.34) and assuming V small enough to justify a linear approximation to $T(V)$, one obtains a modulated transmission around $T = \frac{1}{2}$:

$$\Delta T = T - T_{1/2} = \frac{FV_m \sin\Omega t}{2V_\pi} \tag{7.39}$$

The amplitude ***modulation depth*** is

$$\frac{\Delta T}{\frac{1}{2}} = \frac{FV_m}{V_\pi} \tag{7.40}$$

which represents an enhancement by a factor F compared with a conventional modulator.

In spite of the enhanced performance of the Fabry–Perot device, an important drawback is the high wavelength sensitivity. In fact, the bias condition $T = \frac{1}{2}$ requires an adequate field bias, which depends on wavelength.

7.9.2. Phase modulators

A pure phase modulator can also be built using a Fabry–Perot geometry. Let us consider an asymmetrical etalon with the back mirror having a 100% reflectance (*Gires–Tournois etalon*). The overall amplitude reflectance r ($|r| = 1$) of the etalon can be written [13] as

$$r = e^{i\Phi} \tag{7.41}$$

with

$$\Phi = -2\arctan\left\{\left(\frac{1 + R^{1/2}}{1 - R^{1/2}}\right)\tan\Delta\phi\right\} \tag{7.42}$$

R being the front mirror reflectivity and $\Delta\phi$ the single-pass dephasage. In the presence of a voltage V and assuming that $\Delta\phi(0) = m\pi$,

$$\Phi = 2\arctan\left\{\left(\frac{1 + R^{1/2}}{1 - R^{1/2}}\right)\tan\left(\frac{\pi V}{V_\pi}\right)\right\} \tag{7.43}$$

that for $V \ll V_\pi$ (small applied voltage) simplifies to

$$\Phi = 2\pi\left(\frac{1 + R^{1/2}}{1 - R^{1/2}}\right)\frac{V}{V_\pi} \tag{7.44}$$

Therefore, the Fabry–Perot configuration enhances the single-pass field-induced dephasage by the factor $(1 + R^{1/2})/(1 - R^{1/2})$, which easily reaches values higher than 10.

The device has the drawback of a reduced spectral range already remarked for amplitude modulators.

7.10. Performance parameters for modulators

The main performance parameters characterizing the operation of modulators and commonly appearing in the commercial literature are:

- *Spectral range* or *optical bandwidth:* the wavelength interval that can be efficiently handled by the modulator.
- *Modulation depth:* the ratio $(I_{max} - I_0)/I_0$, where I_{max} and I_0, respectively, stand for the maximum and average transmitted intensities, under a periodic applied field.
- *Half-wave voltage* (V_π)*:* the applied voltage required for a retardation $\delta = \pi/2$. Values for several Pockels and Kerr materials are given in Table 7.1.
- *Peak optical transmission:* the maximum transmittance of the device, which does not reach unity because of reflection, absorption and scattering losses.
- *Extinction ratio:* this applies to shutter operation and is the ratio between maximum and minimum transmittance. Values for commercial devices range typically from 100 : 1 to 1000 : 1.
- *Bandwidth* (Δf)*:* the frequency range over which modulation depth remains at a minimum of half the maximum value. It is closely related to the response time τ of the device (including material and associated electronic circuitry) through

$$\tau = \frac{2\pi}{\Delta f} \tag{7.45}$$

- *Rise* and *fall times:* the time needed to reach the steady-state transmission after switching the field on and off respectively.
- *Specific energy:* the power required per unit bandwidth.
- *Acceptance angle:* the external angle for which transmittance (modulation depth) reaches 10% of the maximum value. This parameter is larger for longitudinal than for transverse modulators. As an example, the value in longitudinal configuration is 51 mrads cm$^{1/2}$ for KD*P and 53 mrads cm$^{1/2}$ for $LiNbO_3$.
- *Useful aperture:* the transverse size of the light beam keeping the modulator specifications.

Table 7.2 *Damage thresholds for electrooptic materials [27].*

Material	Damage threshold (GW/cm^2)
KDP	0.20
$LiIO_3$	0.50
$LiNbO_3$: Mg	0.05
BBO [28]	5.0, 15
KTP	1.00
LBO [28]	25.0
$AgGaS_2$	0.015
Urea	1.5
m-NA	0.20
POM	2.0
MAP	3.0

The values refer to $\lambda = 1.06\,\mu m$. The data for BBO and LBO taken from reference [28] correspond to excitation by 0.1 ns pulses.

- *Damage threshold:* the maximum light power that can be handled by the modulator material without being damaged. This threshold markedly depends on excitation conditions, such as light wavelength and pulse duration. It is a useful parameter when dealing with high power applications. Threshold values for some inorganic and organic materials are listed in Table 7.2. One should note the comparable (or even superior) performance of organic crystals and the very high laser resistance of BBO and LBO.

7.10.1. Figures of merit

Several parameters are generally used as figures of merit, to evaluate the performance of a modulator. Although some of them have been already mentioned, they are here summarized:

- $n^3 r_{\text{eff}}$ or equivalently $V_\pi \approx 1/n^3 r_{\text{eff}}$. They measure the amount of phase retardation that can be achieved. Note that high refractive indices contribute to a high figure of merit.
- $n^3 r_{\text{eff}}/\epsilon$, which takes into account the deleterious effect of the dielectric constant on the response time. Therefore, low dielectric constant materials (e.g. organic crystals) have a definite advantage on most inorganic counterparts.
- $F = n^7 r_{\text{eff}}^2/\epsilon$. This figure of merit introduced by Kaminow and Turner [29] serves to evaluate the reactive power requirements of a modulator. In fact, the power from the modulating source per unit bandwidth, $P/\Delta f$, is inversely proportional to F.

7.11. Spatial light modulators (SLMs)

SLMs are devices that impose a two-dimensional amplitude modulation on a light wave. In other words, they write an image onto the optical beam. There is a large variety of operative devices, using different physical principles. After a brief general presentation and a discussion of the relevant performance parameters, we will focus here on those devices based on the electrooptic effect. For a more complete description of SLMs the reader should consult appropriate reviews [30–32] and books [33,34].

The main application areas of SLMs are *image storage* and *projection displays*. Recently, much effort is being devoted to their role as input stage in optical image processors to provide an interface between the real world and the device.

SLMs are generally classified into *optically addressed* (OASLMs) and *electronically addressed* (EASLMs) devices. The first type uses a light beam to record an image onto the SLM, whereas the EASLM uses a video camera, a frame store or an electron beam. Both classes can be either reflective or transmissive, depending on whether the spatial modulation is imposed on the reflected or transmitted read-out beam.

7.11.1. Performance parameters

The main performance parameters for SLMs are:

- *Threshold sensitivity:* the minimum write beam intensity required to record an image on the read-out beam. For unambiguous specification, other parameters (e.g. resolution) should be specified. Obviously, it only applies to OASLMs.
- *Frame time:* the time required to replace an image by the next one on the read-out beam. The quoted value generally corresponds to the instant when 90% of full response has been reached. This time is limited by the decay or erasure of the recorded image.

- *Resolution:* the minimum spacing between written lines that can be resolved. For a quantitative assessment, the *modulation transfer function* (MTF) of the SLM is first defined as the fringe visibility of an interferometric sinusoidal grating relative to that at low enough spatial frequency. The resolution is then the value corresponding to a 50% MTF.
- *Contrast ratio:* the quotient between maximum and minimum light intensity in the read-out beam.
- *Optical flatness:* this is given by the r.m.s. or peak-to-peak distortion caused by the device on a plane wavefront.

7.12. Electrically addressed spatial light modulators (EASLMs)

These provide an interface between electronics and optics. Their main applications include *displays* (from wrist watches to large information panels), to be specifically discussed in Sections 7.13 to 7.15, programmable *optical interconnections*, and *input* and *output devices* in optical processors.

Many operative philosophies have been implemented. Some of them do not rely on electrooptic effects and will not be considered here, e.g. *deformable mirror devices*, *acoustooptical SLMs* and *magnetooptical SLMs*. Others, such as the *laser diode array* are electrooptic in the broader sense described in Chapter 1 and will not be discussed either. Proper electrooptic devices are *liquid crystal light valves* (LCLV), such as the *CCD-addressed LCLV* and the *PLZT light valve*. Since they are mostly used as display devices, their discussion is deferred until Sections 7.15 and 7.16.

7.13. Optically addressed spatial light modulators (OASLMs)

Here the image imposed on the read-out beam is transferred from an excitation or writing beam. The basic structure of the so-called *hybrid OASLM* consists of a photoactive (photoconductive, photoelectric or photovoltaic) stage converting light into a pattern of electric signals and a modulator based on an electrooptic material. In some cases, the two stages can be merged into a single one by using an appropriate photoactive and electrooptic material (*monolithic approach*). Typical applications include incoherent to coherent converters, image amplifiers, wavelength image converters (e.g. IR to visible or

Table 7.3 *Main characteristics of OASLMs. (Adapted from [33].)*

Device	Size (cm)	Sensitivity ($-/cm^2$)	Frame time (ms)	Resolution (lines/mm)	Contrast ratio
PROM	2.5 dia.	5 μJ	33	6	$10^3:1$
PRIZ	1.8 dia.	5 μJ	100	10	—
LCLV	5×5	8 mW	80	25	$10^2:1$
LCLV (BSO)	1.5×1.5	380 μW	165	31	21 : 1
LCLV (α-Si)	9.7 cm^2	20 μW	50	35	—
Micro-channel SLM	1.6 dia.	30 nJ	150	10	$>10^3:1$

Note: all devices operate in reflection mode except PRIZ which operates in transmission mode.

vice versa), and programmable optical interconnections. Several types have been already fabricated and commercialized, such as the *Pockels read-out memory* (PROM), the *Liquid crystal light valves* (LCLV), and the *microchannel light modulator*. Measured performance parameters for these devices are listed in Table 7.3 for comparison.

Recently, devices using polymeric materials are being developed in several laboratories. They present improved performance parameters with regard to classical OASLMs [35].

7.13.1. Liquid crystal light valves (LCLVs) [36,37]

These are generic hybrid devices that combine an electrooptically active liquid crystal together with a photoconductive material converting a light signal into an electric field. The schematic diagram of the device fabricated by Hughes is shown in Figure 7.10. The recording light enters from the left side, passes through a transparent electrode followed by a photoconductive layer of CdS and is blocked by a CdTe film. Next on the right one finds a dielectric mirror, the electrooptic liquid crystal and the other transparent electrode. The linearly polarized read-out light is incident from the right and reflected back by the mirror.

The operating principle of the device is as follows. In the absence of a recording light, the external voltage is mostly applied to the photoconductive CdS layer. No voltage exists across the liquid crystal and the reflected read-out light is blocked by an analyser crossed at $\pi/2$ with regard to the polarizer. The recording light generates mobile charges in the illuminated spots and most of the voltage is then applied to the liquid crystal. Consequently the birefringence of the liquid crystal is modified through the electrooptic effect. The dephasage between the two linearly polarized components of the read-out beam rotates the polarization plane of the incident light. As a result, light passes through the analyser and yields an optical signal. Generally, the liquid crystal cell operates in the twisted nematic configuration.

7.13.2. Pockels read-out modulators (PROMs) [38,39]

These are classical SLMs, made from BSO as a Pockels material. At the same time BSO acts as a photoconductor to generate an electric field in response to the optical excitation.

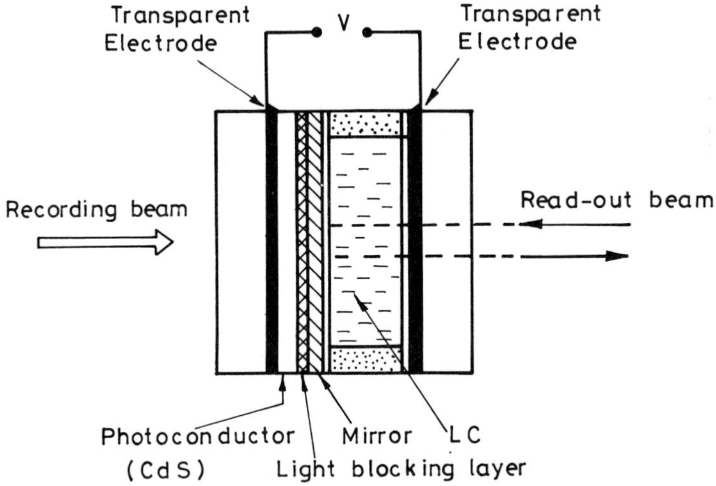

Figure 7.10 *Liquid crystal light valve (LCLV). (Adapted from [34].)*

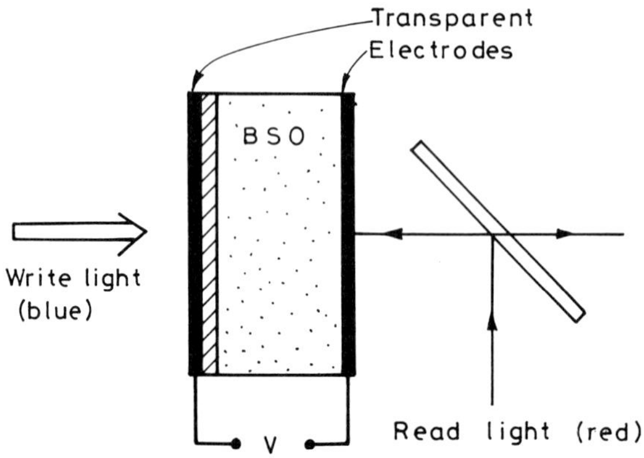

Figure 7.11 *PROM spatial light modulator.*

In other words, one is dealing with a monolithic approach in which BSO behaves as a photoconductive and electrooptic (e.g. photorefractive) material. The operation of a reflective PROM is illustrated schematically in Figure 7.11. First, the crystal is illuminated with a recording UV light beam, and a high voltage is applied. During this stage, photogenerated electrons accumulate at the anode until the applied field is compensated. Second, the applied field is reversed with the illumination off. Since carriers are trapped, the space-charge field is maintained in the regions previously illuminated. This field induces a two-dimensional refractive index pattern which stores the information carried by the recording beam. Then, this information can be read by diffracting a probe beam, whose wavelength is in the red region to avoid any modification of the recorded pattern.

The PROM SLM uses a crystal slab cut along a [100] face. Similar SLMs, using [111] and [110] crystal cuts, have been developed by Russian workers [40,41] and are designated as PRIZ (Russian acronym for image converter).

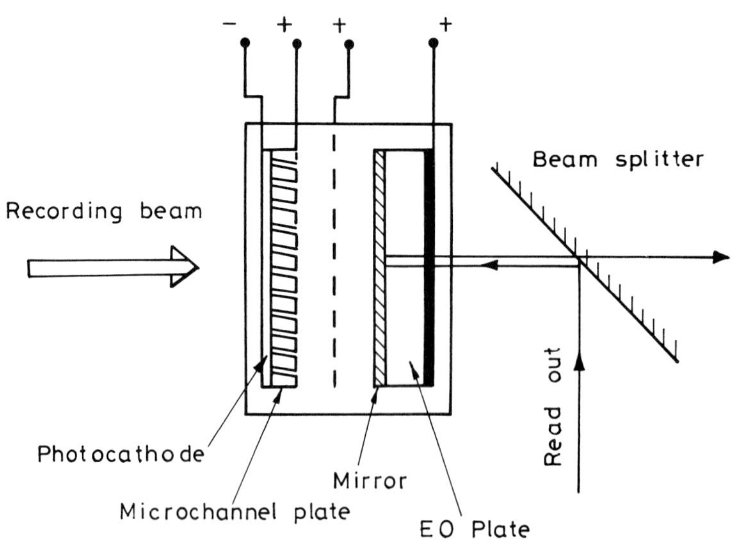

Figure 7.12 *Microchannel spatial light modulator. (Adapted from [43].)*

7.13.3. Microchannel spatial light modulators

The operating principle of these modulators [42] can be discussed with the schematic diagram shown in Figure 7.12. For the recording stage the device uses a photocathode as input element to convert the optical image into a pattern of photoemitted electrons. The electron current is amplified with a microchannel plate multiplier and then deposited onto a slab of electrooptic material, after passing through an accelerating grid. The charge deposited on the electrooptic slab generates an electric field distribution that causes the recorded birefringence pattern. For the reading process light is sent to the electrooptic section of the device through a polarizing beam splitter, as indicated in the figure. After traversing the slab, the light is completely reflected from a dielectric mirror covering the back face. When no information has been recorded, the reflected light passes through the beam splitter and appears as a uniform (homogeneous) beam. When information has been recorded, a phase retardation pattern is imposed on the reflected light which is converted into a spatially intensity-modulated beam by the polarizing element. This modulation is a replica of that carried by the writing beam.

7.14. Displays

Display devices [44,45] convert bidimensional information carried by a pattern of electrical signals into an optical image (*optical information*). The need for displays exists in a vast variety of technologies, both civilian and military, and is one of the fields experiencing a more dramatic growth in the optoelectronic market. Specific application areas include computer monitors, TV screens, projection displays and information panels (sign boards), in addition to small-size consumer applications (watches, pocket calculators, fax machines, audioequipment, etc.). Although in these latter markets, liquid crystal electrooptic displays (LCD) are well established, the situation is different in large-size applications, such as computers and TV. Here, the cathode-ray tube (CRT) continues clearly to occupy a leading position. Therefore, present R&D efforts have centred on the so-called *flat panel* technologies to substitute and improve on CRTs for large area display applications.

There are a number of competing flat panel technologies: (a) electroluminescent displays (ELD), (b) plasma-discharge panels (PDP), (c) vacuum-fluorescence (VF) lamps, (d) light-emitting diodes (LED) and (e) electrooptic devices, including LCDs. Some of these technologies are active, i.e. they use light-emitting devices, whereas others rely on external (ambient) illumination. Electrooptic displays are the subject of the next section.

7.15. Liquid crystal displays (LCDs) [7,44–46]

Essentially, these are electrically addressed spatial light modulators (EASLM), which have been already discussed in Section 7.12. The only difference lies in the specific application to be considered. As mentioned in 7.12, EASLMs refer to devices acting as input stages in optical processors and computers, whereas one speaks of displays when given information has to be visualized. We will refer exclusively to this application in this section. First, the simple small-size LCDs will be considered. In a later subsection the large-area applications will be discussed. Here, the high number of pixels poses a severe problem on the electrical addressing to the LCD, since individual pixel addressing is no longer possible. General features of LCDs are a low operation voltage ($\sim 2\,\text{V}$), and very low power consumption ($\sim 1\,\mu\text{W/cm}^2$). An up-to-date assessment of the LCD market for computing, communications and defense is presented in reference [47].

7.15.1. Simple LCD

A simple *numeric display*, such as those used in wrist watches, is illustrated schematically in Figure 7.13. It consists of a thin LC cell sandwiched between glass plates on which transparent electrodes and conductors are deposited. This sandwich is then placed between crossed polarizers and followed by a back mirror (reflective mode operation). In the absence of a field, the twisted nematic (TN) crystal causes a $\pi/2$ rotation of the light polarization, yielding a bright reflected beam. When the appropriate field is applied, the TN does not rotate the light polarization and no reflected beam is observed. The electrode is patterned into seven bar segments arranged as illustrated in Figure 7.13. By applying a voltage to appropriate combinations of segments all numbers from 0 to 9 are reproduced. The anodes (or cathodes) are wired together, whereas each cathode (anode) has a separate connection, the total number of connections being $N + 1$, N being the number of cathodes ($N = 7$ in the example of Figure 7.13).

7.15.2. Matrix-addressed displays

When the information content increases, requiring a large number of electrode elements, a different approach to the electrical drive has to be applied. A useful scheme is the so-called *matrix-addressed display* illustrated in Figure 7.14. The $N \times M$ electrode elements are organized in a matrix of N rows and M columns. Each row and column has an independent connection, so that the total number of connections is $N + M$. This number is considerably smaller than $N \times M$ for N, $M \gg 1$.

The electrode matrix is scanned according to the following procedure. A voltage V is successively applied to each of the M columns, whereas the voltage is zero for the other

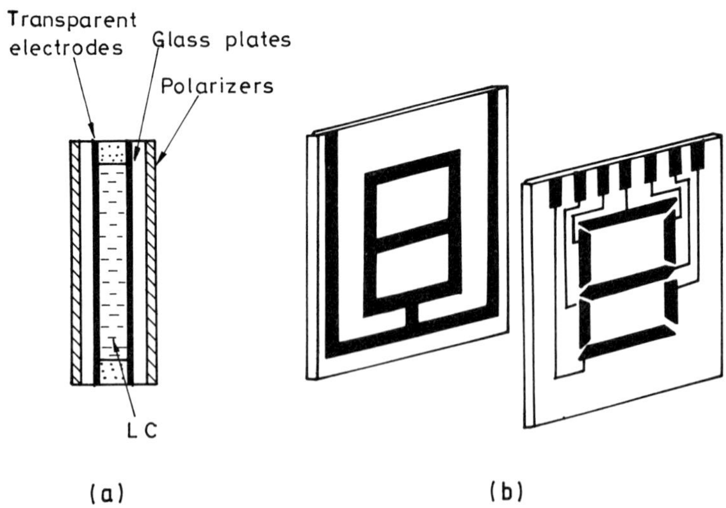

Figure 7.13 *Simple numerical display: (a) cross-sectional view; (b) electrical-addressing layout. (Adapted from [7].)*

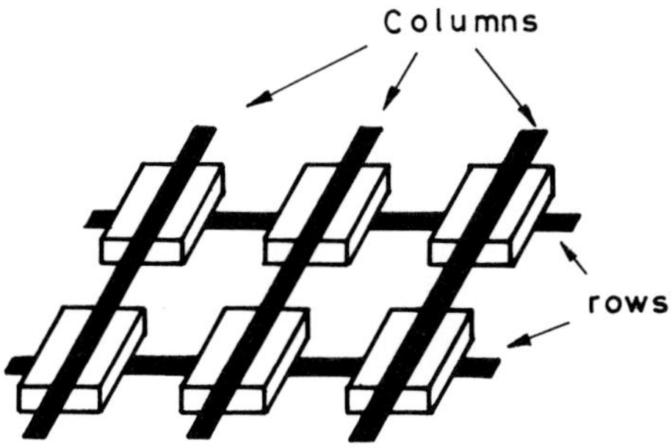

Figure 7.14 *Matrix-addressing scheme for a complex display [45].*

columns. At the same time, a voltage either $+V$ or $-V$ is applied to the rows, depending on the potential that should appear on each node or pixel, either 0 or $2V$. It is programmed so that full response is obtained when the node potential is $2V$. On the other hand, a potential V should not operate the device.

If t_S stands for the time elapsed between the application of voltage to two adjacent columns, the total scanning time is Mt_S. In order to avoid flicker this time has to be lower than the persistence time of the image in the visual system i.e. $Mt_S < \frac{1}{45}$ s. Since t_S should be comparable or longer than the response time τ of the liquid crystal to assure a good signal to noise ratio, the above relation sets an upper bound on τ. Therefore, this severely limits the maximum M compatible with the speed of available liquid crystals.

In accordance with the above considerations, the overall transmission *vs* voltage curve of the device has to be very steep to guarantee efficient performance. Ideally, no transmission should occur for a voltage $2V$, whereas full transmission should be observed at a voltage V. This situation is not substantiated in practice, since available LCs show rather smooth thresholds. Another drawback is that the active voltage is only applied during a fraction $1/M$ of the total scanning time, the voltage being V for the rest of the time. This may lead to a reduced contrast or a poor signal-to-noise ratio when the number M of columns increases. Therefore, in order to enhance this performance, an active element is incorporated into each matrix node (*active matrix display*) [48]. Its function is to keep the voltage applied during the whole addressing cycle. This can be accomplished with a nonlinear element, such as a zinc-oxide varistor, a metal–insulator–metal (MIM) film, a field-effect transistor (FET), or a thin-film transistor (TFT). Displays containing up to 640 × 400 lines, with a diagonal size of 10″ have been fabricated.

Colour can be achieved by subdividing every pixel into three independent subpixels, each one associated with one primary colour (red, green or blue). The images of the subpixels are below the resolving power of the eye. So, by addressing adequately the three subpixels, the device generates three component light signals that merge in the retina to yield the desired colour sensation. New developments in colour display technology are described in references [49,50]. In particular, small colour TV sets are already in the market.

7.15.3. Ferroelectric LCDs

To cope with the high information content of large flat panels, the much faster response of ferroelectric liquid crystals (FLC) offers a very promising alternative to TNLC [51–53]. In fact, the high switching speed (μs) of FLC is well adapted to the high multiplexing rates involved in complex matrix-addressed displays.

The bistability associated with the surface stabilized ferroelectric liquid crystals (SSFLC) can be used to build a binary intensity (or phase) modulator or optical switching device. Let us consider an FLC layer of thickness l placed between crossed linear polarizers. If l is chosen so that $l\Delta n = \lambda n/2$, where Δn is the crystal birefringence and λ the operating wavelength, the phase retardation introduced by the layer amounts to π. For $\Delta n = 0.15$ and $\lambda = 0.633\,\mu$m, one should have $l = 2\,\mu$m. In an initial state, the optic axis of the FLC, which coincides with the ferroelectric axis is oriented parallel to the axis of the entrance polarizer. Reversing the field $\mathbf{E}$ causes the rotation of the optic axis by 2θ to reach the other ferroelectric polarization state (see Figure 6.13). Then, as a consequence of the π retardation, the linear light polarization direction moves to the symmetric position with respect to the new optic axis. This implies a 4θ rotation from the initial polarization direction. For $\theta = 22.5$, this rotation amounts to $\pi/2$ and the light is transmitted through the analyser. Extinction ratios higher than 1000 : 1 have been reached. In the absence of polarizers the device works as a binary polarization modulator.

Some drawbacks of FLCD are related to their complex relaxation behaviour and the persistence of defects introduced by mechanical or electrical disturbances. On the other hand, since FLCs operate as binary elements, they are not adapted to handle grey levels. However these can be achieved through spatial or temporal multiplexing. As an example, if four grey levels have to be reproduced, a 2×2 matrix of binary elements per pixel is required.

Analogue devices can be made by using the electroclinic effect in SmA* LC, or distorted helix ferroelectrics (DHF) [52,54].

7.16. PLZT displays

The operating principle [22,55] is the same as that described for LCDs. A PLZT thin plate plays the role of the liquid crystal layer and it is sandwiched between crossed polarizers. They are commercially available and compete with LCDs. The main applications are in the military and automotive markets. The main drawback with regard to LCDs is the much higher voltage required ($>$100 V).

Active matrix-addressed displays have been also developed using thin-film transistors (TFT) as switching elements. Active addressing is very important for PLZT displays due to the poorly defined and high-speed switching behaviour.

EASLMs, working similarly to LCLVs, have been proposed and fabricated. One-dimensional devices (linear gate arrays) have used direct addressing. On the other hand, bidimensional SLMs built on PLZT offer a number of advantages over LCLVs: high speed (0.5–50 μs), high resolution (40 lines/mm) and good thermal stability. Therefore, they may represent an excellent alternative for future developments [56].

References

1. V. J. Fowler and J. Schafer, *Appl. Optics* **5**, 1675 (1966).
2. M. Gottlieb, C. L. M. Ireland and J. M. Levy, *Electrooptic and Acoustooptic Scanning and Deflection*, Marcel Dekker, New York (1983).
3. A. Yariv, *Quantum Electronics*, 3rd edn, Wiley, New York (1989).
4. Y. Ninomiya, *IEEE J. Quantum Electron.* **QE-9**, 791 (1973).

5. U. J. Schmidt, *Philips Tech. Rev.* **36**, 117 (1976).
6. G. Labrunie and S. Valette, *Appl. Optics* **13**, 1802 (1974).
7. E. P. Raynes, in *Electrooptic and Photorefractive Materials* (ed P. Günter), Springer-Verlag, Berlin (1987).
8. A. F. Fray and D. Jones, *Electron. Lett.* **11**, 358 (1975).
9. C. L. M. Ireland, *Opt. Commun.* **30**, 99 (1979).
10. I. P. Kaminow and E. H. Turner, *Proc. IEEE* **54**, 1374 (1966).
11. A. P. Harvey, *Coherent Light*, Wiley-Interscience, London (1970).
12. I. P. Kaminow, *An Introduction to Electrooptic Devices*, Academic Press, New York (1974).
13. A. Yariv and P. Yeh, *Optical Waves in Crystals*, John Wiley, New York (1984).
14. J. M. Heritier, *Laser Focus Wld* April, 113 (1990).
15. B. E. A. Saleh and M. C. Teich, *Fundamentals of Photonics*, Wiley-Interscience, New York (1991).
16. D. Eimerl, *Ferroelectrics* **72**, 95 (1987).
17. R. Guenther, *Modern Optics*, John Wiley, New York (1990).
18. S. H. Wemple and M. diDomenico, in *Applied Solid State Science. Advances in Materials and Device Research*, Vol. 3 (ed. R. Wolfe), Academic Press, New York (1972).
19. N. Suzuki and K. Tada, *Jpn. J. Appl. Phys.* **23**, 1011 (1984).
20. R. T. Bailey, F. R. Cruickshank, D. Pugh and J. N. Sherwood, *Int. J. Optoelectron.* **5**, 89 (1990).
21. G. Haertling, in *Ceramic Materials for Electronics* (ed. R. C. Buchanan), Marcel Dekker, New York (1986).
22. G. Haertling, *Ferroelectrics* **131**, 1 (1992).
23. T. V. Higgins, *Laser Focus Wld* Dec., 45 (1991).
24. J. W. Doane, *MRS Bull.* **16**, 22 (1991).
25. T. Day, *Laser Focus Wld* May, 183 (1992).
26. G. White and G. M. Chin, *Opt. Commun.* **5**, 374 (1972).
27. J. T. Lin and C. Chen, *Lasers Optron.* Nov., 59 (1987).
28. Ch. Chen, Y. Wu, A. Jiang, B. Wu, G. You, R. Li and S. Lin, *J. Opt. Soc. Am.* **6B**, 616 (1989).
29. I. P. Kaminow and E. H. Turner, in *Handbook of Lasers* (ed. R. J. Pressley), Chemical Rubber, Cleveland (1971).
30. A. D. Fisher and J. N. Lee, *Opt. Hybrid Comput.* **634**, 352 (1986).
31. U. Efron, SPIE **700** (10CC), 132 (1986).
32. J. A. Neff, R. A. Athale and S. H. Lee, *Proc. IEEE* **78**, 826 (1990).
33. N. Collings, *Optical Pattern Recognition Using Holographic Techniques*, Addison-Wesley, Wokingham (1988).
34. A. D. McAulay, *Optical Computer Architectures*, Wiley-Interscience, New York (1991).
35. D. Fichou, J. M. Nimzi, F. Charra and N. Pfeffer, *Adv. Mater.* **6**, 64 (1994).
36. M. L. Blinov, *Electrooptical and Magnetooptical Properties of Liquid Crystals*, Wiley-Interscience (1983).
37. D. Armitage, J. I. Thackara and W. D. Eades, *Appl. Optics* **28**, 4763 (1989).
38. S. G. Lipson and P. Nisenson, *Appl. Optics* **13**, 2052 (1974).
39. Y. Owechko and A. R. Tanguay Jr, *J. Opt. Soc. Am.* **1A**, 635 (1984).
40. M. Petrov, *Sov. Phys. Tech. Phys.* **25**, 752 (1980).
41. D. Casasent, F. Caimi and A. Khomenko, *Appl. Optics* **20**, 4215 (1981).
42. C. Warde, A. M. Weiss, A. D. Fisher and J. J. Thakara, *Appl. Optics* **20**, 2066 (1981).
43. S. A. Collins Jr., in *Optical Computing* (eds. B. S. Wheret and F. A. Tooley), Scottish University Summer School in Physics, pp. 23–53 (1989).
44. J. I. Pankove (ed.), *Display Devices*, Topics in Applied Physics, Vol. 40, Springer-Verlag, Berlin (1980).
45. J. Wilson and J. F. B. Hawkes, *Optoelectronics. An Introduction*, Prentice-Hall, Englewood Cliffs (1983).
46. J. Funfschilling, *Condensed Matter News* **1**, 12 (1991).
47. W. O'Mara, *Phys. Wld* June, 36 (1992).

48. H. J. Plach, P. A. Breddels and B. Rieger, *Solid State Technol.* June, 186 (1992).
49. J. E. Midwinter, *Phys. Technol.* **19**, 101 (1988).
50. L. E. Tannas Jr, *Phys. Today* Dec., 52 (1992).
51. J. S. Patel, *Opt. Engng* **26**, 129 (1987).
52. L. A. Beresnev and L. M. Blinov, *Ferroelectrics* **92**, 335 (1989).
53. S. T. Lagerwall, N. A. Clark, J. Dijon and J. F. Clerc, *Ferroelectrics* **94**, 3 (1989).
54. K. Johnson, *Phys. Wld* Sep., 37 (1992).
55. G. H. Haertling, *Ferroelectrics* **75**, 25 (1987).
56. S. H. Lee, S. C. Esener, M. A. Title and T. J. Drabnic, *Opt. Engng* **25**, 250 (1986).

Chapter Eight

Electrooptics and Integrated Optics

8.1. Introduction

The subject of *integrated optics* arose from the need for miniaturization of optical components required by communication systems based on fibre optics. Later on, other emergent applications such as signal processing, sensing and optical computing, started (and continue) to demand further developments in integrated optics. As with micro-electronics, miniaturization of optical components is an essential route for increasing reliability and performances whilst reducing size, weight, power requirements and cost on large-scale production. In the last decade enormous progress has been made in materials, fabrication processes, component development and device demonstrations. The extra-high frequency of the light as carrier wave (in comparison with microwaves used previously) makes possible massive data transmission at speeds several orders of magnitude greater than those available a decade ago. A rough estimate shows that a present optical-fibre/integrated-optical system may transmit the contents of a 30-volume encylopaedia in just a tenth of a second. Also, the system needs no repeaters up to a distance of 200 km, and the typical error is of the order of 1 in 10^9, which is equivalent to mistaking one letter in the whole process.

The basic elements of integrated optical circuits are optical waveguides. They consist of a region (of a thickness approximate to a wavelength) of "high" refractive index bounded by media of lower refractive index. Within the simple ray-optics picture, they confine light by means of total internal reflections. According to the number of spatial dimensions that provide confinement they are called: *planar* (or *slab*) waveguides when confinement only occurs in one dimension, and *channel* (or *strip*) waveguides when dealing with two-dimensional confinement. Fibre optics are cylindrical guides to be included within the last type. Either of the two kinds of waveguides can also be categorized as: *step-index* waveguides when the index of refraction is constant within the confinement region, and *graded-index* (or *grin*) waveguides when the index varies inside and/or outside of the confinement region. Step-index and grin planar waveguides form the basic substrates for the fabrication of optical circuits. The different types of channel waveguides form the individual elements of those circuits.

In addition to guiding the light in a transparent material, integrated optical devices must generate, manipulate and detect light waves. They must also transfer optical waves to and from guiding layers of the same or other integrated optical devices. An essential mechanism for many of these manipulation and control devices is the electrooptic effect Advantage is taken of the ultra-short response time ($\sim 10^{-14}$ s) associated with the electronic processes. In fact, bandwidth limitations found in bulk electrooptic devices are overcome in a waveguide geometry. This permits better exploitation of optical communication capabilities. Although the fundamental concepts of electrooptics discussed in Chapter 3 are applicable to optical waveguides, the technological relevance and the peculiar features these devices present deserve a separate chapter.

The main objective of this chapter is to emphasize the important role played by electrooptics in practical integrated optical devices. Basic theoretical concepts of optical waveguides are also included in order to make the chapter self-contained. There are many textbooks on the theory and applications of optical waveguides, references [1–5] being good examples. Here, the convenient notation of the popular Kogelnik's article [6] will be followed.

In addition to electrooptic effects, other nonlinear processes such as second-harmonic generation or optical Kerr effect can be more efficiently performed in a waveguide configuration in comparison to bulk samples. Although this expanding field will not be covered in the chapter, the interested reader may consult some appropriate references [7–9].

8.2. Light propagation in optical waveguides: modes

To introduce the basic concepts, let us consider the step-index planar (or slab) dielectric waveguide. It consists of a thin film with uniform refractive index n_f and thickness h, lying on a substrate with refractive index n_s and covered by a medium of refractive index n_c, as shown in Figure 8.1. Under the condition $n_f > n_s, n_c$, a light ray inside the film forming an angle with the normal to the film greater than the critical angles of the substrate and cover ($\theta > \theta_s, \theta_c$) will bounce by total internal reflections between the substrate and the cover, and propagate along the film. Very often the cover is simply air so that $n_c = 1$. Substrate and film indices commonly range between 1.46 (silica) and 3.6 (GaAs). Typical thicknesses are of the order of 1 μm.

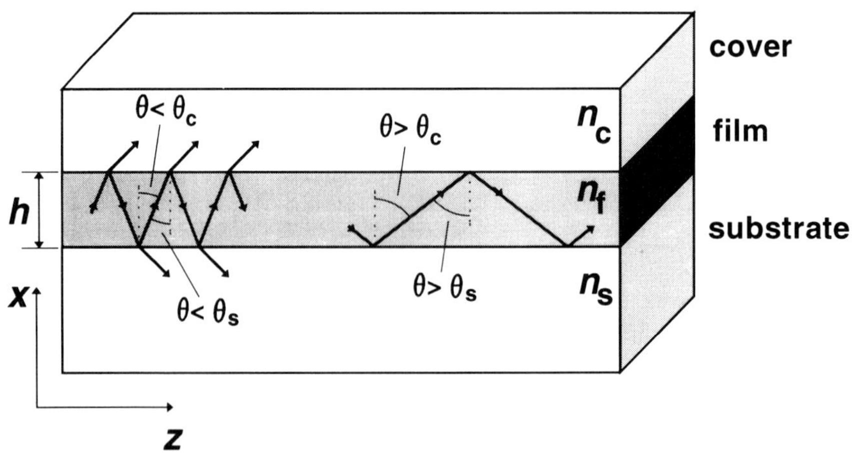

Figure 8.1 *Unguided and guided light beams in a planar waveguide.*

8.2.1. Dispersion equation of the step-index planar guide

With the choice of the coordinate system made in Figure 8.1, light propagates in the Z direction, confinement occurs in the X direction and the film and the light are uniform in the Y direction. Let us assume a monochromatic and plane wave with angular frequency ω, free-space wavelength λ and wavevector $\mathbf{k}n_{\rm f}$, the absolute value of $\mathbf{k}$ being

$$k = \frac{2\pi}{\lambda} = \frac{\omega}{c} \tag{8.1}$$

with c the velocity of light in vacuum. The space dependence of this wave is written

$$e^{in_{\rm f}(k_x x + k_z z)} = e^{in_{\rm f}k(x\cos\theta + z\sin\theta)} \tag{8.2}$$

The *propagation constant* β* (and the associated phase velocity $v_{\rm p}$) predicted by this simple model for a guided wave is the z-component of the wavevector $\mathbf{k}n_{\rm f}$, i.e.

$$\beta = n_{\rm f} k_z = \frac{\omega}{v_{\rm f}} = kn_{\rm f}\sin\theta = kN \tag{8.3}$$

where the "effective guide index" $N = n_{\rm f}\sin\theta$ has been introduced. However, not all propagation constants β are allowed, since only a discrete set of angles (and sometimes none) are possible for a stable configuration of the optical waves to occur. These correspond to the so-called *guided modes*. They are determined by imposing the self-consistency condition that a wave reproduces itself after each round trip along the X direction (giving rise to constructive interference). The optical path in a round trip is $2n_{\rm f}h$ and the corresponding phase shift $2n_{\rm f}hk_x$. Additional phase shifts are introduced on total reflections from the film-cover, $2\phi_{\rm c}$, and the film-substrate, $2\phi_{\rm s}$. The self-consistency (or "transverse resonance") condition implies that the total phase shift be zero or a multiple of 2π, so that it is written

$$kn_{\rm f}h\cos\theta + \phi_{\rm s}(\theta) + \phi_{\rm c}(\theta) = v\pi, \qquad v = 0, 1, 2, \ldots \tag{8.4}$$

where v is the mode number. Expressions of $\phi_{\rm c}(\theta)$ and $\phi_{\rm s}(\theta)$ as functions of θ depend on the light polarization and can be obtained from the Fresnel's coefficients for total reflection (e.g. [10], p. 49, or [11], p. 207) and write

$$\tan\phi^{\rm TE}_{\rm c,s} = -\frac{\sqrt{n_{\rm f}^2\sin^2\theta - n_{\rm c,s}^2}}{n_{\rm f}\cos\theta} < 0 \tag{8.5}$$

for TE polarization (i.e. electric field perpendicular to the incidence plane), and

$$\tan\phi^{\rm TM}_{\rm c,s} = \frac{n_{\rm f}^2}{n_{\rm c,s}^2}\tan\phi^{\rm TE}_{\rm c,s} < 0 \tag{8.6}$$

for TM polarization (i.e. magnetic field perpendicular to the incidence plane). In equation (8.5) the sign of the square root must be taken as positive for consistency with the sign chosen in the phase of (8.2), so that $\phi^{\rm TE}_{\rm c,s}$ and $\phi^{\rm TM}_{\rm c,s}$ are negative.

Equation (8.4) yields a modal angle θ_v for each mode number v, from which the propagation constant β_v is obtained as a function of frequency $\omega = ck$ and film thickness h.

---

* Not to be confused with the second-order molecular polarizability.

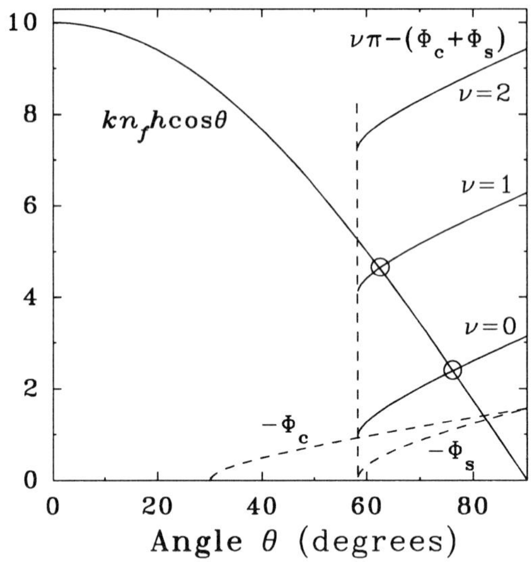

Figure 8.2 *Graphical solution of the dispersion equation of asymmetrical planar waveguides.*

This is, then, the *dispersion equation* of the guide. A graphical solution of this equation is sketched in Figure 8.2 with the condition $n_f > n_s > n_c$, i.e. $(\theta > \theta_s > \theta_c)$ for guided modes. The different terms of the equation have been drawn as functions of the angle θ, namely minus half the phase shifts at the film boundaries ϕ_s and ϕ_c, the phase shift introduced on crossing the film $(2\pi/\lambda)n_f h \cos\theta$, as well as the sum $\nu\pi - \phi_c - \phi_s$. The intersections of the two last functions determine the mode angles θ_ν. Note that θ_ν gets smaller (steeper zigzags) as $h\lambda$ gets smaller, and when $\theta_\nu < \theta_s$ that particular mode can never be guided. When this situation occurs for the fundamental mode, $\theta_0 < \theta_s$, no mode can be guided by the film and the *cut-off* condition is reached. If the guide is symmetrical, the value $\nu\pi - \phi_c - \phi_s = 2\phi_c$ for the fundamental mode can be as small as needed (since $\phi_c(\theta_c) = 0$), so that there is always a solution (no cut-off) even for very small film thickness or very large wavelength.

The range of possible values of effective indices and propagation constants for guided modes is easily found. Since $N_\nu = n_f \sin\theta_\nu < n_f$ and, for guided modes, $n_f \sin\theta_\nu > n_s$ (for total reflection to occur), the effective index of guided modes is bounded by the substrate and film indices

$$n_s < N_\nu < n_f \tag{8.7}$$

and β_ν is bounded by the propagation constants of plane waves propagating along the Z-axis in substrate and film

$$kn_s < \beta_\nu < kn_f \tag{8.8}$$

8.2.2. The Goos–Hänchen shift and the effective guide thickness

The phase shifts introduced on total reflection from the film-cover, $2\phi_c$, and film-substrate, $2\phi_s$, are important elements in the understanding of the energy flow through the guide. Energy flow is determined by the Poynting vector and its direction is usually referred to as the light ray. Although propagation through the cladding does not occur

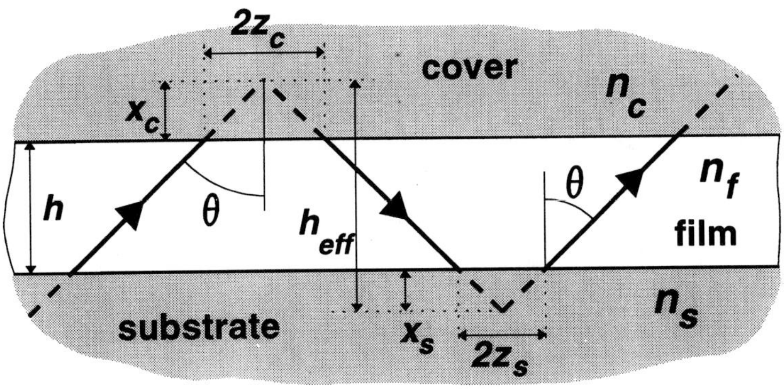

Figure 8.3 *Ray picture of the Goos–Hänchen shifts, z_s and z_c, and the effective guide thickness, h_{eff}, in the slab waveguide.*

on total reflection, there always exists a ray or energy penetration into it. This penetration produces a lateral shift z_c or z_s of the ray on each total reflection, known as *Goos–Hänchen shift*, and gives rise to an apparent or *effective guide thickness* h_{eff} (see Figure 8.3). It can be determined by considering a wave packet instead of a pure plane wave (e.g. [6], p. 17):

for TE waves:
$$z_{c,s}^{TE} = \frac{\tan\theta_v}{k\sqrt{N_v^2 - n_{c,s}^2}} \tag{8.9}$$

for TM waves:
$$z_{c,s}^{TM} = \frac{z_{c,s}^{TE}}{\dfrac{N_v^2}{n_{c,s}^2} + \dfrac{N_v^2}{n_f^2} - 1} \tag{8.10}$$

$$h_{\text{eff}} = h + x_s + x_c = h + \frac{z_s + z_c}{\tan\theta_v} \tag{8.11}$$

8.2.3. Field distributions for the step-index planar guide

To obtain the field distributions corresponding to the guided modes of a step-index planar waveguide, the wave equation with the appropriate boundary conditions must be solved. The wave equation is obtained from Maxwell's equations and, for TE polarization with $\mathbf{E} \equiv (0, E_y, 0)$ (TM polarization is treated in a similar way and analogous formulae are obtained), can be written [6]

$$\frac{\partial^2 E_y}{\partial x^2} = -n^2 k_x^2 E_y = (\beta^2 - n^2 k^2) E_y \tag{8.12}$$

The sign of $n^2 k_x^2$ is positive inside the film and this gives rise to an oscillatory solution of equation (8.12). On the contrary the sign is negative inside the cover and substrate so that evanescent waves are obtained as solutions in both media. Taking

$$\begin{aligned} n_c^2 k_x^2 &= n_c^2 k^2 - \beta^2 = -\gamma_c^2 < 0 \\ n_s^2 k_x^2 &= n_s^2 k^2 - \beta^2 = -\gamma_s^2 < 0 \end{aligned} \tag{8.13}$$

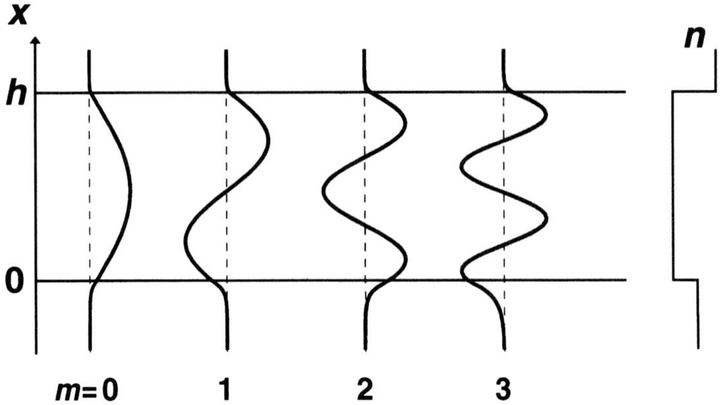

Figure 8.4 *Field distributions of TE modes in an asymmetrical planar dielectric waveguide.*

the solutions of equation (8.12) write

$$
\begin{aligned}
E_y(x) &= E_c\, e^{-\gamma_c (x-h)} && \text{for } x > h \text{ (cover)} \\
E_y(x) &= E_f \cos(n_f k_x x + \phi_s) && \text{for } 0 < x < h \text{ (film)} \\
E_y(x) &= E_s\, e^{\gamma_s x} && \text{for } x < 0 \text{ (substrate)}
\end{aligned}
\tag{8.14}
$$

The boundary conditions require that E_y and $\partial E_y/\partial x$ be continuous across the film boundaries at $x = 0$ and $x = h$. On applying these conditions to equation (8.14) the following results are obtained: (i) ϕ_s is found to coincide with definition (8.5), (ii) the dispersion equation (8.4) is obtained again, and (iii) the following relations between the field amplitudes appearing in (8.14) are found:

$$E_f^2(n_f^2 - N_\nu^2) = E_s^2(n_f^2 - n_s^2) = E_c^2(n_f^2 - n_c^2) \tag{8.15}$$

Figure 8.4 sketches the field distributions of the first few modes as given by equation (8.14). Notice that, unlike a beam in free space, guided light does not spread in the transverse direction as it propagates, since mode configurations are stable along the propagation direction. This important property of guided light makes waveguides particularly suitable as nonlinear optical devices for low power semiconductor lasers, since the light wave of high intensity is kept strictly confined over a long length of material where nonlinear interaction can take place.

Light waves with an arbitrary transverse field distribution (and not only individual modes) can also be guided by a structure like that shown in Figure 8.1. The unique limitation is that of satisfying the appropriate boundary conditions. However, the optical power is divided among available modes, which travel with different propagation constants and group velocities. As a consequence, the transverse distribution of the field changes as it travels through the waveguide.

8.2.4. Other waveguide geometries

Planar waveguides

Besides the simple step-index planar waveguide considered in former sections, practical devices involve other geometries such as graded-index planar and channel waveguides. For a graded-index planar waveguide, where the refractive index $n(x)$ varies gradually over the cross-section of the guide, equation (8.12) can be written

$$\frac{\partial^2 E_y(x)}{\partial x^2} = [\beta^2 - k^2 n^2(x)] E_y(x) \tag{8.16}$$

which has the same form as the Schrödinger equation of quantum mechanics, $-N^2 = -\beta^2/k^2$ corresponding to the energy level and $-n^2(x)$ to the potential energy well. Thus, the extensive literature of quantum mechanics can be used both for specific solutions and for methods of analysis. Perturbative and variational methods, as well as multilayer and Wentzel, Kramers and Brillouin (WKB) approaches have been applied. With some modifications that improve its applicability [12], the WKB method is fast and yields satisfactory results for slowly varying refractive index profiles, as is the case with Ti–$LiNbO_3$ and other waveguides prepared by diffusion. For sharper index profiles such as those obtained in proton-exchanged $LiNbO_3$ guides, or more complicated profiles as those produced by ion implantation, a multilayer approach is better suited for experimentalists. To simulate the guide reflectivity it is considered as a stack of very thin layers, each with $n(x) \approx$ constant, together with the appropriate boundary conditions for neighbouring layers [13].

A further degree of complexity appears when dealing with the anisotropic behaviour of the film or substrate material. Most waveguides are optically anisotropic to a certain degree since, even when the starting material is not anisotropic, the structural, chemical or stress changes induced by the guide fabrication procedure make it so. For instance, $LiNbO_3$ is an uniaxial substrate material where all common guide fabrication procedures result in approximately uniaxial guides. These guides may support modes on ordinary and/or extraordinary indices. Confinement for extraordinary polarization may even occur simultaneously with radiation for ordinary polarization (*hybrid leaky modes*) [14]. Simulations of wave propagation in anisotropic structures involve solving Maxwell equations with numerical methods. Although the finite element method can be used for planar guides [15], the *transfer-matrix* formalism [16] is preferred in this case because it is exact and faster.

Channel waveguides

Channel waveguides confine the light in two dimensions of the order of the wavelength, and, in general, the refractive index $n(x, y)$ is a function of both transverse coordinates x, y. These guides form the basic elements for many integrated optical devices and can present several structures and designs. Figure 8.5 shows three important types of optical channel

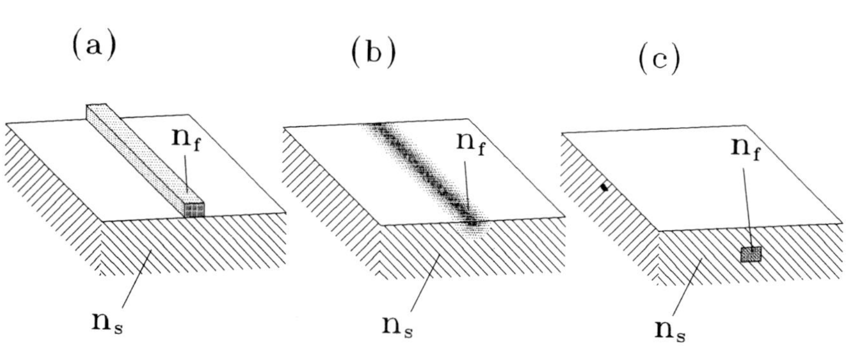

Figure 8.5 *Some types of channel waveguides: (a) raised; (b) graded-index embedded; (c) buried.*

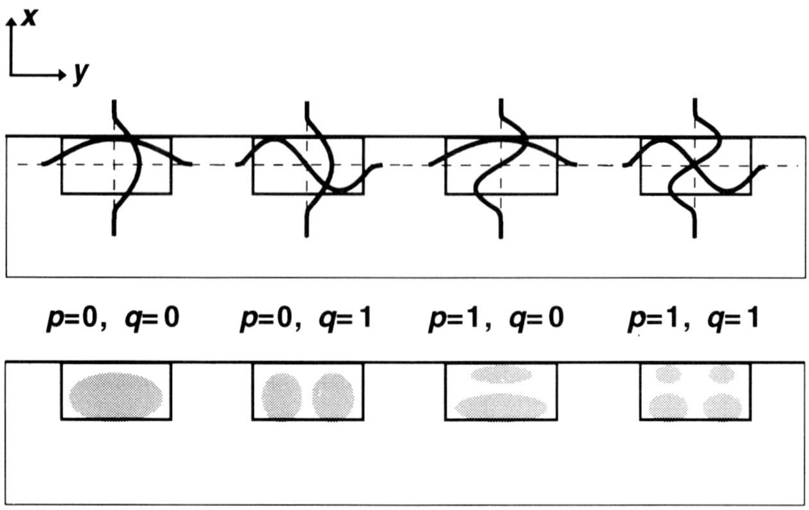

Figure 8.6 *Spatial variations of field and intensity for a few modes in an embedded channel waveguide.*

guides. For a rigorous theoretical study of these guides, a full three-dimensional vector wave equation is required that includes propagation along the guide direction, confinement in the two transverse directions and polarization effects. A derivation of this equation, as well as approximate methods of solving it and original references, can be found in reference [6] for example. In many practical cases, the *effective index method* that approaches the two-dimensional problem as two one-dimensional problems is very useful. Numerical methods already quoted [15] have to be used when higher precision is required. Notice that even the simple optical fibre made of amorphous silica presents geometry- and/or stress-induced anisotropy that varies along the fibre and complicates the desirable maintenance of the polarization [17]. However, the two-dimensional confinement gives rise to two mode numbers (or quantum numbers) that involve the corresponding modes in the field distribution. Figure 8.6 schematically illustrates the field distributions of a few modes in an embedded channel guide.

8.3. Inorganic electrooptic waveguides

Useful waveguides have been fabricated on a number of inorganic electrooptic materials, both dielectrics and semiconductors. They include $LiNbO_3$ and $LiTaO_3$ [18,19], $KNbO_3$ [20], KTP [21], GaAs [22] and InP [23]. Presently, the leading material is $LiNbO_3$ which overwhelmingly dominates the electrooptic market. Therefore, the various techniques to fabricate waveguides in dielectric materials will next be described with reference to $LiNbO_3$. A short section on semiconductor waveguides will be also included.

8.3.1. $LiNbO_3$ waveguides

Lithium niobate exhibits two important characteristics that make it attractive for waveguide devices: good electrooptic coefficients and commercial availability in large substrates (wafers of up to 7.5 cm in diameter) at a reasonable cost. It is also relatively easy to process and its material properties are extensively documented. The knowledge of the phenomenological aspects of $LiNbO_3$ waveguide technology is sufficient to fabricate devices of satisfactory overall performance (low losses, low drive voltages, large bandwidths). However, the relation between understanding of material properties, processing knowhow and guide performance is still evolving and further progress is expected. In the infrared region, semiconductor materials offering the possibility of monolithic integration threaten the present leading position of $LiNbO_3$ (in spite of the efficient $LiNbO_3$ waveguide lasers reported [24,25] which promise almost monolithic integration in this material). For visible applications, such as signal processing devices [26] or second harmonic generation [27,28], the leading position of $LiNbO_3$ is not presently threatened.

A number of techniques have been successfully used to produce waveguides on $LiNbO_3$ substrates, most of which have also been extended to $LiTaO_3$. Since the first success with thermal out-diffusion of Li_2O in a vacuum [29], waveguides have been formed by sputtering [30], epitaxial growth [31], thermal in-diffusion [32], ion implantation [33], ion exchange [34], molecular beam epitaxy (MBE) [35] (see Section 9.2 on semiconductors) and laser ablation [36]. However, only three of them have produced useful devices: ion implantation, proton exchange and, the most extended and the only one used in commercial devices, titanium in-diffusion. The following sections are devoted to describe these three methods.

Titanium in-diffusion

Titanium in-diffusion is the most popular method for producing optical waveguides in $LiNbO_3$ substrates and the only one actually used in commercial devices. A thorough description of the method and its practical applications is given in references [18] and [37]. Figure 8.7 shows the basic steps commonly followed for preparing a strip Ti–$LiNbO_3$ waveguide. After spin-coating with photoresist the polished crystal surface, it is exposed to UV light through the mask with the desired guide pattern, and developed to remove the exposed part. Then, a titanium film of a thickness ranging between 10 nm and 100 nm is deposited over the entire surface by any common deposition method, and the sample

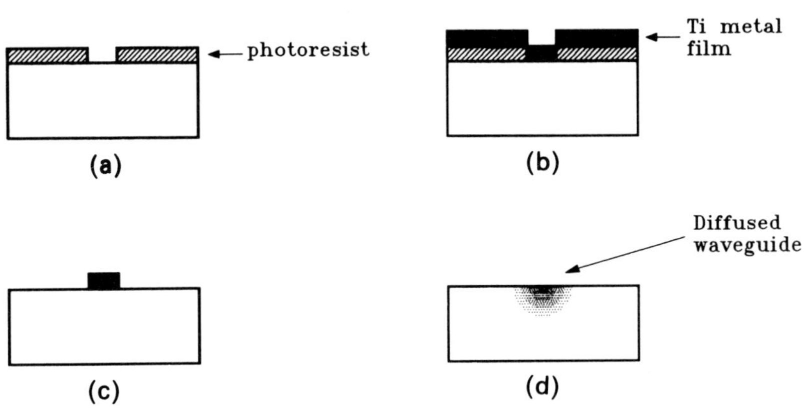

Figure 8.7 *Basic steps for fabricating strip Ti–$LiNbO_3$ waveguides. (a) Exposure of waveguide pattern. (b) Deposition of titanium dopant. (c) Lift-off. (d) Diffusion.*

is immersed in a photoresist solvent to remove the unwanted titanium layer. Titanium in-diffusion is carried out by heating the sample at temperatures ranging from 900 °C to 1100 °C for typical diffusion times of 0.5–30 h. The treatment temperature must be well below the Curie temperature (~1150 °C) to avoid the need of repoling the crystal after diffusion. To prevent Li_2O out-diffusion, which induces a planar waveguide in the whole surface, and to improve the optical damage resistance, the heating and diffusion treatments are performed under an oxygen flow with a relative humidity of about 80%. The physical mechanisms involved in the diffusion/doping and refractive index change are not yet fully understood, although Ti-induced strain seems to be a significant factor via the photoelastic effect. Other metals have been used as diffusant in $LiNbO_3$ and the method has also been applied to other substrates, particularly to $LiTaO_3$ where niobium is used instead of titanium.

Waveguide parameters can be independently controlled through the fabrication parameters. The guide depth d depends on the diffusion time t and temperature T, and to a lesser extent on the titanium strip width. The waveguide width, w, is defined in the photolithographic process, provided that correction for lateral diffusion is taken into account. Also, the guide index change Δn for fixed t and T can be controlled with the titanium thickness and density (density depends on the deposition method and ranges from bulk density to 70% of this value). The extraordinary index increase is proportional to the titanium concentration, whilst the ordinary index increase deviates from linearity for large changes. In the simple diffusion model, the dependence of Δn with the depth x in the crystal can be written

$$\Delta n(x) = \Delta n(0)\, e^{-(x/d)^2} \tag{8.17}$$

where $\Delta n(0)$ depends on the deposited titanium thickness and width and on d. Here d is the diffusion depth that is related to the diffusion time t and diffusion coefficient

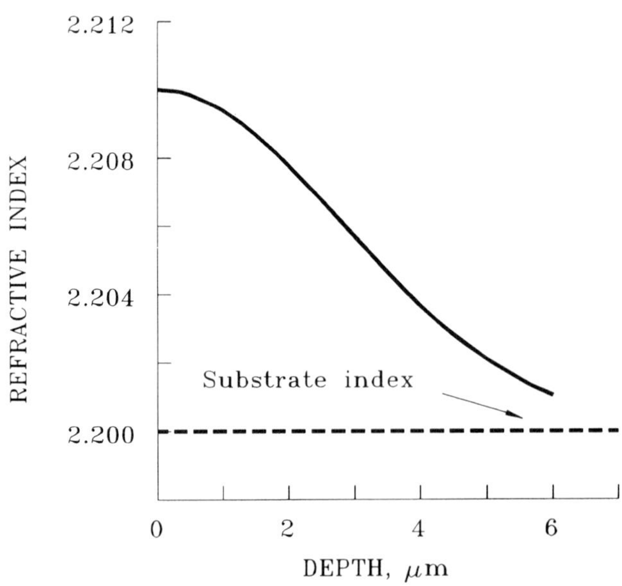

Figure 8.8 *Extraordinary refractive index profile of a Ti–$LiNbO_3$ waveguide.*

$D = D_0 \exp(\Delta\epsilon/k_B T)$ by

$$d = 2\sqrt{Dt} \tag{8.18}$$

Numerical values for these parameters are the following: $\Delta n(0) < 0.01$, $D_0 = 2.5\times 10^{-4}\,cm^2/s$, $\Delta\epsilon \approx 2.2\,eV$. Figure 8.8 illustrates a typical index profile of a Ti–$LiNbO_3$ waveguide. Losses as low as 0.2 dB/cm can be reached. A negative feature of these guides is the strong optical damage (photorefractive effect) they present in the visible and near infrared regions that make them useless for visible applications. Mg doping or an additional proton exchange treatment substantially improves their optical damage resistance.

Proton exchange

The method of proton exchange to form waveguides in $LiNbO_3$ substrates is a variant of the ion exchange method commonly employed to produce guides in glass and other dielectric substrates. It was first reported in 1982 [34] and since then has developed to a considerable degree. A detailed description of the technique can be found in references [38] and [39]. Apart from lithographic procedures (illustrated in Figure 8.7) to form channel waveguides, the essential of this method is the index changes induced by the proton-lithium exchange that can be obtained on immersing a $LiNbO_3$ substrate in a hot acid. Although many different acids have been employed, benzoic acid continues to be the most popular. The treatment temperature ranges from 160 °C to 350 °C depending on the desired index profile characteristics. Above the boiling temperature of benzoic acid (249 °C) a sealed ampoule has to be used to prevent loss of acid. The treatment time ranges from a few minutes to several hours according to the desired profile, the temperature and the melt composition. Typical diffusion coefficients are of the order of 1 $\mu m^2/h$.

Early proton exchanged guides exhibited mode degradation with time, as well as with the application of a d.c. electric field, and a substantial reduction of the electrooptic coefficients with respect to the bulk value. All these problems are now overcome by fabricating the guide at temperatures above 300 °C and/or adding an appropriate lithium benzoate amount (between 0.5% and 5%) to the benzoic acid melt that makes the process less aggressive [39]. Thermal annealing treatments posterior to the guide fabrication also improve the guiding quality and the recovery of the electrooptic coefficient up to the bulk value.

Figure 8.9 shows, as dotted lines, typical refractive index profiles induced by proton exchange. Step (Fermi-like) profiles are found for both extraordinary and ordinary indices, although index changes are different. Whilst the extraordinary index increases ($\Delta n_e \approx 0.11$) with respect to the substrate value, the ordinary index decreases ($\Delta n_o \approx -0.04$). These guides, then, can only support modes with extraordinary polarization, which is an advantage in polarization selective devices, but a disadvantage when both polarizations have to be handled. By a reverse exchange process in melts very rich in lithium, guides with ordinary polarization can also be prepared, together with a buried extraordinary guide [39,40]. Figure 8.9 shows the index profiles (solid line for ordinary and dashed line for extraordinary index) for this case where parabolic-like shapes are found. In either case, losses are found to be under ~0.5 dB/cm. The origin of index changes is not clear as yet and most likely several mechanisms are simultaneously contributing, e.g. changes in density, ionic site preference and changes in lattice constant [41].

At present, proton exchange is the best method to produce large and sharp index changes in $LiNbO_3$. Thus, these guides are the best choice for applications requiring tight wave confinement such as ring resonators [42] and other structures with a small radius of

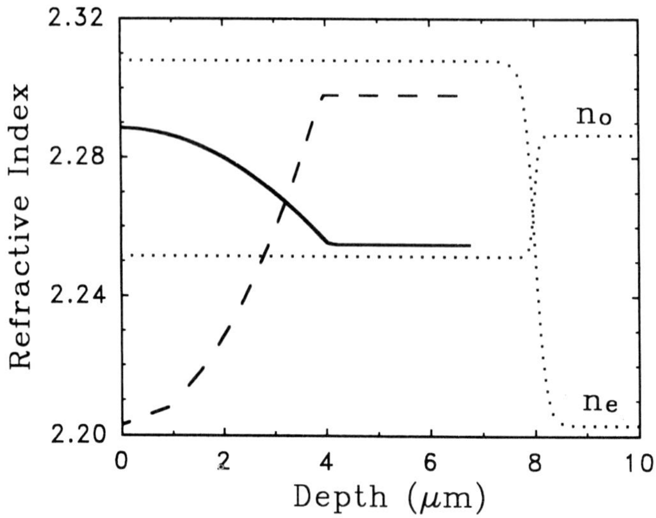

Figure 8.9 *Extraordinary and ordinary refractive index profiles of a proton exchanged waveguide (dotted lines). Extraordinary (dashed line) and ordinary (solid line) index profiles of a reverse exchanged waveguide.*

curvature, planar lenses [43], gratings and others. A major advantage of proton exchanged guides is their much smaller sensitivity to optical damage, down to several orders of magnitude smaller than that of Ti-guides [44,45]. Proton exchange has also been applied to Ti-guides to produce the so-called TIPE-guides [46] with reduced optical damage sensitivity and other specific properties. Low optical damage is particularly advantageous for applications with visible light, where the damage sensitivity presents its maximum. An important example is second harmonic generation (SHG) of blue-green light [47], that also takes advantage of an improved efficiency of the nonlinear interaction. Efficient operation of CW and mode-locked lasers have also been demonstrated in proton exchanged guides on Nd–MgO–$LiNbO_3$ substrates [48,49].

Ion implantation

Ion implantation is a general technique that has been used to form waveguides in many dielectric materials as well as in some semiconductors [50]. In $LiNbO_3$, by an appropriate handling of the lattice damage produced by the implanted ions, good waveguides and some devices have been reported. A detailed description of the method and results are reviewed in references [50,51]. The main features of the method are the following. Bombardment with high energy (of the order of 1 MeV) ions of light mass, e.g. He^+ ions, produces two effects along their track within the lattice before stopping. Near the surface, where the ion velocity is very high, ions mainly interact with electrons and produce ionization, i.e. lattice defects, and slight index changes are produced. Deeper into the sample, where the ion velocity is low, nuclear collisions predominate and ion displacements produce lattice disordering and density decrease, although in $LiNbO_3$ amorphization is not reached. This second process induces a strong decrease of both ordinary and extraordinary indices. Thus, a low index barrier is produced that confines the light in the outer layer, whose refractive index can be smaller than that of the substrate, although greater than in the barrier. The depth of the barrier is controlled with the energy of the ions, and the refractive index change (the barrier height) with the dose.

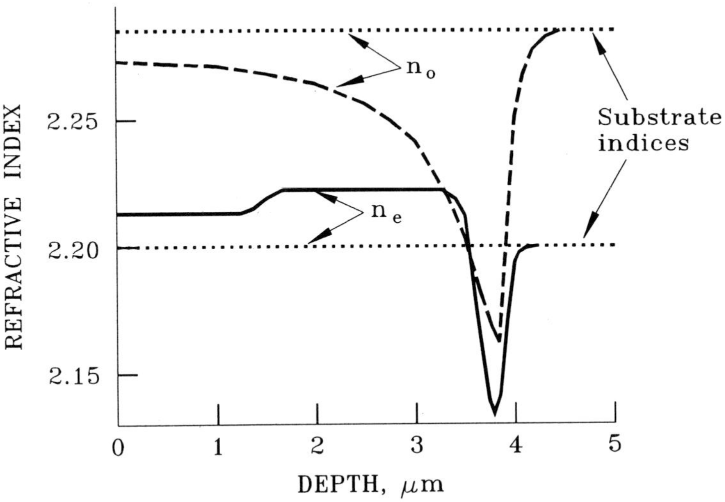

Figure 8.10 ***Extraordinary (solid line) and ordinary (dashed line) refractive index profile of an implanted $LiNbO_3$ waveguide.***

Figure 8.10 shows typical ordinary and extraordinary index profiles of an implanted $LiNbO_3$ waveguide. The ordinary index is seen to decrease, slightly at the electronic damage region and strongly at the nuclear damage region near the ion stopping distance. The decrease can be approximately fitted to an exponential function, closely resembling that followed by the nuclear stopping damage profile with a point defect tail extending to the surface. Then, the ordinary index sharply increases to the substrate value forming the index barrier. The extraordinary index first increases slightly and then decreases to form a barrier sharper than that of the ordinary index. It is believed [51] that during ion bombardment lithium ions diffuse towards the highly disordered nuclear damage peak where ionic mobility is enhanced.

Ion dose can range from $\sim 10^{14}$ ions/cm^2 to the saturation value of $\sim 10^{17}$ ions/cm^2 at which a 15% refractive index change can be obtained. Most often a moderate dose of $\sim 10^{16}$ ions/cm^2 is employed that produces a barrier of 2% or 3%. Guide depths can range from tenths of a μm to several μm, depending on the ion projected range (i.e. on the ion mass and energy). The barrier width can be broadened with multiple energy implants. The electronic damage in the guiding region produces considerable guide losses and reduction of electrooptic coefficients. As with proton exchanged guides these problems are overcome by post-implantation thermal annealing at temperatures between 200 °C and 300 °C. Losses below 1 dB/cm have been reported. The sensitivity to optical damage of these guides is lower than that of Ti-diffused but substantially greater than that of proton exchanged guides [44].

Some devices for use with visible light have been reported for $LiNbO_3$ implanted guides, e.g. second harmonic generators [28] and waveguide lasers [25], although Ti diffusion and proton exchange are efficient, cheap and well established fabrication methods with which, at present, competition is hard. Nevertheless, light ion implantations can easily produce double well structures [52], and heavier ion (as Ti [53]) implantations on Ti in-diffused or proton exchanged guides may be a good complementary tool. Of much more interest is the easy application of ion implantation to a large variety of different materials as discussed in reference [50].

8.3.2. Semiconductor waveguides

For optoelectronic applications, leading materials are the alloys $Al_xGa_{1-x}As$ and $In_xGa_{1-x}As_{1-y}P_y$. To fabricate waveguides in them, one takes advantage of the dependence of refractive index on alloy composition, i.e. on x and y. As an example, for AlGaAs, the refractive index decreases with increasing Al content. Therefore, a waveguide can be formed by depositing a layer of GaAs on a $Al_xGa_{1-x}As$ ($x \neq 0$) substrate. Film preparation techniques include: liquid-phase epitaxy (LPE), vapour-phase epitaxy (VPE), metal-organic chemical vapour deposition (MOCVD) and molecular beam epitaxy (MBE). Some of them will be discussed in Chapter 9, when dealing with the fabrication of semiconductor nanostructures. All these techniques have reached a high degree of maturity because of their application to microelectronic devices.

8.4. Organic electrooptic waveguides

The very promising electrooptic and nonlinear optical properties of organic materials described in Chapter 6 are promoting an extensive research effort directed towards the preparation and characterization of waveguides and integrated optical devices [54]. A number of fabrication procedures have been explored and in a number of cases have yielded good performance waveguides. However, much more work is needed in this area before those procedures can actually be incorporated by industry.

One group of techniques uses *deposition* of the organic guiding layer onto an organic or inorganic substrate. It includes: solution deposition, vacuum evaporation, plasma polymerization and the Langmuir–Blodgett (LB) technique.

Among solution techniques, the *spinning* or *spin-coating method* has been most often used. Thin-films of soluble polymers have been often deposited by this technique. The thickness of the deposited layer may range from $\sim$0.1 to $\sim$1 μm. It can be controlled to a certain extent through the viscosity of the solution and the spinning velocity. Standard micro-lithography can be used to prepare channel waveguides.

Vacuum evaporation (or better sublimation) is a relatively simple technique frequently used with inorganic compounds. It also works for many organic substances, although one is faced with the problem of thermal molecular decomposition. The microcrystalline structures common for evaporated films leads to large scattering losses. Improved performance can be achieved by playing with the substrate temperature.

In the plasma polymerization technique [55], a radiofrequency discharge is established in an enclosure containing the monomers in vapour phase. As a consequence of the discharge, bonding of the monomers occurs by cross-linking and a polymer layer is deposited on the substrate.

The LB technique has been already described in Chapter 6. It has the advantage of a very precise control of layer thickness (at molecular level) and also allows the preparation of ordered structures appropriated for second-order nonlinear effects. However, the thickness obtained appears to be limited so far up to $\sim$0.5 μm. Moreover, the deposited layer shows a large number of structural defects which lead to poor performance. Promising results have been recently obtained by avoiding crystallization on the water–air monolayer. Films containing 500 monolayers, exhibiting an attenuation as low as 2 dB/cm, have been prepared [56].

One should here mention epitaxial growth methods, from the liquid or vapour phases. Some recent experiments have produced single-crystal layers of active organic compounds with suitable orientation for efficient second-harmonic generation [57].

Another group of waveguide preparation techniques involves the modification of a given organic substrate by adequate physical or chemical surface treatments, e.g. *photo-sensitization* (or electron beam sensitization), *photolocking* and *ion implantation.* In photon- [58] or electron-induced [59] sensitization, a given material is sensitized by illumination or electron-beam irradiation respectively, and is subsequently developed. This is a simple extension of the standard phorotesist technique used in microelectronics. As an example [60], photomethylmethacrylate (PMMA) films doped with an initiator have led after illumination to refractive index changes as high as 5×10^{-2}. Photolocking [61] uses a polymer doped with a suitable dopant, so that illumination induces cross-linking of the dopant which becomes locked to the polymer structure. Ion implantation has also been recently applied to organic materials. Monomode and multimode guides have been produced as reported for PMMA [62].

8.5. Electrooptic modulation in waveguides

There are several characteristics of electrooptic switches and modulators of particular relevance when operating in waveguides. In fact, optical waveguides constitute a very convenient configuration for the electrooptic effect and present superior performance parameters in comparison with bulk devices. A waveguide modulator, such as the single-channel modulator illustrated in Figure 8.11, uses a transverse geometry. Its small interelectrode gap ($\sim \mu$m) *vs* electrode length, leads to *half-wave voltages* as low as a few volts, in accordance with equation (3.68). This advantage is further substantiated by the absence of any beam divergence (diffraction) of the waveguide modes, which allows long light

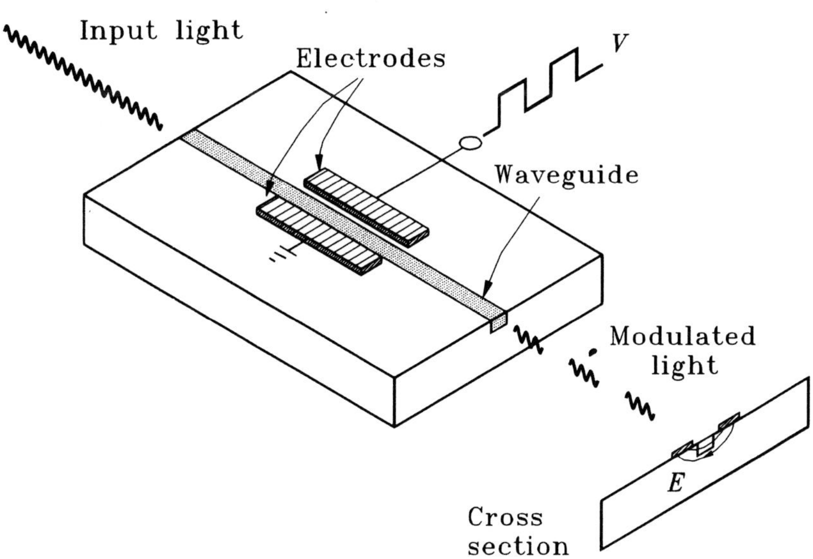

Figure 8.11 *Single waveguide electrooptic modulator.*

paths. The much reduced V_π simplifies the electrical source and implies a low *power per unit bandwidth*. However, losses of optical power are necessarily increased by insertion into the waveguide modulator.

Bandwidths can be also very large and modulation speeds over 70 Gb/s have been reached. Wide bandwidth is a critical factor when many signals are to be multiplexed onto the same optical beam. On the other hand, the fast switching speed (switching time $\tau \approx 1/\Delta f$) is most important on large-scale arrays of switches used to address light waves to desired routes.

In principle, one may think of modulating the waveguide laser itself as the preferable modulation method (*monolithic approach*). However, direct modulation of lasers produces "chirping" (see Chapter 7) and this affects dispersion during fiber propagation and so impairs performance of the communication links. Main types of waveguide modulators will now be described in more detail.

8.5.1. Single-waveguide electrooptic modulators

Electrooptic switches and modulators can be fabricated either on planar, channel, single or double waveguides, and they can function as either amplitude (intensity) modulators, phase modulators or as optical switches. The single channel waveguide modulator shown in Figure 8.11 can work as a *phase modulator*. Upon applying an electric field E to the guide, a refractive index change will be induced by electrooptic effect, (equation (3.11)),

$$\Delta n_{\mathrm{EO}} = -\tfrac{1}{2} n^3 r E \tag{8.19}$$

where r is the appropriate electrooptic coefficient. Then, the corresponding change of the propagation constant of the guide is

$$\Delta\beta = k\,\Delta n_{\mathrm{EO}} = -\frac{\pi}{\lambda_0} n^3 r E \tag{8.20}$$

so that the phase modulation is

$$\Delta\phi_{\mathrm{EO}} = L\,\Delta\beta = -\frac{\pi}{\lambda_0} n^3 r E L \tag{8.21}$$

where L is the length of the modulator in the Z (propagation) direction. Obviously, a phase coherent detection system must be used with phase modulators.

The same design of Figure 8.11 can be used as a *polarization modulator* by introducing the optical beam linearly polarized at 45° to the X- and Y-axes. Since the phase change mainly occurs for Y polarization in this figure, a rotation of the polarization vector will take place as the waves propagate in the Z direction. Detection of the change of polarization can be achieved by placing an appropriate analyser ahead of the detector.

The device of Figure 8.11 can also be used as *intensity modulator*. Intensity modulation is easier to detect than phase or polarization modulation. However, the difference in refractive index at the waveguide-substrate interface must be critically controlled during the guide fabrication process. It must be just at the threshold for guiding the lowest order mode (*cut-off*, see Section 8.2.1) and smaller than the index change induced via electrooptic effect. The layer will guide or not according to the magnitude and sign of the applied electric field.

There are good examples of the above mentioned modulators fabricated with Ti in-diffused $LiNbO_3$ waveguides. Typical guide widths are between 2 μm and 5 μm, and applied voltages between electrodes within the range 1–10 V (fields from 2 to 50 kV/cm).

Phase modulators require about 1 μW/(MHz rad) and achieve up to 1 rad with a voltage of about 1 V, bandwidths between 1 GHz and 10 GHz, and insertion losses between 2 dB and 5 dB. Intensity modulators easily reach extinction ratios between 15 dB and 20 dB, which is normally sufficient, for modulation voltages of $\pm$10 V between electrodes. Further details and references for Ti–$LiNbO_3$ can be found in reference [18].

The Franz–Keldysh effect (see Section 5.10.3) can also be used in semiconductor waveguides to produce a different type of device: the *electroabsorption modulator*. Modulation is based on the electric-field-enhanced absorption near the semiconductor band edge. This involves photocurrent and an associated heat dissipation problem not present in modulators based on the electrooptic effect. Heat dissipation becomes important especially for applications in which high power has to be handled. Operation near the semiconductor band edge can also lead to high insertion loss, and short ($\sim$1 mm) devices have to be used. Other waveguide effects associated with the enhanced absorption are the index change, which affects the modes and produces chirping effects in modulation, and the dichroism induced by electroabsorption. Achievements in InGaAsP have shown a 20 dB extinction ratio and bandwidth of $\sim$1.6 GHz. In AlGaAs/GaAs, a 30 dB extinction ratio and 3 GHz bandwidth have been proved, with driving voltages $<$10 V and power requirements of the order of 0.1 mW/MHz. Insertion losses are of the order of 10 dB in both cases. For further details and references see reference [63].

8.5.2. Mach–Zehnder type modulators

An important two-waveguide modulator is based on the waveguide version of the Mach–Zehnder interferometer. The basic features of this modulator are shown schematically in Figure 8.12. Light is coupled into a single-mode waveguide and divided into two equal beams by a waveguide beamsplitter. These beams travel through two single-mode guides which recombine into an output single-mode guide. The output intensity arises from interference of phase coherent light waves that have travelled over different paths. Ideally, the path lengths and guide characteristics are identical, so that the output will consist of the same unique mode as in the input guide. When an appropriate electric field is applied to one of the paths, a phase change of π radians is introduced between the two arms. The recombination results in a field distribution which is zero at the centre of the output guide

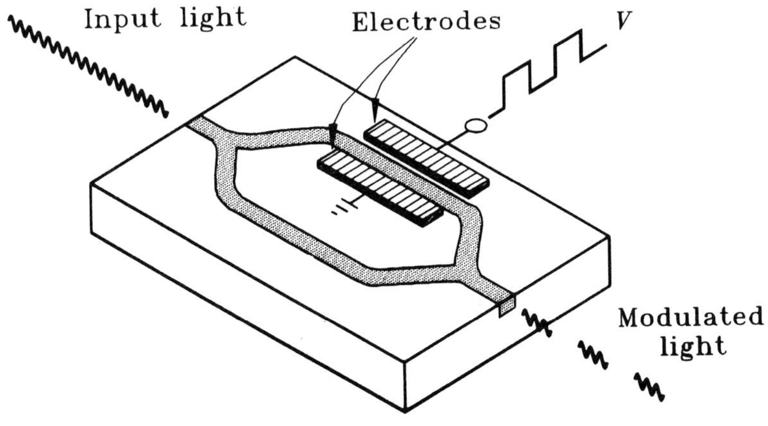

Figure 8.12 *Basic structure of Mach–Zehnder type modulator.*

and corresponds to the first order $m = 1$ mode. In a single-mode guide this mode quickly dissipates by substrate radiation. The modulator can be, then, switched from a transmitting to a nontransmitting state by a voltage application.

In practice the two arms of the interferometer cannot be fabricated identically. Even minute modifications of fabrication parameters result in modulators which are not in the pure *on* state. Thus, a bias voltage has to be supplied for compensation. Most successful Mach–Zehnder modulators have been prepared with Ti–$LiNbO_3$ guides [18]. There are commercially available modulators with on/off ratios over 20 dB for applied voltages of about 15 V and speeds of a few GHz. In laboratory experiments speeds up to 25 GHz have been reported.

8.5.3. Electrooptic Bragg modulators

Electrooptic control of either reflection or diffraction of a guided beam has been proved useful for fabricating waveguide modulators and switches. Most common diffraction modulators use Bragg diffraction from an optical grating produced inside the waveguide. In order to induce the grating a voltage is applied to interdigitated surface electrodes (see Figure 8.13) which via the electrooptic effect produce a refractive index modulation (phase grating). The light is diffracted with maximum efficiency if the incidence angle in the waveguide is adjusted to the Bragg angle θ_B. The Bragg angle is given by

$$\sin \theta_B = \frac{\lambda_0}{2N_\nu \Lambda} \tag{8.22}$$

where N_ν is the effective guide index (β_ν / k), and Λ the grating spacing. In planar guides of Ti–$LiNbO_3$, electrooptic Bragg modulators with diffraction efficiencies over 98% and extinction ratios about 300 : 1 (24.7 dB) were soon achieved [64]. The spatial separation of the diffracted beam from the original beam is an additional advantage of Bragg modulators that allows them to be used directly in switching applications.

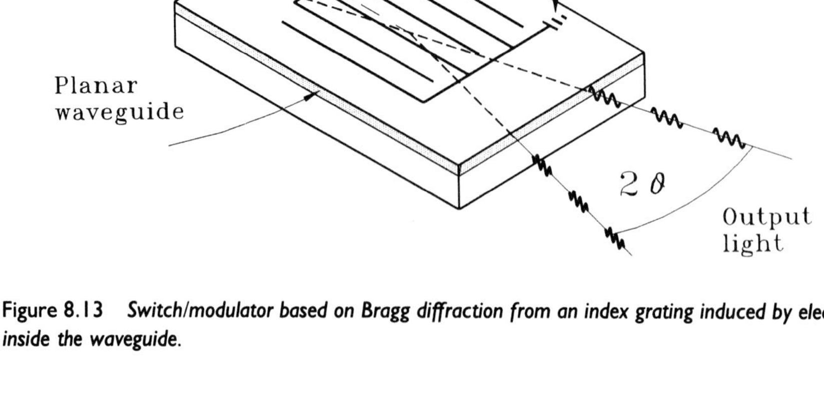

Figure 8.13 ***Switch/modulator based on Bragg diffraction from an index grating induced by electrooptic effect inside the waveguide.***

8.6. Coupled-mode formalism

Modes in a waveguide can interact with each other through a variety of mechanisms, giving rise to energy exchanges among them. The same type of interaction also takes place when two waveguides are sufficiently close so that their evanescent fields overlap. Thus, energy transfer from one guide to the other occurs through interaction of the modes of the two-guide structure, which are different from those of each of the isolated waveguides. More complex guiding structures formed of several coupled waveguides show the same phenomenon as well. Mode coupling is of particular relevance in electrooptic devices which use an applied electric field to control energy transfer between waveguides (through a change of the mode interaction). For an adequate analysis of these situations, Maxwell's equations and appropriate boundary conditions must be used. However, generating a solution to Maxwell's equations from a computer for each particular geometry of interest is a tedious task, which usually does not provide guidance to the designer. A simplified approximation has thus been developed that usually gives satisfactory results for the common case of weak coupling. This is the so-called *coupled-mode formalism* that has become the most widely used theoretical tool for studying mode interactions. Whilst the exact analysis involves computing the normal modes of the whole structure using Maxwell's equations, the coupled-mode approximation involves obtaining those "perturbed" modes (Figure 8.14(b)) by summation of normal modes of the "unperturbed" system (uncoupled waveguides, Figure 8.14(a)).

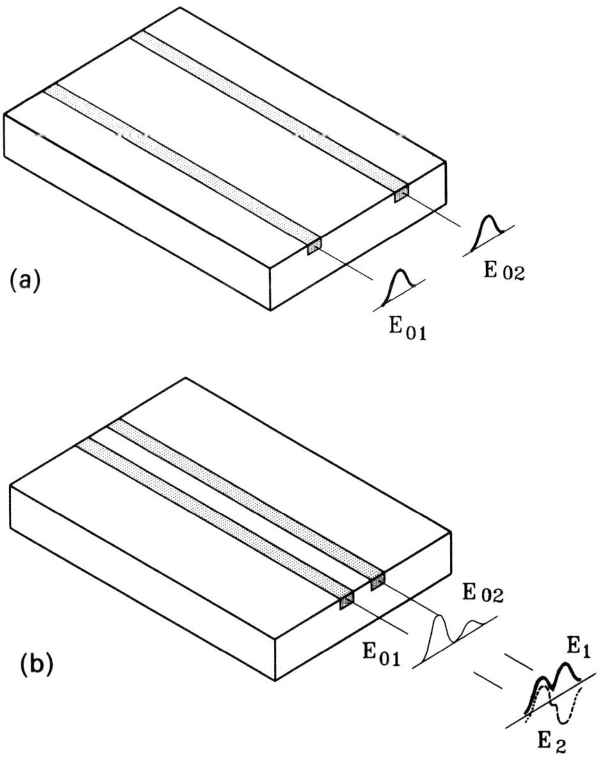

Figure 8.14 *Field distributions of normal modes of two identical waveguides: (a) uncoupled; (b) coupled.*

We will introduce the basic concepts and results of the coupled-mode approach through the simplest example of two weakly coupled waveguides. Modes are considered to propagate in the same direction in both waveguides (co-directional interaction) and, when uncoupled, they are assumed to guide only the fundamental mode. A more detailed treatment, including a derivation of the coupled-mode equation to be used below, can be seen in references [5] and [6]. Let the field distributions of the considered modes for the uncoupled situation be:

$$E_{0j}(x,y,z) = A_{0j}\,\mathrm{e}^{i\beta_j z} f_j(x,y), \qquad j = 1,2 \tag{8.23}$$

where A_{0j} is a constant and $f_j(x,y)$ is the transverse distribution normalized to give unit power flow. When the waveguides are weakly coupled, it is assumed that the modes of equation (8.23) keep the same propagation constants β_j and transverse distributions $f_j(x,y)$, but the amplitudes vary slowly along the propagation direction. Then, the perturbed modes are

$$E_j(x,y,z) = A_j(z)\,\mathrm{e}^{i\beta_j z} f_j(x,y) \tag{8.24}$$

The evolution of $A_j(z)$ is governed by the following coupled-mode equations, derived from Maxwell equations [5,6]:

$$\begin{aligned}\frac{\mathrm{d}A_1(z)}{\mathrm{d}z} &= i\kappa A_2(z)\,\mathrm{e}^{i(\beta_2-\beta_1)z}\\ \frac{\mathrm{d}A_2(z)}{\mathrm{d}z} &= i\kappa A_1(z)\,\mathrm{e}^{-i(\beta_2-\beta_1)z}\end{aligned} \tag{8.25}$$

where $\Delta\beta = \beta_2 - \beta_1$ corresponds to the phase mismatch of the modes (it is zero when both propagation constants are the same), and coefficient κ is the coupling per unit length, essentially determined by the overlap integral of the two involved modes. This constant produces a rate of change of $A_1(z)$ proportional to $A_2(z)$ and *vice versa*. In the absence of guide coupling ($\kappa = 0$), $A_j(z)$ reduces to the constant A_{0j} of the unperturbed modes. Assuming that light only enters waveguide 1, i.e. with boundary conditions $A_1(0) = 1$ and $A_2(0) = 0$, the solution of equation (8.25) can be written

$$\begin{aligned}A_1(z) &= \left(\cos\beta_c z - i\frac{\Delta\beta}{2\beta_c}\sin\beta_c z\right)\exp\left(i\frac{\Delta\beta}{2}z\right)\\ A_2(z) &= \left(i\frac{\kappa}{\beta_c}\sin\beta_c z\right)\exp\left(-i\frac{\Delta\beta}{2}z\right)\end{aligned} \tag{8.26}$$

where

$$\beta_c = \sqrt{\kappa^2 + \left(\frac{\Delta\beta}{2}\right)^2} \tag{8.27}$$

which, for identical waveguides ($\beta_1 = \beta_2$, phase-matched modes), reduces to $\beta_c = \kappa$. From these equations the power flow can be written

$$\begin{aligned}\frac{P_1(z)}{P_1(0)} &= \frac{|A_1(z)|^2}{|A_1(0)|^2} = 1 - \frac{\kappa^2}{\beta_c^2}\sin^2\beta_c z\\ \frac{P_2(z)}{P_1(0)} &= \frac{|A_2(z)|^2}{|A_1(0)|^2} = \frac{\kappa^2}{\beta_c^2}\sin^2\beta_c z\end{aligned} \tag{8.28}$$

Equations (8.28) show that power is exchanged periodically between the two guides with a period $2\pi/\beta_c$. The maximum transfer of optical power from guide 1 to guide 2 is determined by κ^2/β_c^2, which is equal to 1 only for $\beta_1 = \beta_2$ (phase-matched modes), but smaller than 1 for $\beta_1 \neq \beta_2$ (phase-mismatched modes). The maximum energy transfer occurs at a distance

$$L = \frac{\pi}{2\beta_c}$$
$$\left(L = \frac{\pi}{2\kappa} \quad \text{for } \beta_1 = \beta_2\right) \tag{8.29}$$

called *coupling* or *transfer length*. Notice that for phase-matched modes with $\Delta\beta = 0$, in equation (8.28) $\kappa^2/\beta_c^2 = 1$ and the coupling coefficient κ only affects the coupling length but not the maximum power transfer, which is 100% irrespective the value of κ.

8.7. Directional couplers

An important application of the electrooptic effect is the electrical control of the coupling efficiency of waveguide couplers. Considers a waveguide coupler made up of two identical waveguides with phase-matched modes ($\beta_1 = \beta_2$) interacting along a coupling length given by equation (8.29), $L = \pi/2\kappa$. The device will produce a complete energy transfer from one guide to the other. However, if a small phase mismatch $\Delta\beta = \beta_2 - \beta_1$ is introduced via the electrooptic effect, using appropriate electrodes, the power-transfer ratio of the device can be changed in a controlled manner. Then, the coupling efficiency, $\eta = P_2(L)/P_1(0)$, from equations (8.28) and (8.29) can be written

$$\eta(\Delta\beta) = \left(\frac{\pi}{2}\frac{\sin\alpha}{\alpha}\right)^2 \qquad \text{with} \qquad \alpha = \frac{\pi}{2}\sqrt{1 + \left(\frac{L\,\Delta\beta}{\pi}\right)^2} \tag{8.30}$$

As shown in Figure 8.15, this function exhibits a maximum value of unity at $L\,\Delta\beta = 0$ (notice that $\alpha \geq \pi/2$), which decreases with increasing $\Delta\beta$, and then vanishes (no energy is transferred at waveguide 2) when $L\,\Delta\beta = \sqrt{3}\pi$. The geometry of a typical electrooptic directional coupler, with coupling length L and distance between guides d, is illustrated in Figure 8.16. The applied voltage V generates an electric field $E \approx V/d$ whose lines go downward at one waveguide and upward at the other. Thus, the refractive index increases in one waveguide and decreases in the other, and the total difference is given by equation (8.19): $2|\Delta n| \approx n^3 r(V/d)$. This corresponds to a phase mismatch factor of

$$L\,\Delta\beta \approx \frac{2\pi}{\lambda_0} n^3 r L \frac{V}{d} \tag{8.31}$$

which is proportional to the applied voltage V. The voltage V_s necessary for a complete switching of the optical power is determined by the vanishing condition of equation (8.30), $L\,\Delta\beta = \sqrt{3}\pi$, i.e.

$$V_s = \frac{\sqrt{3}\lambda_0 d}{2n^3 r L} \tag{8.32}$$

This is called *switching voltage*. Then, from relation $L\,\Delta\beta = \sqrt{3}\pi V/V_s$, the argument α of

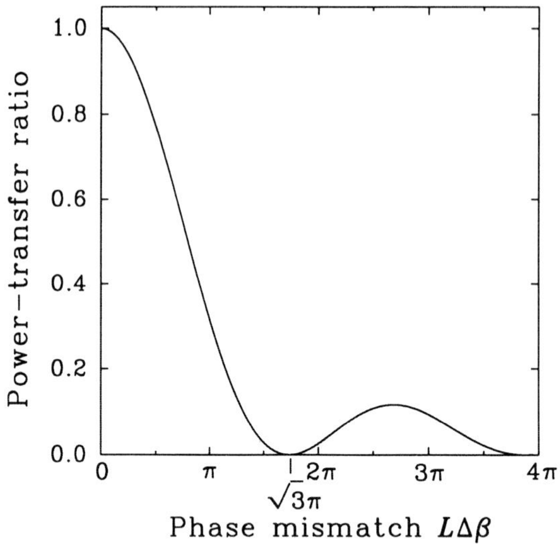

Figure 8.15 *Coupling efficiency of a directional coupler as a function of the phase mismatch* $\Delta\beta$.

function (8.30) is

$$\alpha = \frac{\pi}{2}\sqrt{1 + 3\left(\frac{V}{V_s}\right)^2} \tag{8.33}$$

so that the plot of Figure 8.15 can be drawn in terms of the applied voltage V. The key parameters determining V_s are the electrooptic figure of merit n^3r, and the coupling

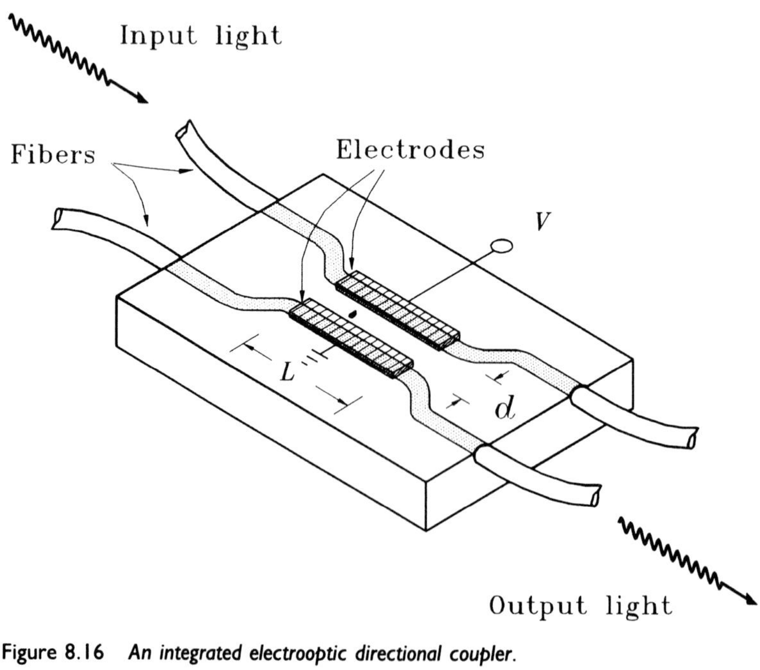

Figure 8.16 *An integrated electrooptic directional coupler.*

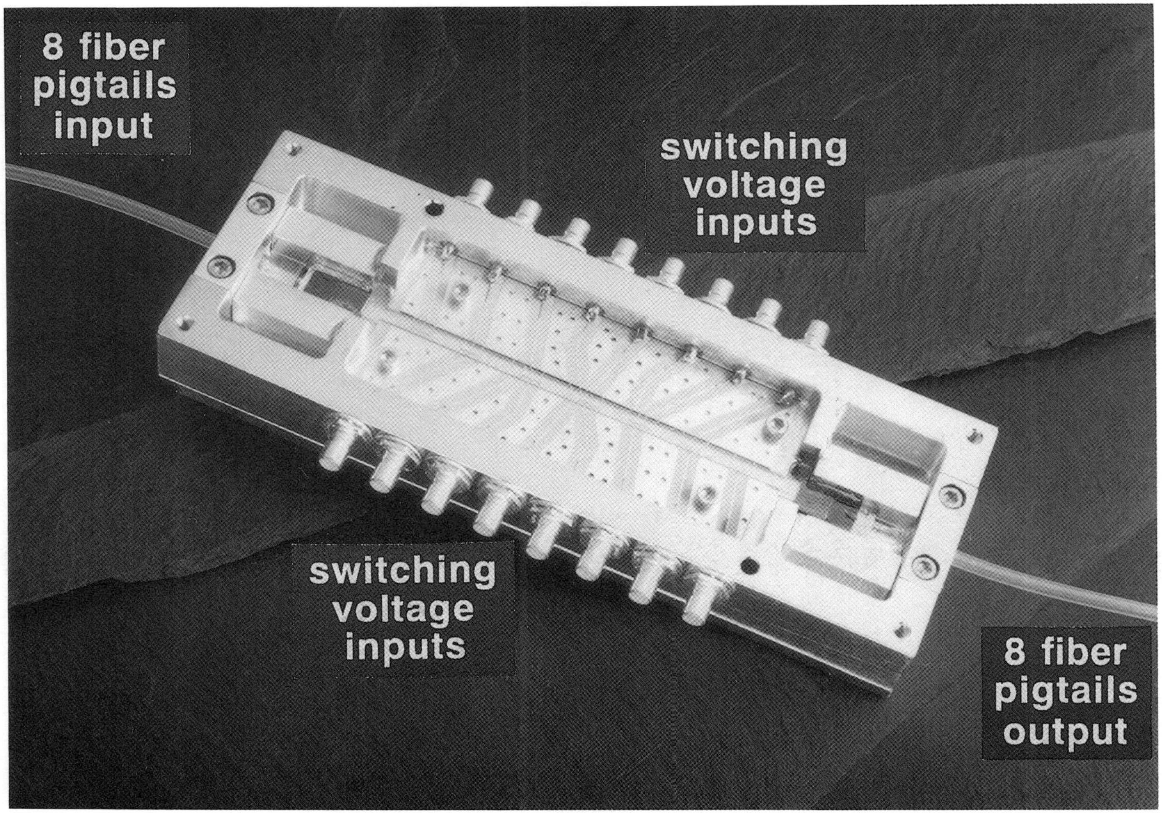

Figure 8.17 *A commercial pig-tailed 8 × 8 electrooptic switch based on an integrated optics Ti–LiNbO$_3$ circuit. (Courtesy of Crystal Technology.)*

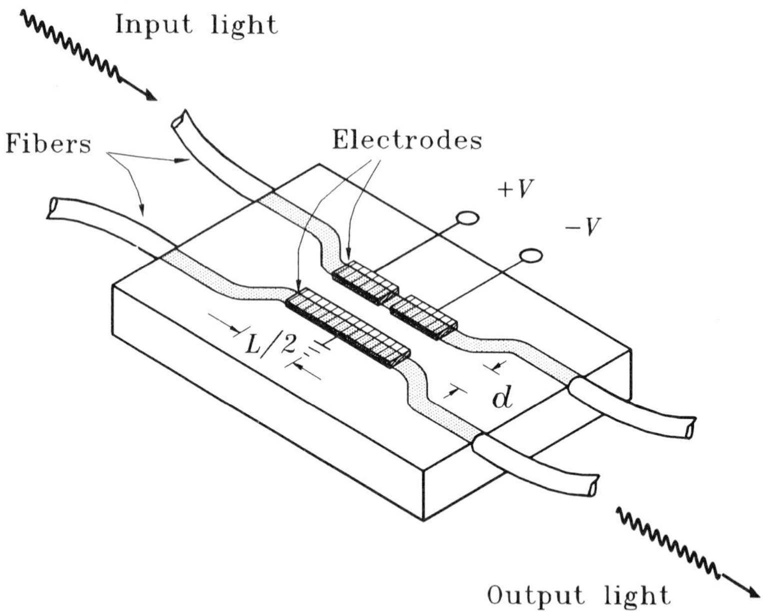

Figure 8.18 *Directional coupler with reversed $\Delta\beta$ control.*

length L. According to (8.29), L is determined by the coupling coefficient κ, which in turn depends on the refractive index and the coupler geometry.

Coupling lengths from $\sim$200 μm to $\sim$1 cm have been demonstrated in couplers based on Ti-diffused lithium niobate waveguides [65]. These devices can provide a practically complete *crossover* or *cross state* in which the energy is completely transferred from guide I to guide II. In the *straight through* or *bar state* all optical power comes out from guide I. Almost 100% modulation is achieved with applied voltage V under 10 V. With periodically reversed travelling-wave electrodes (to accomplish velocity matching of both the electrical and optical waves, a key requirement at high speeds) operating speeds of 72 Gb/s for a 2 × 2 Ti–$LiNbO_3$ switch have been reached [66]. Figure 8.17 shows a photograph of a commercial 8 × 8 switch based on Ti–$LiNbO_3$ technology able to work at 4 Gb/s.

The standard switched directional coupler just described requires careful control of the device length during the fabrication process in order to establish maximum coupling with no applied bias. Although this may be achievable, a split-electrode coupler, called *reversed $\Delta\beta$ directional coupler*, makes the fabrication tolerance much less strict. Figure 8.18 shows a simple example of such device with two coupled waveguides. The polarity of the applied voltage is reversed in the two halves in order to obtain two sections with $\Delta\beta$ of equal magnitude, but opposite sign. This makes possible an electrical adjust of both the *on* and *off* states for a relative wide range of lengths, $L < l < 3L$, maximizing the extinction ratio and minimizing the cross-talk. Crosstalk values over 40 dB have been achieved in Ti–$LiNbO_3$ waveguide couplers [67].

8.8. Other electrooptic devices

Another important application of electrooptic directional couplers is as wavelength multiplexers in transmission systems with single mode fibres, in order to exploit their enormous potential transmission capacity. This is achieved with the so-called *tunable directional coupler filter*. It consists of two asymmetrical ($\beta_1 \neq \beta_2$) coupled waveguides of different widths and refractive indices, the wider guide having lower index (Figure 8.19). Because of those differences between the two guides, they show distinct modal dispersion curves $N_{\text{eff}}(\lambda)$. The guides are designed so that the effective indices are equal at the chosen filter wavelength $\bar{\lambda}$. Thus, complete light transfer between the guides can be obtained by phase-matching at this wavelength. Other wavelengths $\bar{\lambda} \pm \Delta\lambda$, sufficiently different from $\bar{\lambda}$, cannot be transferred and demultiplexing is achieved. Tunability is reached by applying an adequate voltage that modifies the modal dispersion curve and shifts phase-matching to a different wavelength $\bar{\lambda}'$. The device has been demonstrated [68] in Ti–$LiNbO_3$ waveguides

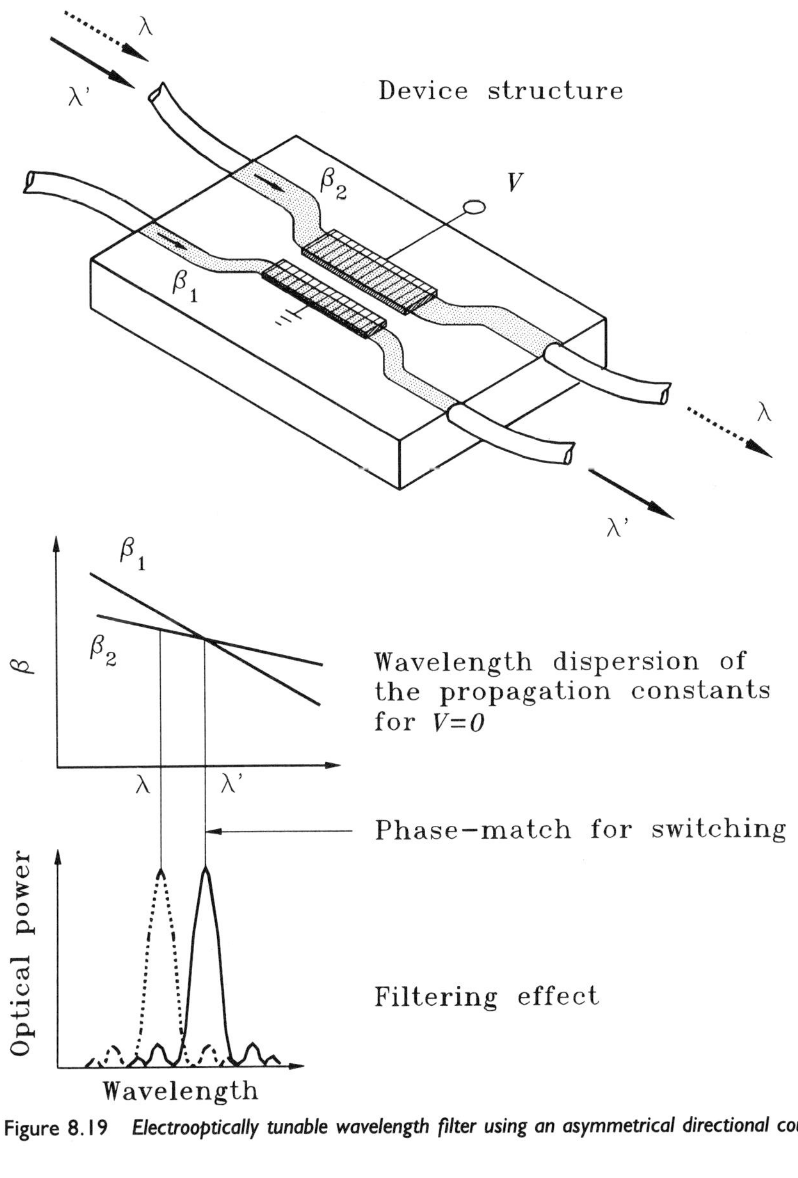

Figure 8.19 *Electrooptically tunable wavelength filter using an asymmetrical directional coupler.*

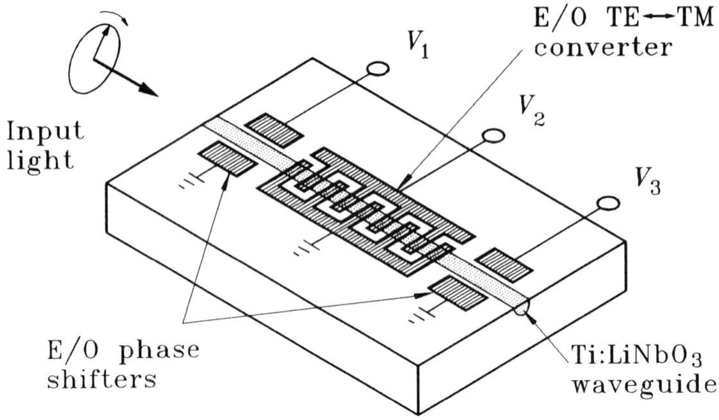

Figure 8.20 *Integrated optical circuit for electrooptic control of the polarization.*

with 1.5 cm interaction length, 70 nm bandwidth, a peak crossover efficiency of ~100% and a tuning range over 160 nm at a rate of 8.5 nm/V. Since the device can also work as a switch, multiplexing and switching can all be carried out at the same time.

Another relevant application of electrooptics to integrated optics is the polarization control of guided beams. Polarization control is important because coherent communication systems require a specified signal polarization, whilst typical single-mode fibres do not maintain polarization, rather they provide unknown elliptically polarized light.

Figure 8.20 sketches a device based on a Ti–$LiNbO_3$ substrate that converts an input signal of arbitrarily given polarization into a specified output polarization. The central part of this figure consists of a periodic electrode that induces TE–TM mode conversion. Despite the very large phase-mismatch between the two types of modes, a coupling efficiency close to 100% can be achieved with periodic electrodes, provided that the phase-match condition is satisfied [18]:

$$\beta_{\text{TM}} - \beta_{\text{TE}} = \frac{2\pi}{\Lambda} \tag{8.34}$$

where Λ is the period of the electrode. For a total TE–TM conversion in Ti–$LiNbO_3$ a period of 18 μm is required for $\lambda = 1.3\,\mu$m. Notice that, in order to modify coupling by the applied voltage with the geometry of Figure 8.20, the off-diagonal electrooptic coefficient r_{51} must be used in equation (8.31) so that an x-directed electric field can couple the x- and z-polarization components. In Figure 8.20 the TE–TM mode converter is placed between two electrooptic phase shifters to provide electrical control over the relative TE/TM phase angle.

Apart from the examples discussed above, many other integrated optical devices, which exploit the electrooptic effect to control guided lightwaves, have been demonstrated in $LiNbO_3$ as well as in electrooptic III–V semiconductors. A larger and more detailed description of the subject is given in references [3]–[5].

8.9. Measurement of electrooptic coefficients in waveguides

For a reliable assessment of the electrooptic properties of a waveguide material, it is highly desirable whenever possible that measurements be performed in those waveguide configurations to be actually used in real devices or as close to it as possible. There are two major reasons that justify this procedure. On one hand, most fabrication techniques of integrated optical components produce structural and/or physicochemical changes in the waveguide material with respect to the starting bulk or powder material. It has been mentioned in Section 8.3.1 that any of the Ti in-diffusion, proton-exchange or ion-implantation processes without post-annealing reduce the electrooptic coefficient of bulk $LiNbO_3$ to a considerable extent. Thus, in general, bulk measurements of electrooptic coefficients should only be taken as rough indicative values. On the other hand, the actual value and distribution of the applied electric field in an integrated optical device are usually difficult to determine. It also happens that the waveguide modes are affected by both the index change induced by the electrooptic effect and the thickness change induced by piezoelectricity or electrostriction.

Therefore, a full characterization of the effects induced by an electric field should ideally be able to separate the various contributions and to measure all coefficients. Also the modulating frequency at which coefficients are measured is an important factor. At frequencies above the mechanical resonance frequency of the layer ($\sim$1 GHz for a 1 μm clamped configuration), the strain-induced electrooptic effect disappears. Several common methods of measuring electrooptic coefficients in waveguides are reviewed in reference [69] in connection with organic waveguides. Those general methods based in nonguiding configurations are also applicable to thin films and have already been discussed in Chapter 3. Here, the two most familiar methods using guiding configurations, namely the Mach–Zenhder interferometer and the attenuated total reflection (ATR) technique will be discussed.

8.9.1. The Mach–Zenhder interferometer method

In Section 8.5.2 (see also Figure 8.12) the principle of light modulation based on an integrated optics Mach–Zenhder interferometer has been described. On the other hand, in Section 3.13.2 a discussion has been presented of the Mach–Zenhder method for measuring electrooptic coefficients in a nonguiding configuration. Both ideas can be put together in the guiding configuration of the Mach–Zenhder method to measure electrooptic coefficients of a guiding layer. The main difference in this case, as already mentioned above, is that the electric field is not homogeneous and affects in a different way each guide mode. This makes the analysis much more difficult, particularly in graded-index guides [70], and the accuracy of the electrooptic coefficient value is rather limited. An important advantage of this configuration is that it is much closer to real devices.

8.9.2. Attenuated total reflection (ATR)

This method is based on the popular prism coupler technique to observe the "dark-mode" pattern, which is commonly used to determine the thickness and refractive index of dielectric waveguides. A detailed discussion of the application of this method to the determination of electrooptic coefficients of polymer waveguides is given in reference [69].

Once the film is sandwiched between two electrodes, one of them being transparent, light is coupled into the film with a prism coupler (Figure 8.21). Light will actually couple only when the incidence angle φ (experimentally measured by the goniometer angle Φ) corresponds to a guide mode. Thus, an intensity dip (or dark line) will be exhibited by the reflectivity $R(\varphi)$ for each mode angle, i.e. when the tangential component k_x of the

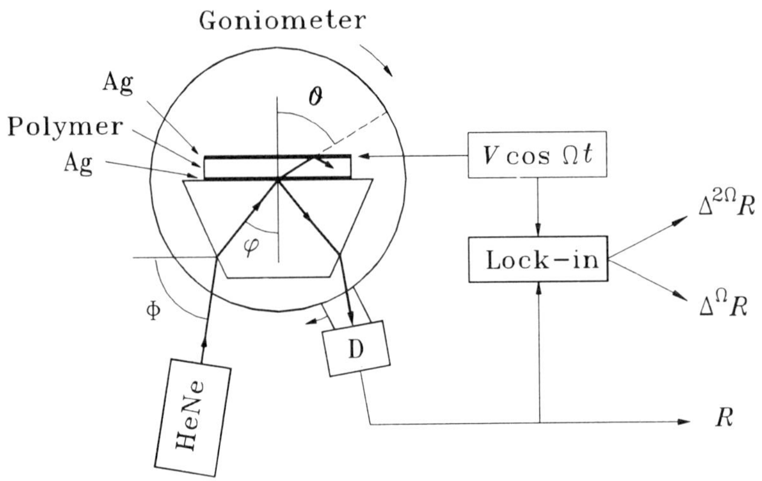

Figure 8.21 *Measurement of electrooptic coefficients of a guide with the attenuated total reflection (ATR) method.*

incident wavevector equals that of a guided mode:

$$kn_{\mathrm{p}}\sin\varphi_v = k_{xv} = kn_{\mathrm{f}}\sin\theta_v = k\beta_v \tag{8.35}$$

where n_{p} is the refractive index of the prism. The dip shape is determined by the reflectivity curve $\mathrm{R}(\varphi)$ which must be computed and is a rather complicated function of the guide losses. However, the position of the dip is not very sensitive to losses and the isolated mode condition (8.4) may be used in first approximation:

$$kn_{\mathrm{f}}h\cos\theta_v + \phi_{\mathrm{s}}(n_{\mathrm{f}},\theta_v) + \phi_{\mathrm{c}}(n_{\mathrm{f}},\theta_v) = v\pi \tag{8.36}$$

Applying a low-frequency voltage $V\cos\Omega t$ and using a lock-in amplifier tuned at Ω and 2Ω, the linear and quadratic electrooptic coefficients can be obtained as follows. If losses can be neglected, the electric field $E = V/h$ will move the angular position of the dips without changing their profiles. Then, if the detector is placed near an angle φ_v of a dip minimum, the reflectivity change can be written

$$\Delta\mathrm{R}_v(\varphi) = \frac{\partial\mathrm{R}_v}{\partial\varphi}(-\Delta\varphi_v) = -\frac{\partial\mathrm{R}_v}{\partial\varphi}\left(\frac{\partial\varphi_v}{\partial h}\Delta h + \frac{\partial\varphi_v}{\partial n_{\mathrm{f}}}\Delta n_{\mathrm{f}}\right) \tag{8.37}$$

Now, using equation (8.37), or the equivalent for quadratic effects, a relationship is obtained between the experimental data $\Delta\mathrm{R}_v$ and $\partial\mathrm{R}_v/\partial\varphi$. At least two different measurements have to be taken in order to obtain r and the piezoelectric coefficient d^{E} (or s and the electrostriction coefficient Q^{E}). This method presents the advantage of being a guiding method, and the above equation applies both at TE-modes and TM-modes so that ordinary and extraordinary index changes can be determined in anisotropic guides. Also, by using exact computed dip profiles and comparing with experimental ones, additional information can be obtained on the imaginary parts of refractive indices.

The method requires that the waveguide film is sandwiched between electrodes. In many practical cases, as in most guides prepared on a substrate by diffusion or ion exchange, the film cannot be sandwiched and other geometrical configurations have to be used. One of these involves applying the electric field through the substrate and/or an air gap such as in reference [71], where a careful estimate of the actual electrical field inside the guiding layer has been taken into account. Changes of the substrate index induced by electrooptic effect also may affect the accuracy of the measured guide parameters.

References

1. R. G. Hunsperger, *Integrated Optics: Theory and Technology*, 2nd edn, Springer-Verlag, New York (1984).
2. D. L. Lee, *Electromagnetic Principles of Integrated Optics*, Wiley, New York (1986).
3. L. D. Hutcheson (ed.), *Integrated Optical Circuits and Components*, Marcel Dekker, New York (1987).
4. T. Tamir (ed.), *Guided Wave Optoelectronics*, Springer-Verlag, New York (1990).
5. H. Nishihara, M. Haruna and T. Suhara, *Optical Integrated Circuits*, New York (1989).
6. H. Kogelnik, *Theory of Optical Waveguides*, Chapter 2 of reference [4].
7. G. I. Stegeman, C. T. Seaton, W. M. Hetherington III, A. D. Boardman and P. Egan, in *Nonlinear Optics: Materials and Devices* (eds C. Flytzanys and J. L. Oudar), Springer, Berlin (1986), pp. 31–64.
8. G. I. Stegeman, E. M. Wright, N. Finlayson, R. Zanoni and C. T. Seaton, *J. Lightwave Technol.* **6**, 953 (1985).
9. G. I. Stegeman and E. M. Wright, *Opt. Quantum Electron.* **22**, 95 (1990).
10. M. Born and E. Wolf, *Principles of Optics*, Pergamon Press, Oxford (1980).
11. B. E. A. Saleh and M. C. Teich, *Fundamentals of Photonics*, John Wiley, New York (1991).
12. K. S. Chiang, *J. Lightwave Techn.* **LT-3**(2), 385 (1985).
13. P. J. Chandler and F. L. Lama, *Opt. Acta* **33**, 127 (1986).
14. J. Olivares, M. A. Diaz-García and J. M. Cabrera, *Optics Commun.* **92**, 40 (1992).
15. M. Zoboli and P. Bassi, in *Anisotropic and Nonlinear Optical Waveguides* (eds G. C. Someda and G. Stegeman), Elsevier, Amsterdam (1992), pp. 77–116.
16. L. Torner, J. Recolons and J. P. Torres, *J. Lightwave Technol.*, **11**, 1592 (1993).
17. F. Matera and G. C. Someda in *Anisotropic and Nonlinear Optical Waveguides* (eds G. C. Someda and G. Stegeman), Elsevier, Amsterdam (1992), pp. 1–38.
18. R. C. Alferness, *Titanium-Diffused Lithium Niobate Waveguide Devices*, Chapter 4 of reference [4].
19. J. Jackel, *Integrated Optical Circuits*, SPIE Vol. 1583, 54 (1991).
20. T. Bremer, W. Heiland, B. Hellerman, P. Hertel, E. Krätzig and D. Kollewe, *Ferroelectr. Lett.* **9**, 11 (1988).
21. J. D. Bierlein, A. Ferretti, L. H. Brixner and M. Y. Hsu, *Appl. Phys. Lett.* **50**, 1216 (1987).
22. S. Y. Wang, S. H. Lin and Y. M. Houng, *Appl. Phys. Lett.* **51**, 83 (1987).
23. V. Koren, B. I. Miller, T. L. Koch, G. Eisenstein, R. S. Tucker, I. Bar-Joseph and D. S. Chemla, *Appl. Phys. Lett.* **51**, 1132 (1987).
24. E. Lallier, J. P. Pochelle, M. Papuchon, C. Grezes-Besset, E. Pelletier, M. De Micheli, Q. He and D. B. Ostrowsky, *Electron. Lett.* **27**, 11 (1991).
25. S. J. Field, D. C. Hanna, D. P. Shepherd, A. C. Tropper, P. J. Chandler, P. D. Townsend and L. Zhang, *Optics Lett.* **16**, 481 (1991).
26. T. Suhara and H. Nishihara, *IEEE J. Quantum Electron.* **QE-22**, 845 (1986).
27. M. De Micheli, J. Botineau, S. Neven, F. Sibillot, D. B. Ostrowsky and M. Papuchon, *Optics Lett.* **8**, 116 (1983).
28. R. W. Keys, A. Loni, R. M. de la Rue, C. N. Ironside, J. H. Marsh, B. J. Luff and P. D. Townsend, *Electron. Lett.* **26**, 1193 (1990).
29. I. P. Kaminow and J. R. Carruthers, *Appl. Phys. Lett.* **22**(7), 326 (1973).
30. D. P. Gia Russo and C. S. Kumar, *Appl. Phys. Lett.* **23**(5), 229 (1973).
31. S. Fukunishi, N. Uchida, S. Miyazawa and J. Noda, *Appl. Phys. Lett.* **24**(9), 424 (1973).

32. R. V. Schmidt and I. P. Kaminow, *Appl. Phys. Lett.* **25**(18), 458 (1974).
33. G. L. Destefanis, P. D. Townsend and J. P. Gailliard, *Appl. Phys. Lett.* **32**, 293 (1978).
34. J. L. Jackel, C. E. Rice and J. J. Veselka, *Appl. Phys. Lett.* **41**(7), 607 (1982).
35. R. A. Betts and C. W. Pitt, *Electron. Lett.* **21**, 960 (1985).
36. C. N. Alfonso, F. Vega, J. Gonzalo and C. Zaldo, *Appl. Surf. Sci.* **69**, 149 (1993).
37. M. N. Armenise, in *Properties of Lithium Niobate*, EMIS Datareviews Series No. 5, INSPEC, The Institution of Electrical Engineers, London (1989), p. 253.
38. K. K. Wong, in *Properties of Lithium Niobate*, EMIS Datareviews Series No. 5, Sections 8.1 and 8.2. INSPEC, The Institution of Electrical Engineers, London (1989).
39. M. J. Li, M. de Micheli, D. Ostrowsky and M. Papuchon, *Ann. Télécommun.* **43**(1–2), 73 (1988).
39a. V. A. Gan'shin, V. Sh. Ivanov, Yu. N. Korkishko and V. Z. Petrova, *Sov. Phys. Tech. Phys.* **31**, 794 (1986).
40. J. Olivares and J. M. Cabrera, *Appl. Phys. Lett.* **62**, 2468 (1993); *Fiber Integ. Optics*, **12**, 277 (1993).
41. M. M. Abouelleil and F. Leonberger, *J. Am. Ceram. Soc.* **72**(8), 1311 (1989).
42. A. Mahapatra and W. C. Robinson, *Appl. Optics* **24**, 2285 (1985).
43. D. Y. Zang and C. S. Tsai, *Appl. Phys. Lett.* **46**, 703 (1985).
44. E. Glavas, J. M. Cabrera and P. D. Townsend, *J. Phys.: Appl. Phys.* **D22**, 611 (1989).
45. J. Olivares, E. Diéguez, F. J. López and J. M. Cabrera, *Appl. Phys. Lett.* **61**(6), 624 (1992).
46. M. De Micheli, J. Botineau, F. Sibillot, D. B. Ostrowsky and M. Papuchon, *Optics Commun.* **42**, 101 (1982).
47. M. De Micheli, J. Botineau, S. Neven, F. Sibillot, D. B. Ostrowsky and M. Papuchon, *Optics Lett.* **8**, 116 (1983).
48. E. Lallier, J. P. Pocholle, M. Papuchon, C. Grezes-Besset, E. Pelletier, M. De Micheli, M. J. Li, Q. He and D. B. Ostrowsky, *Electron. Lett.* **26**(13), 927 (1990).
49. E. Lallier, J. P. Pocholle, M. Papuchon, C. Grezes-Besset, E. Pelletier, M. De Micheli, Q. He and D. B. Ostrowsky, *Electron. Lett.* **27**, 11 (1991).
50. P. D. Townsend, P. J. Chandler and L. Zhang, *Optical Effects of Ion Implantation*, Cambridge University Press, Cambridge (1993).
51. P. J. Chandler, L. Zhang and P. D. Townsend, *Solid State Phenom.* **27**, 129 (1992).
52. P. J. Chandler, L. Zhang and P. D. Townsend, *Appl. Phys. Lett.* **55**(17), 1710 (1989).
53. S. Fouchet, A. Carenco and N. Duhamel, *Proc. 4th European Conf. on Integrated Optics*, Glasgow, Scotland, May 1987, p. 54.
54. J. Zyss, *J. Molec. Electron.* **1**, 25 (1985).
55. P. K. Tien, G. Smolinsky and R. J. Martin, *Appl. Opt.* **11**, 637 (1972).
56. U. Falk, W. Hickel, D. Lupo, W. Pross and U. Schevnemann, in *Nonlinear Optics. Fundamentals, Materials and Devices* (ed. S. Miyata), Elsevier, Amsterdam (1992), pp. 272–282.
57. S. Miyata (ed.), *Nonlinear Optics. Fundamentals, Materials and Devices*, Elsevier, Amsterdam (1992), pp. 219–224 and 501–506.
58. H. P. Weber, R. Ubrich, E. A. Chandross and W. J. Tomlinson, *Appl. Phys. Lett.* **20**, 143 (1972).
59. J. C. Dubois, M. Gazard and D. B. Ostrowsky, *Opt. Commun.* **7**, 237 (1973).
60. R. Schriever, H. Franke, H. G. Festl and E. Krätzig, *Polymer* **26**, 1423 (1985).
61. E. A. Chandross, C. A. Pryde, W. J. Tomlinson and H. P. Weber, *Appl. Phys. Lett.* **24**, 72 (1974).
62. J. R. Kulisch, H. Franke, R. Irmscher and Ch. Buchal, *J. Appl. Phys.* **71**, 3123 (1992).
63. F. J. Leonberger and J. P. Donnelly, *Semiconductor Integrated Optic Devices*, Chapter 6 of reference [4].
64. G. L. Tangonan, L. Persechini, J. F. Lotspeich and M. K. Barnosky, *Appl. Optics* **17**, 3259 (1978).
65. R. C. Alferness, R. V. Schmidt and E. H. Turner, *Appl. Optics* **18**, 4012 (1979).
66. S. K. Korotky and J. J. Veselka, *Digest Conf. Optical Fiber Communications*, Optical Society of America, TUH2 (1990).

67. G. A. Bogert, E. J. Murphy and R. T. Ku, *IEEE Lightwave Technol.* **4**, 1542 (1986).
68. R. C. Alferness and R. V. Schmidt, *Appl. Phys. Lett.* **33**, 161 (1978); see also R. C. Alferness and J. J. Veselka, *Electron. Lett.* **21**, 466 (1985).
69. M. Dumont, Y. Levy and D. Morichère, in *Organic Molecules for Nonlinear Optics and Photonics* (ed. J. Messier), NATO ASI Series, Kluwer Academic Publishers (1990).
70. M. Rottschalk, A. Rasch and W. Karthe, *J. Opt. Commun.* **9**, 19 (1988).
71. P. G. Suchoski, T. K. Findakly and F. J. Leonberger, *Optics. Lett.* **13**(11), 1050 (1988).

Chapter Nine

Semiconductor Quantum Wells and Superlattices

9.1. Introduction

In Section 5.9 we discussed the electrooptic properties of homogeneous crystalline bulk semiconductors. Their electronic properties ultimately depend on the constituent material and therefore are limited by the discrete number of natural semiconductors. In recent years a new class of man-made semiconductors has emerged which allows a fine tuning of the electronic and optical properties starting from a few basic substances.

These new semiconductors are made by altering the material characteristics, like the composition or the doping, over distances of a few nanometers or less, that is, close to the lattice constant or interatomic distances. The movement of free carriers (electrons in the conduction band or holes in the valence band) through these artificial structures, due to its reduced dimensions, will show new quantum effects not present in the bulk material. These effects have a strong influence on the electronic properties, which can be tailored by appropriately modifying the structure.

The field of the so-called nanostructured materials started thanks to the development of epitaxial growth techniques that allowed the deposition of atom-thick layers. The first materials were layered semiconductors such as *quantum wells* and *superlattices* [1–6], formed by sandwiching thin layers of a semiconductor between layers of other semiconductors with larger bandgaps. Hence the carriers are confined in the small-gap layers. Nowadays a strong effort is being made to tailor structures in all the three directions of space by means of photolithography, electron beam lithography, growth on low symmetry orientations, and others. With these techniques other small-scale structures can be built, such as *quantum wires*, where the carriers are confined in "wires" with a cross-section of a few square nanometres, or *quantum dots*, where the carriers are confined in a volume of a few cubic nanometres (for an introduction see, e.g. references [5] (Chapter 6), [7] and [8]).

Apart from the intrinsic tunability of their properties, semiconductor quantum wells and superlattices have other interesting advantages when compared with bulk semiconductors [8a]. For example, they have strong excitonic effects, even at room temperature or at high electric fields, and show a large quadratic electrooptic effect, which makes them very attractive for electrooptic applications. Furthermore, they could be easily integrated with semiconductor lasers and other microelectronic components to build optoelectronic integrated devices.

This chapter begins with a description of the growth techniques, since they determine a great deal of the possibilities and limitations of semiconductor nanostructures. Then it gives a brief introduction to their electronic and optical properties and the effect of an externally applied electric field or electrooptic response. Throughout the chapter we will use the GaAs/(Al,Ga)As system as an example because it is the most widely studied. Finally, some electrooptic devices based on semiconductor nanostructures are discussed.

9.2. Fabrication of quantum wells and superlattices

The epitaxial layer-by-layer growth techniques allow the deposition of layers with thicknesses down to one atom (called *monolayers*), i.e. with a thickness of a few angstroms. The growth is done on a good quality wafer of a semiconductor (the *substrate*) whose structure and lattice parameter are similar to those of the material to be grown.

Table 9.1 lists some of the systems which have been employed to fabricate quantum wells. Among the most important are GaAs/(Al,Ga)As grown on GaAs, because of its applications in optoelectronics, and (In,Ga)As/(Al,Ga)As, whose bandgap energy lies at one of the transmission windows of the optical fibres used in telecommunications. Both systems are *lattice-matched* heterostructures because the lattice parameters of the constituent semiconductors are very close. Nowadays it is also possible to grow a thin layer of a semiconductor on a substrate with a different lattice constant (*lattice-mismatched* heterostructures). In this case the lattice constant of the thin layer accommodates to that of the substrate by elastic strain. The so-called *strained* heterostructures multiply the available combinations of materials.

The most widely used techniques for the growth of semiconductor quantum wells and superlattices are molecular beam epitaxy (MBE) and metal-organic chemical vapour deposition (MOVCD).

In MBE [25,26] atomic beams of the desired elements are deposited onto the substrate in an ultrahigh vacuum environment. A typical system is shown schematically in Figure 9.1. The growth chamber is pumped down to pressures of 10^{-11} torr while its walls are kept at liquid-nitrogen temperature to avoid outgassing. Heated effusion cells, one for each element, evaporate the atomic beams onto the substrate. Additional cells are included for the *n*-type and *p*-type dopants. In front of each cell there is a shutter which allows a rapid termination of the flux.

The substrate is heated up to a temperature that permits the migration of the deposited atoms over the surface to the right crystal sites, but not so high that atoms of deeper layers diffuse to the surface. For GaAs the growth temperature is of the order of 600 °C.

Different techniques can be added to control the growth process. RHEED (Reflection High Energy Electron Diffraction) [27] is one of them. The thickness of the layers can be controlled by the change in the RHEED intensity every time that an atomic monolayer is completed or, after a careful calibration of the growth rate, by the deposition time. In MBE the growth rate is very slow (of the order of a few angstroms per second or, equivalently, a micrometre per hour) and allows very good control of the interface sharpness.

As an example, to make a GaAs/(Al,Ga)As quantum well, first the Ga, the As and the Al shutters are opened to grow the bottom $Al_xGa_{1-x}As$ layer (the Al mole fraction is determined by the Al cell relative temperature). Next the Al shutter is closed to grow the intermediate GaAs layer. Finally the Al shutter is opened again to deposit the upper (Al,Ga)As layer. Be and Si can be used as *p*- or *n*-type dopants, respectively.

In MOCVD [28,29] the elements are introduced in the growth chamber, which is at low pressure (0.1–1.0 atm), in the form of organo-metallic gases that decompose on the hot

Table 9.1 *Some examples of semiconductor quantum well systems and their electrooptical properties ($\Delta\mathscr{E} = \mathscr{E}_{meas} - \mathscr{E}_0$, $\mathscr{E}_{meas}$ and $\mathscr{E}_0$ are the energies of the light and the quantum well excitonic gap, respectively).*

GaAs/$Al_xGa_{1-x}As$ on GaAs	$L_w = 50$ Å, $x = 0.3$ $s_{33} = -2.2 \times 10^{-19}\ m^2/V^2$	[9]
	$L_w = 122$ Å, $x = 0.24$ $\mathscr{E}_0 - 1.445$ eV, $\Delta\mathscr{E} = -49$ meV $r_{63} = -1.6 \times 10^{-12}$ m/V $s_{33} - (s_{12} + s_{13})/2 = 6 \times 10^{-20}\ m^2/V^2$	[10]
	$L_w = 94$ Å, $x = 0.30$ $\mathscr{E}_0 = 1.470$ eV, $\Delta\mathscr{E} = -12$ meV $\lvert s_{33} \rvert = 4.6 \times 10^{-17}\ m^2/V^2$	[11]
	$L_w = 85$ Å $\mathscr{E}_0 = 1.467$ eV, $\Delta\mathscr{E} = -34$ meV $r_{63} = -1.75 \times 10^{-12}$ m/V $s_{13} = 1.40 \times 10^{-18}\ m^2/V^2$	[12]
		[12a]
(In,Ga)(As,P)/InP	$L_w = 85$ Å $\mathscr{E}_0 = 0.849$ eV, $\Delta\mathscr{E} = -14$ meV $r_{63} = 5.4 \times 10^{-12}$ m/V $s_{13} = -3.01 \times 10^{-18}\ m^2/V^2$	[13]
	$L_w = 70$ Å $\mathscr{E}_0 = 0.992$ eV, $\Delta\mathscr{E} = -42$ meV $r_{63} = -2.11 \times 10^{-12}$ m/V $s_{13} = -3.36 \times 10^{-18}\ m^2/V^2$	[13]
GaAs/AlAs on GaAs		[14]
(In,Ga)As/InP on InP		[15]
In(As,P)/InP on InP		[16]
(In,Ga)As/(In,Al)As on InP		[17]
(In,Ga)(As,P)/(In,Ga)(As,P) on InP		[18]
GaSb/(Al,Ga)Sb		[19]
InAs/GaAs on GaAs	$L_w = 60$ Å $\mathscr{E}_0 = 1.272$ eV, $\Delta\mathscr{E} = -6.6$ meV $r_{63} = -5.55 \times 10^{-11}$ m/V $s_{13} = -3.96 \times 10^{-17}\ m^2/V^2$	[20]
ZnSe/(Zn,Mn)Se		[21]
ZnTe/(Cd,Zn)Te on GaAs		[22]

substrate, depositing the desired elements while the remaining gas is removed from the chamber. For instance, to grow (Al,Ga)As the starting components are $Ga(CH_3)_2$ and $Al(CH_3)_2$ and AsH_3. After the chemical reaction on the substrate an (Al,Ga)As layer is deposited and CH_4 is formed as a waste material.

Other growth methods are liquid-phase epitaxy (LPE) [23] and vapour-phase epitaxy (VPE) [23]. The former is widely used in the fabrication of heterojunction laser diodes.

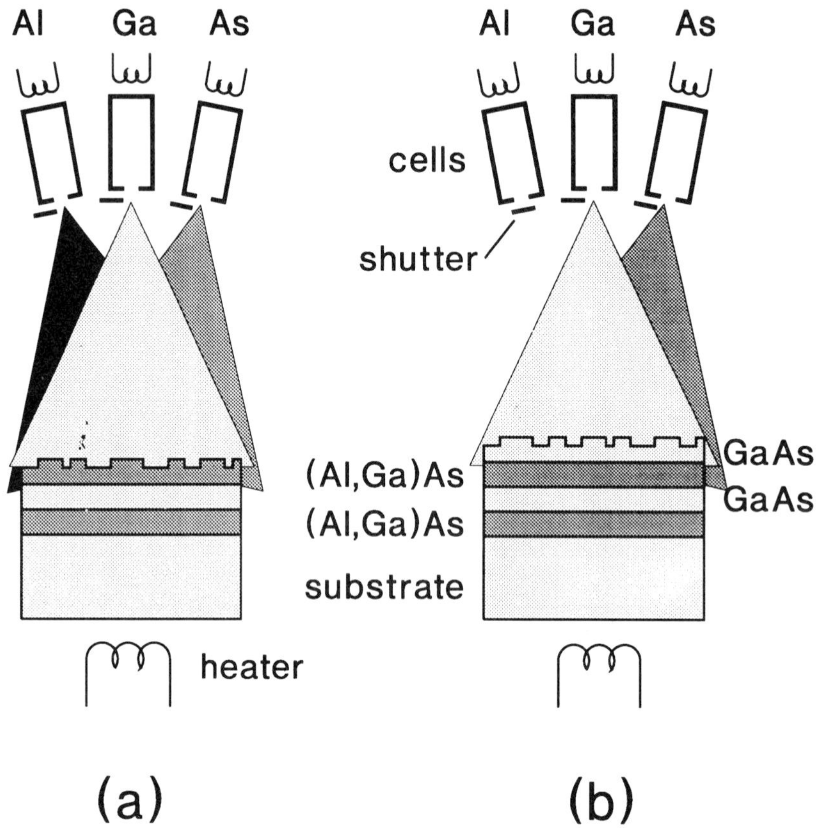

Figure 9.1 *Schematic diagram illustrating the growth of a GaAs/(Al,Ga)As multiple quantum well by molecular beam epitaxy. Deposition of (a) (Al,Ga)As; (b) GaAs. The system is under high vacuum.*

9.3. Fabrication of quantum well electrooptic devices

For electrooptic applications an electric field is applied to the heterostructure by means of an external bias voltage. With this purpose electrooptic devices are fabricated as p–i–n diode structures [24] on a doped substrate, with the heterostructure in the intrinsic region (see Figure 9.2). After growth the device can be etched down to the substrate forming many individual mesa structures from a single wafer. In reverse bias, for voltages smaller than the breakdown voltage, the current through the devices is very small and one can apply very high and uniform electric fields. The approximate relationship between the internal electric field E and the applied external voltage V (for $V < V_{bi}$, reverse bias) is

$$E = \frac{V_{bi} - V}{W} \tag{9.1}$$

where V_{bi} is the built-in voltage of the diode and W is the depletion-region width (or approximately the intrinsic-region width). One can typically get field values of up to a few hundred kV/cm.

When the growth technique does not allow for the deposition of doped layers one can use a Schottky diode structure, although in this case the applied electric field is not

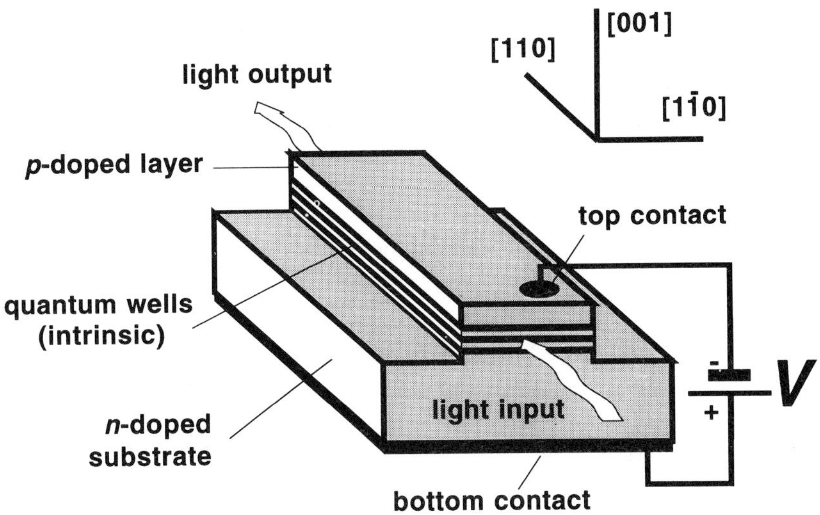

Figure 9.2 *Stripe-waveguide quantum-well modulator built in a* p–i–n *diode. Axes give the crystallographic directions.*

so homogeneous. If the electric field has to be applied in both directions one can resort to a *n–i–n* structure, which allows the application of small fields around flat band.

An important part of the devices are the contacts between the semiconductor structure and the external wires that provide the electrical connection. Good ohmic contacts [30] are obtained by evaporating a few metal layers on the semiconductor surface and then alloying them at a certain temperature in an inert atmosphere to diffuse the metals into the semiconductor. The wire is bonded to the contact using, for instance, an ultrasonic bonder. The detailed metal composition of the contact and the processing depend very much on the specific semiconductor and its doping.

Finally, to handle the light in the device one can incorporate waveguides, mirrors, etalons and other optical elements (see Chapter 8). In integrated electrooptic devices [23,31] most of these elements can be made from the same semiconductor materials by modifying the refractive index through proper choice of the composition, doping, or layer structure, and by etching [32] the appropriate geometries. For example, for device applications GaAs/(Al,Ga)As quantum wells are usually grown in the middle of an $Al_xGa_{1-x}As$ thick layer surrounded by $Al_yGa_{1-y}As$ outer layers with $x < y$. Since the $Al_xGa_{1-x}As$ layer has a refractive index higher than the $Al_yGa_{1-y}As$ layer, it confines the light, forming a leaky slab waveguide that enhances the electrooptic effects of the quantum well.

9.4. Electronic structure

The electronic properties of semiconductor nanostructures are determined both by the bulk properties of the component materials and by the geometry of the artificial structure itself. The former permits a broad selection from the available semiconductors of bandgaps (direct or indirect), lattice constant, effective masses, band offsets and so on. The latter permits fine tuning by choosing the appropriate parameters of the structure, such as well widths, mole fractions and others. In this section we describe the electronic properties of free electrons and holes in a semiconductor quantum well and how they depend on the different parameters. We are not going to consider free carriers other than those generated by the light itself, e.g. we are not going to consider carrier injection or the effects of doping. For the moment, we will use the single-particle approximation, that is, we will assume that there is either one single electron in the conduction band or one hole in the valence band. In the next section we will consider the case when there are simultaneously an electron in the conduction band and a hole in the valence band, bound by the electrostatic interaction. This has important consequences in the optical properties.

9.4.1. Electronic properties of a single quantum well

We are going to deal only with quantum wells of type I, because they are the most studied. In this case the quantum well is formed by three semiconductor layers. The centre layer (e.g. GaAs) has smaller bandgap than the outer layers (e.g. Al,GaAs). Evenmore, the conduction band and the valence band edges of the middle layer are lower and higher, respectively (Figure 9.3). The absolute value of the difference between the conduction (valence) band edges is known as the conduction (valence) band offset $\Delta\mathscr{E}_c(\Delta\mathscr{E}_v)$. A convenient parameter for a pair of semiconductors is the relative conduction band offset $Q_c = \Delta\mathscr{E}_c/\Delta\mathscr{E}_g$, where $\Delta\mathscr{E}_g$ is the bandgap difference. Note that the relative valence band offset is $Q_v = 1 - Q_c$. There has been a lot of work, both experimental and theoretical, on the determination of band offsets [33] because they define many of the electronic properties. For the $GaAs/Al_xGa_{1-x}As$ heterojunction Q_c does not depend on x (Table 9.2).

According to classical mechanics, if an electron in the conduction band has an energy below the band edge of the larger bandgap material it will be confined to move inside the central layer. Therefore this layer receives the name *well* whereas the outer layers are called *barriers*. When the well thickness L_w is smaller than a few hundred angstroms, quantum effects become very important for movement in the direction perpendicular to the well plane. In this case the kinetic energy in that direction can only have a few discrete values or quantum *energy levels* [38] $\mathscr{E}_i$, where i is the *quantum number*. For any quantum well these energies and the corresponding wavefunctions can be obtained by solving the time-independent Schrödinger equation in the envelope function approximation [3], in which the atomic-scale details of the potential and the wavefunction are neglected. Instead, values averaged over a few atoms are considered. The effect of the real crystal atomic structure enters into the model through only two parameters: the effective mass m and the conduction (valence) band edge energy, which represent the mass and the potential energy of an electron (hole) in the conduction (valence) band moving through the crystal. Note that for a hole the energy increases downwards in the scheme of Figure 9.3. Values for (Al,Ga)As are given in Table 9.2. The total kinetic energy of a carrier in a quantum well will be the sum of the parallel and perpendicular contributions:

$$\mathscr{E}(k_\parallel, i) = \frac{\hbar^2 k_\parallel^2}{2m_\parallel} + \mathscr{E}_i \tag{9.2}$$

where $k_\parallel$ and $m_\parallel$ are the crystal momentum and the effective mass, respectively, for movement in the well plane, and $\mathscr{E}_i$ is the energy of the perpendicular level with quantum number i. The allowed values of the total kinetic energy form semi-infinite bands with the minima at the $\mathscr{E}_i$.

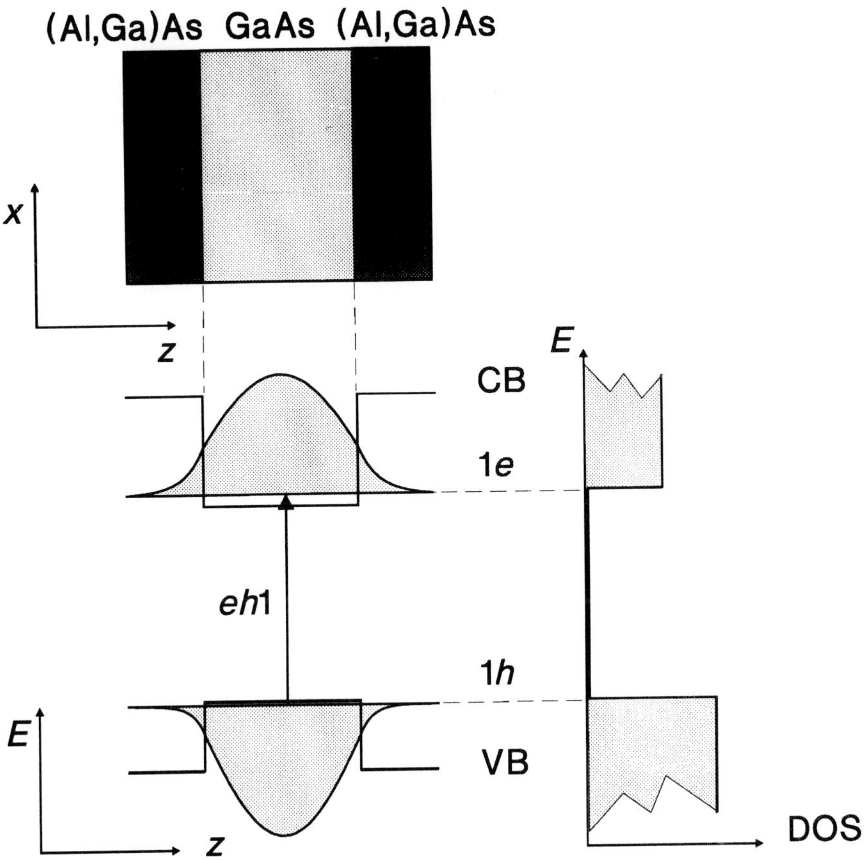

Figure 9.3 *Cross-section of a GaAs/(Al,Ga)As quantum well along the growth direction z (top). Potential energy U(z) for an electron in the conduction band (CB) and a hole in the valence band (VB) (solid line). Note that the energy of a hole increases downwards. Shadowed areas represent the wavefunctions for the fundamental states of electrons (1e) and heavy-holes (1h), plotted on the corresponding energy level. The vertical arrow signals the lowest optical transition (bottom left). Density of states (DOS) as a function of energy, without considering excited states (bottom right).*

Table 9.2 *Relevant parameters for a GaAs/Al_xGa_{1-x}As quantum well grown on a (001)-oriented GaAs substrate (300 K).*

Lattice constant (Å)	$a = 5.6532 - 0.0068x$	[34]
Refractive index (@1.38 eV)	$n(x) = 3.590 - 0.710x + 0.091x^2$	[23]
Bandgap (eV)	$\mathscr{E}_g = 1.4227 + 1.36x + 0.22x^2$	[35]
Relative conduction-band offset	$Q_c = 0.62$	
Conduction band offset (meV)	$\Delta\mathscr{E}_c = 843x = 136x^2$	[35,36]
Valence band offset (meV)	$\Delta\mathscr{E}_v = 517x + 84x^2$	[35,36]
Electron effective mass	$m_\perp = (0.0665 + 0.083x)m_0$	[37]
Heavy-hole effective mass	$m_\perp = (0.034 + 0.138x)m_0$	[37]
Light-hole effective mass	$m_\perp = (0.094 + 0.117x)m_0$	[37]

In the parabolic band approximation, valid for energies close to the bottom (top) of the bulk conduction (valence) band, the effective mass is taken as energy independent. Then, the eigenenergies $\mathscr{E}_i$ and the associated envelope wavefunctions ψ_i are given by the solutions to the Schrödinger equation

$$\frac{\hbar^2}{2m_\perp}\frac{d^2}{dz^2}\psi(z) + [\mathscr{E} - U(z)]\psi(z) = 0 \tag{9.3}$$

where $U(z)$ is the confining potential (Figure 9.3).

For the particular case of a square quantum well with barriers of infinite height, the energies and the normalized wavefunctions in the perpendicular direction are given analytically [38]:

$$\mathscr{E}_i = i^2\frac{\pi^2\hbar^2}{2m_\perp L_w^2}, \quad i = 1, 2, 3, \ldots \tag{9.4}$$

$$\psi_i(z) = \sqrt{\frac{2}{L_w}}\sin\frac{i\pi}{L_w}z, \quad i = 1, 3, 5, \ldots \tag{9.5}$$

$$\psi_i(z) = \sqrt{\frac{2}{L_w}}\cos\frac{i\pi}{L_w}z, \quad i = 2, 4, 6, \ldots \tag{9.6}$$

For finite-barrier wells there are only a finite number of confined states and the wavefunctions decay exponentially inside the barriers (Figure 9.3); but for deep square wells these expressions are still a good approximation for the lowest levels.

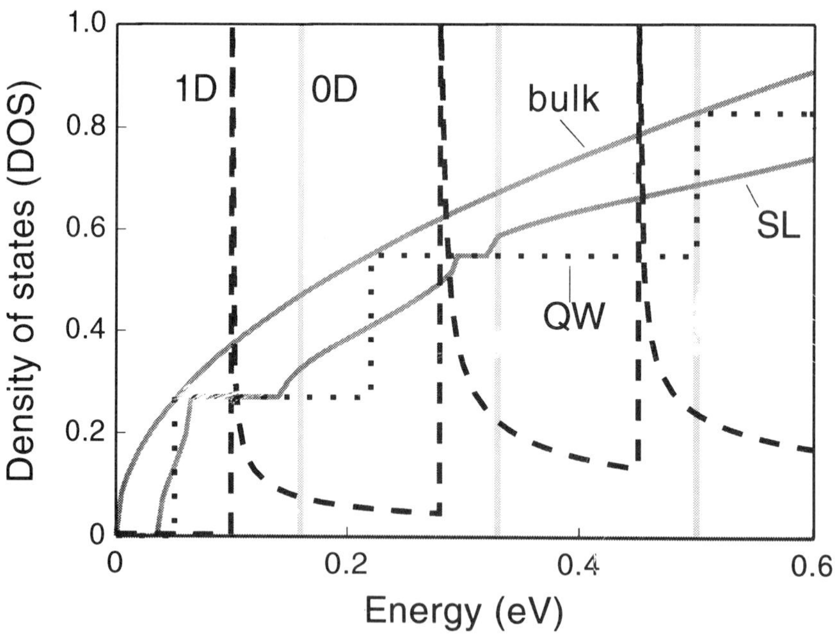

Figure 9.4 *Conduction band density of states (units of $10^{20}/(\text{eV cm}^3)$) for bulk, superlattice (SL), quantum well (QW), quantum wire (1D) and quantum dot (0D) semiconductors. Confining dimensions in all cases are 100 Å. The well material is GaAs and the barriers are considered of infinite height, except for the superlattice, which has 100 Å-thick AlAs barriers. (Adapted from [6].)*

Although the conduction band is quite parabolic for the energies of interest and the effective mass is very isotropic ($m_\perp = m_\parallel$), for the valence band the situation is somewhat more complicated. In III–V bulk semiconductors like GaAs there are two hole bands which are degenerated at the centre of the Brillouin zone ($k = 0$) [3]. However, they have different dispersion relations (different effective masses). According to the relative value of $m_\perp$ they are called the heavy- (hh) and the light-hole (lh) bands. Since the confinement energy in a quantum well depends on the effective mass they will produce two different sets of levels. When the heavy- and light-hole levels are very close they can interact very strongly causing large nonparabolicity and heavy–light hole mixing. In such a case the simple picture given here is no longer valid.

As an example let us consider a 100 Å-thick GaAs/$Al_{0.35}Ga_{0.65}As$ quantum well. In the infinite barrier approximation the first levels for electrons, heavy holes and light holes are $\mathscr{E}_{1e} = 57$, $\mathscr{E}_{1h} = 11$, and $\mathscr{E}_{1l} = 40\,\text{meV}$, respectively. In fact the conduction and valence band barrier heights are only 312 and 191 meV, respectively, and the analytical solution of the Schrödinger equation (9.3) taking the real barrier heights gives $\mathscr{E}_{1e} = 32$, $\mathscr{E}_{1h} = 8$, and $\mathscr{E}_{1l} = 22\,\text{meV}$.

The density of states of a quantum well in the parabolic approximation, in contrast to the square root dependence of the bulk, will have, due to the contribution of the motion in the well plane, a staircase-like dependence [5], with the position of the steps given by the $\mathscr{E}_i$ ($n = 1, 2, 3, \ldots$) (Figures 9.3 and 9.4). Note that the band edge has shifted by $\mathscr{E}_1$ with respect to the bulk value and that the thinner the well is the larger is this shift. These results will have a strong influence on the optical properties.

9.4.2. Electronic properties of coupled quantum wells and superlattices

An important quantum effect is the penetration of the wavefunction inside finite barriers [38] (see for example Figure 9.3). When two or more wells are separated by narrow barriers the wavefunctions of their electronic states overlap. If two states of different wells have similar energies they resonate with each other, producing two states which are extended over the two wells. One then speaks of *coupled quantum wells* (see below, Figure 9.9). In multiple quantum wells a large number of identical quantum wells are grown one after the other forming a periodic array. When the barriers between the wells are so thin that the wells are coupled, the structure is called a *superlattice* (see below, Figure 9.10(a)). In this case the resonance of all the equivalent quantum wells states produces a miniband of allowed levels, in the same manner that atomic levels in a crystal produce crystal bands of allowed electronic states. Since the electronic states extend over many wells, the electronic properties of a superlattice are between those of a two-dimensional system and those of a highly anisotropic bulk system. Similarly, the density of states of a superlattice is between that of a quantum well and that of the bulk (Figure 9.4).

9.4.3. Electronic properties of quantum wires and quantum dots

For a quantum wire the electron is confined in two directions and can move freely only in the remaining direction. They are electronic quasi-one-dimensional systems and their density of states is concentrated in a few peaks (Figure 9.4).

For quantum dots the movement is confined in the three directions of space and the electron energy can only have discrete values. The density of states is zero outside these energies (Figure 9.4). Quantum dots are quasi-zero-dimensional electronic systems.

The fact that the density of states for systems of reduced dimensionality is important only for some energies looks very promising for optical and electrooptic applications.

9.5. Optical properties of quantum wells

From the various optical properties of a material we will concentrate on the absorption spectrum and on the index of refraction because they are the most important for electrooptic applications. Surprisingly most of the experimental work on quantum well systems has avoided direct absorption experiments, since usually the substrate has a lower bandgap than the quantum wells and transmission measurements would require either the removal of the substrate or the use of a waveguide structure. However, a great deal can be learned with related techniques like photoluminescence excitation and photocurrent spectroscopy. The latter is more convenient for electrooptic characterization because the photoluminescence is quenched at high electric fields due to exciton ionization.

We are going to discuss only the effects near the absorption edge, which are the most interesting for device applications. They correspond to optical interband transitions from the highest states of the valence band to the lowest states of the conduction band. For symmetrical rectangular quantum wells there are some important *selection rules* that allow only transitions between hole and electron states of the same quantum number i [38]. Moreover, heavy-hole transitions are forbidden when the electric field vector of the light is

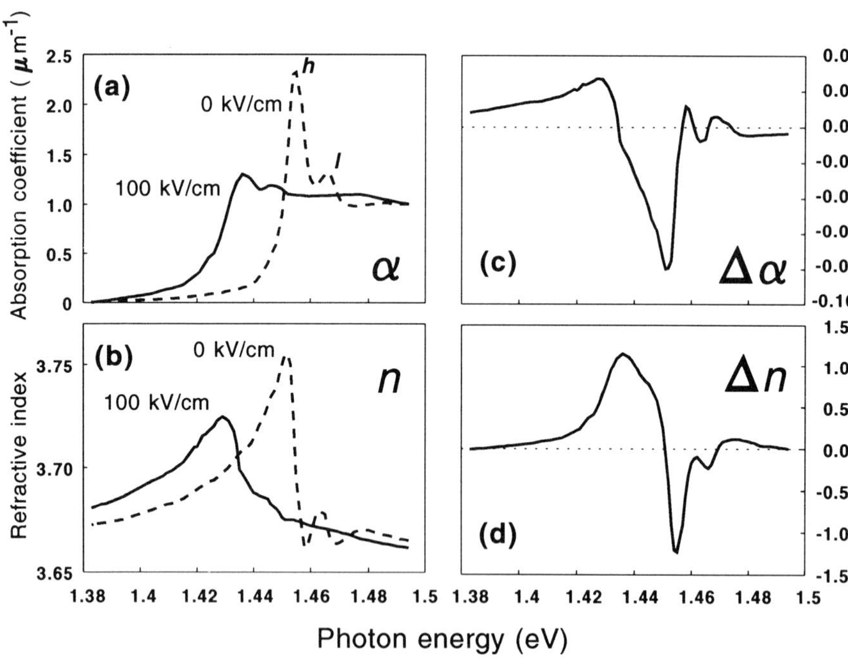

Figure 9.5 *Electrooptic properties of a 100 Å GaAs/$Al_{0.35}Ga_{0.65}As$ quantum well. The left side shows (a) the idealized absorption coefficient α and (b) the refractive index* n *for electric fields of 0 and 100 kV/cm. The right side shows the increase of (c) α and (d)* n *from 0 to 100 kV/cm. h and l refer to the heavy- and light-hole exciton peak, respectively.*

perpendicular to the well plane [5]. This is important when the light beam propagates in the plane of the quantum well, as in slab waveguides (Figure 9.2).

The absorption spectrum will depend on the states available in the conduction and the valence bands. Therefore, in principle it will resemble the staircase-like shape of the joint density of states, with an absorption edge that will be shifted from the bulk one by $\mathscr{E}_{1e} + \mathscr{E}_{1h}$, where $\mathscr{E}_{1e}$ and $\mathscr{E}_{1h}$ are the electron and heavy-hole confinement energies, respectively (Figure 9.5(a)).

9.5.1. Excitonic effects

When a photon is absorbed, an electron in the conduction band and a hole in the valence band, respectively, are excited. Due to their opposite charges, the electron and the hole interact with each other forming a hydrogen-like bound electron–hole pair called an *exciton* [39]. Therefore the absorption spectrum has a strong exciton peak (representing the formation of an exciton in its lowest state) for every step at a slightly lower energy

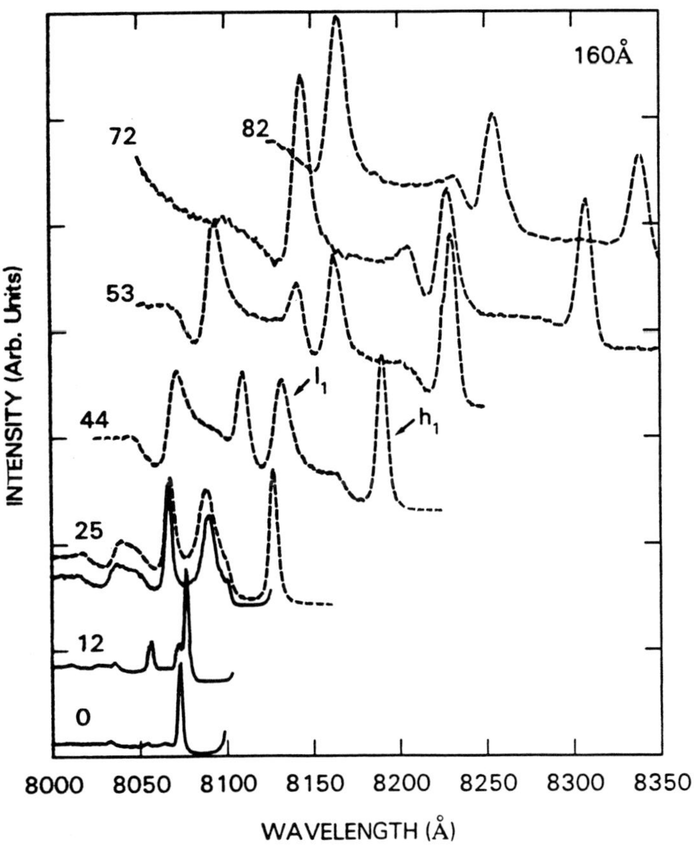

Figure 9.6 *Photocurrent (dashed lines) and photoluminescence excitation (continuous lines) spectra for a 160 Å thick GaAs/$Al_{0.35}Ga_{0.65}As$ quantum well measured at 10 K and at 5 K, respectively, for various electric field values (given in kV/cm). These spectra very closely resemble the shape of the absorption spectra, although the absolute intensities depend on other factors. Note the red shift of the lowest-energy exciton transition (h_1). Spectra have been offset for clarity. (From [40].)*

(Figures 9.5(a) and 9.6). The excitonic absorption edge, which corresponds to the first heavy-hole exciton, appears at an energy $\mathscr{E}_{eh1} = \mathscr{E}_g + \mathscr{E}_{1e} + \mathscr{E}_{1h} - \mathscr{E}_B$, where $\mathscr{E}_B$ is the exciton binding energy.

In a bulk semiconductor the exciton binding energy is of a few meV (4.7 meV for GaAs [34]), and excitons are mostly ionized at room temperature (Section 5.8.1). Therefore the excitonic features of the absorption spectrum in the bulk are smeared out at room temperature. On the other hand, in a semiconductor quantum well the exciton is squeezed because its Bohr diameter in the bulk is a few hundred angstroms (230 Å in GaAs [34]), larger than the usual well thickness. Hence the binding energy is larger than in the bulk and increases as the well thickness decreases. For example, in a GaAs/$Al_{0.3}Ga_{0.7}As$ quantum well the binding energy of the heavy-hole exciton reaches a maximum of around 9.5 meV for $L_w \approx 30$ Å and then decreases again due to the penetration of the wavefunctions into the barriers [41].

Summing up, the absorption spectrum of a semiconductor quantum well, in contrast to the bulk, will have a staircase shape with sharp excitonic peaks even at room temperature, what is very attractive for device applications. Also, the absorption edge is blue-shifted with respect to the bulk and can be tuned by an appropriate design of well width and/or barrier height.

The index of refraction can be obtained from the absorption spectrum through the Kramers–Kronig relations (see Section 2.5). It has been measured directly in some systems [42], and a typical dispersion relation is given in Figure 9.5(b).

9.6. Electrooptic effects in a quantum well

An electric field adds an additional electrostatic potential to the electrons and holes and can have a significant effect on the electronic properties and, therefore, on the optical properties (for a review see, for example, [43,44]). We will not discuss band filling effects here, which can be important in doped quantum wells or under high-power light excitation. Finally we will limit ourselves to energies of the incident light close to the absorption edge of the quantum well, in the transition from the transparent to the absorbing spectral regions.

Among the electric-field-induced changes in the optical properties we are going to consider those in the absorption coefficient (electroabsorption effects) and those in the index of refraction (properly called electrooptic or electrorefraction effects).

9.6.1. Electroabsorption

We introduce first the effect due to the confined states of single electrons or holes, without considering the Coulomb interaction between the photogenerated electrons and holes (excitonic effects). This is sometimes referred in the literature as the *Quantum Confined Franz–Keldysh Effect* (QCFKE). We will then complete this simple picture by including the effect on the exciton or the *Quantum Confined Stark Effect* (QCSE), which is always very important in quantum wells and yields some of the interesting electrooptic applications of these systems.

The electroabsorption can be easily understood from the variation of the potential profile. We assume that the electric field is applied normal to the plane of the quantum well, because it is in this case that the effects strongly differ from the bulk case. A constant electric field in the z direction adds a term $-eEz$ to the electrostatic potential acting on the

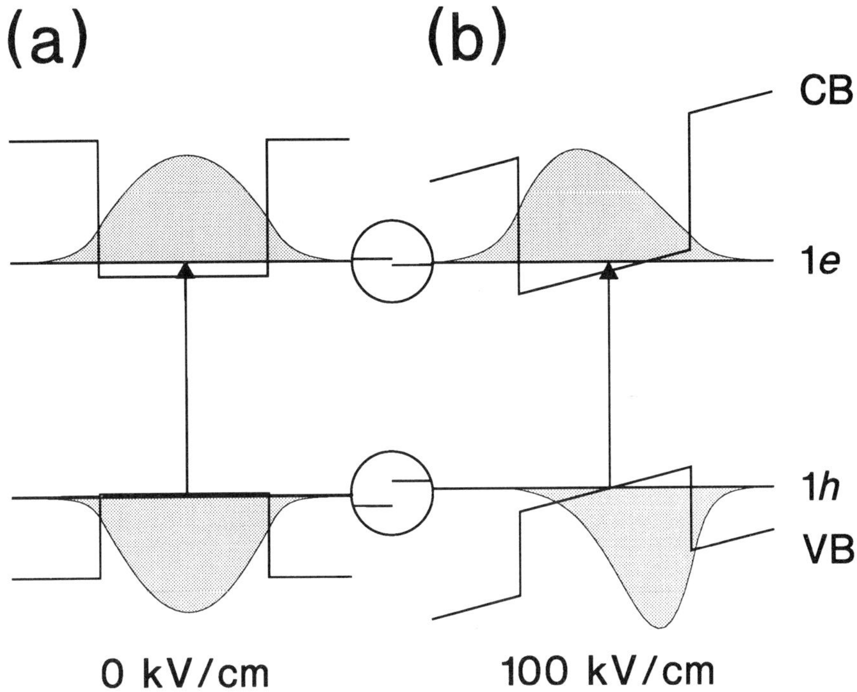

Figure 9.7 *Band profile and fundamental-state wavefunctions for a 100 Å-thick GaAs/$Al_{0.35}Ga_{0.65}As$ quantum well under electric fields of (a) 0 and (b) 100 kV/cm. Circular insets show the Stark shifts of levels amplified three times.*

electrons. For a rectangular well the band edges or the electronic potential profiles become trapezoidal (Figure 9.7(b)). This produces a reduction of the confinement energy of electrons and holes and a red shift of the bandgap with the electric field [45] (see Figures 9.6, 9.7 and 9.8). In the case of a rectangular quantum well with infinite barriers and moderate electric fields, such that the drop of the applied electrostatic potential across the well is much smaller than the confinement energy

$$|e|EL_w \ll \frac{\hbar^2\pi^2}{2m_\perp L_w^2} \tag{9.7}$$

the shift $\Delta\mathscr{E}$ can be obtained analytically from equation (9.3) by considering the applied potential as a second-order perturbation. For the fundamental level [48]

$$\Delta\mathscr{E}_1 \approx -\frac{2.19\times 10^{-3}e^2}{\hbar^2}m_\perp E^2 L_w^4 \tag{9.8}$$

Note that the shift increases sharply with the well width and that it depends quadratically on the electric field (Figure 9.8), being independent on its sign (although this is valid only for symmetrical wells). Moreover, due to the linear dependence with the effective mass, the shift is much larger for the heavy hole than for the electron or the light hole levels, and the electric-field-induced decrease of the quantum well bandgap is due mainly to the heavy-hole contribution.

As an example, for a 100 Å thick GaAs/$Al_{0.35}Ga_{0.65}As$ quantum well under a field of $E = 100\,kV/cm$, equation (9.8) predicts shifts of -10 and -2 meV for the heavy-hole and

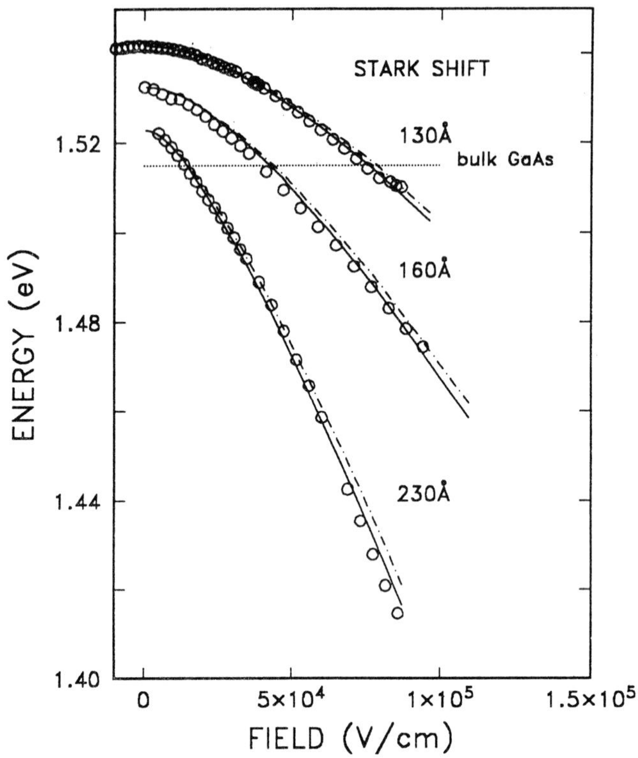

Figure 9.8 *Excitonic bandgap energy as a function of the electric field for GaAs/$Al_{0.35}Ga_{0.65}As$ quantum wells of various thicknesses, obtained from photoluminescence emission experiments at 5 K; lines correspond to different calculations [46]. (From [47].)*

electron levels, respectively. A numerical calculation that takes into account the finite barrier height and the different effective masses in the barriers gives shifts of −15 and −6 meV, respectively.

For higher levels the shift can be positive or negative, depending on the quantum number [49]. Because these levels are not important for electrooptic properties near the band edge they will not be considered here.

Besides the energy shift of the band gap there is a field-induced change in the intensities of the hole-to-electron transitions. The intensities are approximately proportional to the square of the overlap between the electron and the hole wavefunctions. For the allowed transitions at zero field (those involving states with the same quantum number i for electron and hole) the field produces a reduction of the intensity because the electron and the hole wavefunctions are pushed in opposite directions (Figure 9.7(b)).

For example, in a 100 Å-thick GaAs-$Al_{0.35}Ga_{0.65}As$ quantum well the square of the overlap between the fundamental electron and the heavy-hole wavefunctions, which is roughly proportional to the absorption coefficient above the band edge, decreases from 0.98 at zero field to 0.44 at 100 kV/cm. This effect is smaller for narrower wells since the electron–hole separation is limited by the barriers.

For forbidden transitions the field relaxes the selection rules due to the loss of symmetry. For finite fields the wavefunctions are no longer strictly even or odd and the overlap between electron and hole states with different quantum number i becomes different from zero. Therefore forbidden transitions become allowed at high fields. If one considers the sum of the intensities of all the possible heavy-hole to electron or light-hole to electron transitions, respectively, one finds that they are independent of the field. This important result is known as the *sum rule* [50].

It is interesting to note that for very thick quantum wells, where many transitions are possible, the spectrum becomes rather smooth under an electric field, with an exponential tail below the zero-field bandgap and oscillations above. This is characteristic of the bulk Franz–Keldysh effect discussed in Section 5.10.3 [51]. As expected, for very thick quantum wells the bulk limit is recovered.

Finally let us discuss the influence of the electric field on the exciton peak. Since the field pushes electrons and holes in opposite directions, it has a weakening effect on the exciton. With the electric field the exciton peak shifts to lower energies and loses intensity and sharpness [52–55]. The red shift is equal to the shifts of single-particle levels minus the small decrease in the exciton binding energy. The latter is mainly responsible for the weakening of the peak and its broadening. For electric fields $E\mathscr{E} \gg \mathscr{E}_B/ea_B$ (where a_B is the exciton Bohr radius) the excitons are ionized and the exciton peak finally disappears. However, excitonic effects in quantum wells, due to their larger binding energy, withstand larger fields than in the bulk. The sum rules are also verified when excitonic effects are considered [50,56].

Quantum well exciton peaks are very intense and very narrow (typically a few meV) compared with bulk ones. Therefore, spectral regions at energies slightly below the exciton resonance, which are transparent at zero field, become highly absorbing when the field is increased and the exciton peak moves to lower energies. This introduces large electrooptic effects near the bandgap which have application in quantum well electrooptic devices (see Section 9.9).

9.6.2. Electrorefraction. Electrooptic coefficients

Electric-field-induced changes in the refractive index n are related to changes in the absorption coefficient through the Kramers–Kronig relations, and in principle one can be obtained from the other. Figure 9.5(d) shows schematically the change in the refractive index for a typical quantum well.

As we saw in Section 3.2 the variation of the refractive index tensor n_{ij} with electric field can be conveniently expressed in terms of the electrooptic coefficients r and s, where r is the linear electrooptic coefficient (Pockels effect) and s is the quadratic electrooptic coefficient (Kerr effect):

$$\Delta n_{ij}(E) = -\tfrac{1}{2}n_0^3(r_{ijk}E_k + s_{ijkl}E_kE_l) \tag{9.9}$$

(n_0 is the refraction index at zero field, $n_0 = n(0)$.) Higher-order terms are negligible. What makes semiconductor quantum wells very attractive is their large quadratic electrooptic effect near the absorption edge when compared with bulk semiconductors [10]. This is so because in symmetrical quantum wells the quantum confined Stark effect introduces a large electroabsorption, not present in the bulk, which does not depend on the sign of the applied electric field. Therefore, quantum confinement introduces only an additional Kerr effect. In asymmetrical quantum wells the Pockels effect can also differ from that of the bulk.

Before discussing the magnitudes of the electrooptic coefficients let us go briefly through their symmetry properties. For a bulk semiconductor with the zinc-blende structure (group symmetry $\bar{4}3m$ or T_d) the linear electrooptic tensor has three nonzero elements, all equal:

$r_{41} = r_{52} = r_{63}$ (Appendix 3A). In a quantum well grown on a (001)-oriented substrate the z-axis becomes a privileged direction and the cubic symmetry of the bulk reduces to a tetragonal symmetry (group symmetry $\bar{4}2m$ or D_{2d}). Then the three nonzero elements of the first order electrooptic tensor become $r_{41} = r_{52} \neq r_{63}$ (Appendix 3A). For the quadratic electrooptic tensor the argument is similar and the result can be found in the tables of Appendix 3B under the group symmetry D_{2d}.

In the usual geometry (see Figure 9.2) the quantum well is grown in the intrinsic layer of a p–i–n diode, and is perpendicular to the [001] crystal axis. When a reverse bias voltage is applied to the diode the electric field is parallel to the z-axis, and only electrooptic coefficients r_{ij} and s_{ij} with $j = 3$ contribute.

In the simplest geometry the light enters into the device through the top surface, perpendicular to the well plane (along the z-axis). Therefore the polarization of the light lies on this plane, and $r = r_{63}$ and $s = s_{13}$ (Table 9.3). This geometry has two main disadvantages. First, the interaction length of the light with the quantum well is only the well thickness, and a multiple quantum well must be used. Second, the light passes through the substrate. Consequently, the latter must be etched away when its bandgap is smaller than that of the quantum well.

These disadvantages are avoided when the light propagates in the plane of the well and the latter is embedded in a slab or stripe waveguide. In this case the light usually enters perpendicular to the cleavage planes, that is, parallel to [1$\bar{1}$0] or [110] directions. According to the polarization of the light two cases can be distinguished. If the polarization is in the plane of the well (the so-called TE mode of the slab waveguide) then $r = \pm r_{63}$ (the signs + and − stand, respectively, for light propagation along [1$\bar{1}$0] or [110] directions [12]) and $s = s_{13}$. If the polarization is perpendicular to the well plane (the so-called TM mode of the slab waveguide) then $r = 0$ and $s = s_{33}$.

Experimental measurements of the electrooptic coefficients have been done for several quantum well systems, like GaAs/(Al,Ga)As [10–12a,57], (In,Ga)As/InP [58], and (In,Ga)(AsP)/InP [13], but the data are still very scarce. As expected, the linear coefficient r is of the same order of magnitude as in the bulk ($\sim 10^{-12}$ m/V for GaAs/(Al,Ga)As quantum

Table 9.3 ***Electrooptic coefficients for a semiconductor quantum well grown on a (001)-oriented substrate. E and k vectors give, respectively, the directions of polarization and propagation of the light (Figure 9.1).***

	E		
k	[001]	[110]	[1$\bar{1}$0]
[001]	—	$r = r_{63}$ $s = s_{13}$	$r = r_{63}$ $s = s_{13}$
[110]	(TM) $r = 0$ $s = s_{33}$	—	(TE) $r = -r_{63}$ $s = s_{13}$
[1$\bar{1}$0]	(TM) $r = 0$ $s = s_{33}$	(TE) $r = r_{63}$ $s = s_{13}$	—

Measured values for a 85 Å-thick GaAs/$Al_{0.3}Ga_{0.7}As$ quantum well [12] below the exciton peak:

$r_{63} = (-1.75 \pm 0.15) \times 10^{-12}\,\mathrm{mV^{-1}}$

$s_{13} \approx -1.3 \times 10^{-15}/(\mathscr{E}_{1eh} - \mathscr{E})\,\mathrm{meV^2m^2/V^2}$

$s_{33} \approx -1.3 \times 10^{-15}/(\mathscr{E}_{1el} - \mathscr{E})\,\mathrm{meV^2m^2/V^2}$

$n_0 = 3.5$

wells [10,12a,57]). Moreover, it shows a very weak spectral dependence below the bandgap [12,57]. On the contrary, the quadratic coefficient s close to the band gap is much larger than in the bulk (values of $140 \times 10^{-20}\,m^2/V^2$ have been reported for GaAs/(Al,Ga)As quantum wells [12]). Below the bandgap it decreases very sharply. Theoretically, from the sum rules [59] s should decrease inversely proportional to the square of the energy separation $\Delta\mathscr{E}$ from the corresponding exciton resonance ($\mathscr{E}_{1eh}$ for s_{13} and $\mathscr{E}_{1el}$ for s_{33}). Although Zucker *et al.* [13,57] have found an $s \propto 1/\Delta\mathscr{E}$ dependence, the recent results of Jennings *et al.* [12] seem to fit better with the expected $s \propto 1/\Delta\mathscr{E}^2$ dependence.

For phase modulators it is interesting to create a large change in the refractive index n while keeping a low value of the absorption coefficient. For this reason, one has to look for a compromise and find a photon energy well below the absorption edge to avoid a significant increase of the absorption with the electric field, but not too low so that the change of the refractive index becomes very small. Here also quantum wells have some advantages over the bulk because in the latter the Franz–Keldysh effect produces far reaching low energy tails in the absorption. For a quantum well, the largest change in n occurs for the TE mode, with light propagation along the $[1\bar{1}0]$ direction, because the r and s have the same sign (see Table 9.3).

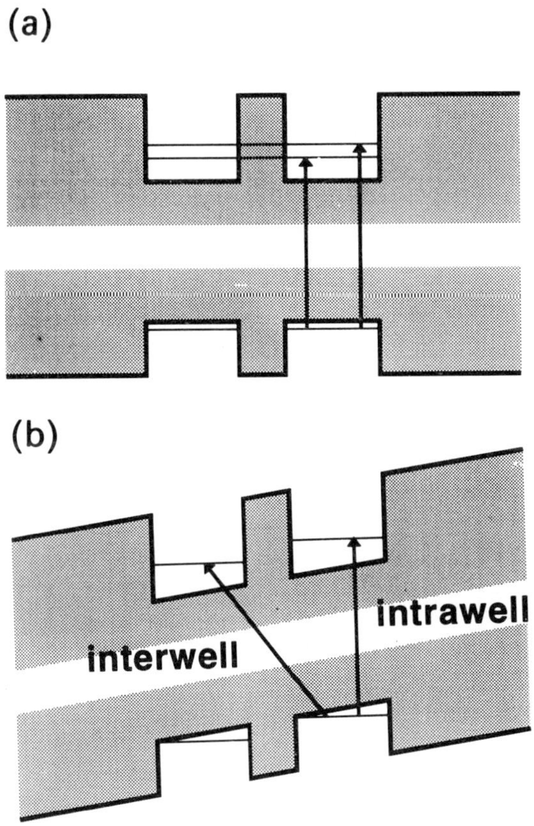

Figure 9.9 *Symmetrical coupled double quantum well showing optical transitions from the right-well hole. At zero electric field (a) the wells are coupled and electron states extend over both wells. At finite electric fields (b) states are localized mostly in either one of the wells and transitions can be classified as intrawell or interwell.*

9.7. Electrooptics of coupled quantum wells and superlattices

In coupled quantum wells the electric field strongly affects the coupling between wells. Since the electrostatic energy of an electron in a uniform electric field depends linearly on the distance, the relative shift of levels originating in different wells is $\Delta\mathscr{E} = eEd$, where d is the distance between the centres of the wells. The resonance between two states varies

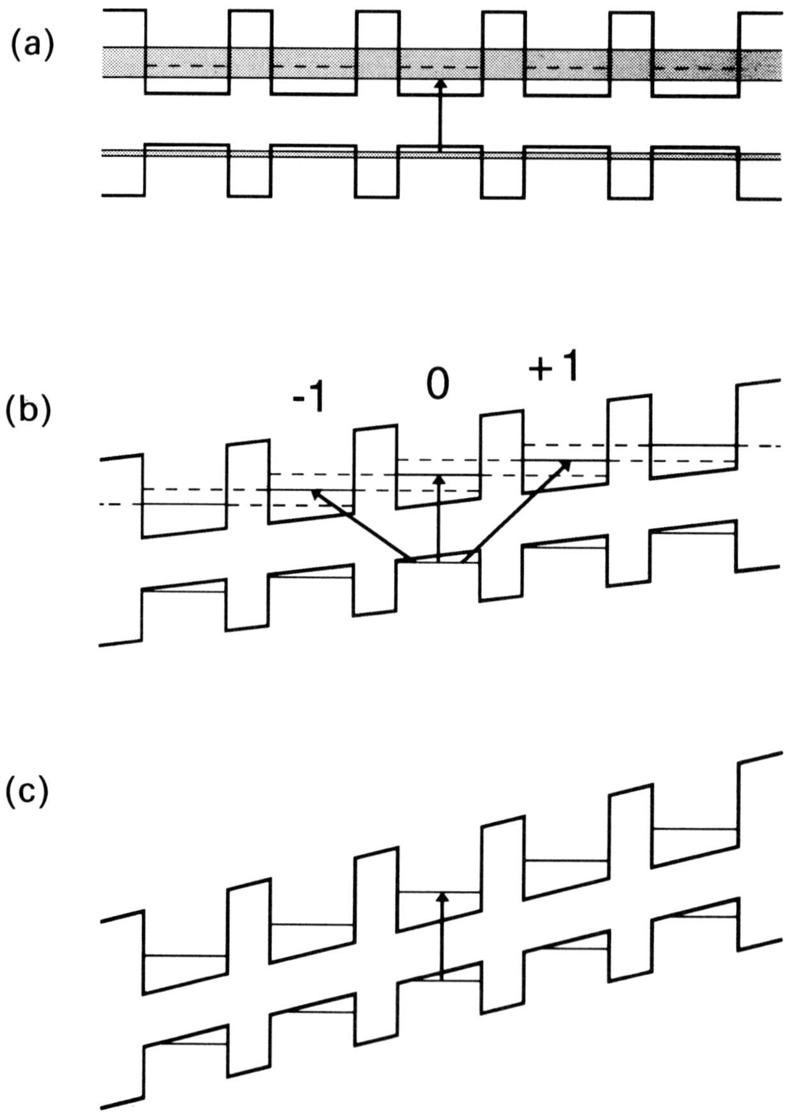

Figure 9.10 *Electric field effects on a superlattice. At zero electric field (a) exciton transition occurs between miniband edges. At moderate electric fields (b), holes are localized and the electron miniband splits into a Stark ladder, producing a number of transitions, intrawell and interwell. At high electric fields (c) electron states localize into single wells, and interwell transitions disappear. From (a) to (c) the absorption edge blue shifts by roughly half the miniband width. For simplicity, transitions are shown only from the central well hole.*

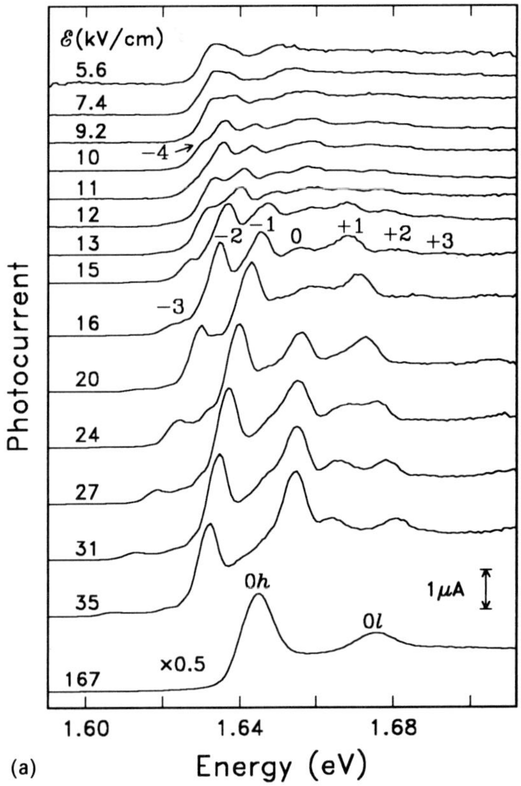

(a)

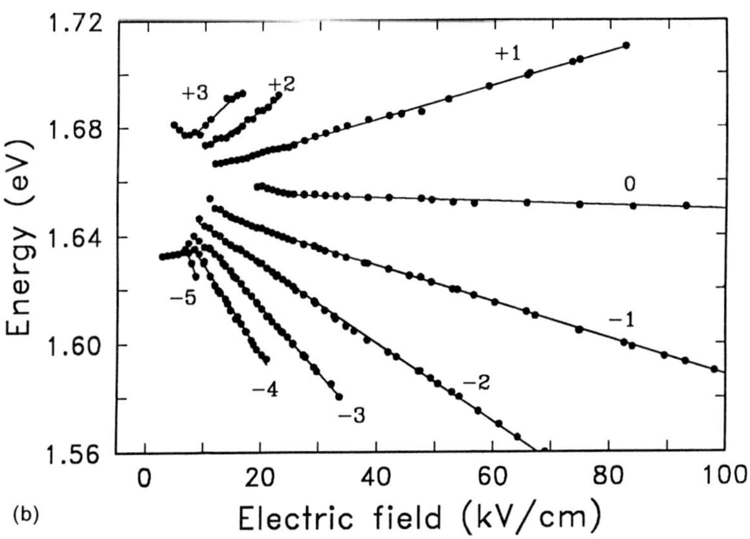

(b)

Figure 9.11 *(a) Photocurrent spectra for a 40–20 Å GaAs/$Al_{0.35}Ga_{0.65}As$ superlattice measured at 5 K for various electric fields. (Spectra have been offset for clarity.) Interwell transitions are labelled according to the number of periods between the centres of the electron and the hole states (see Figure 9.10(b)). Note the blue shift of the absorption edge from 0 to 167 kV/cm. (From [67].) (b) Heavy-hole transition energies vs electric field measured under the same conditions. Straight lines are least-squares fits to the experimental data. (Adapted from [67].)*

roughly inversely proportional to the energy separation of their levels and thus the field can easily tune the coupling among states coming from different wells.

When the coupling is small the electronic states are localized mainly in a certain well and one can classify the optical transitions in *intrawell* transitions, among hole and electron states of the same well, and *interwell* transitions, among hole and electron states of different wells (Figure 9.9(b)). For intrawell transitions the electrooptic effects are similar to those of single quantum wells. For interwell transitions the electric field induces a large linear shift, which can be positive or negative depending on the direction of the field. If the latter points from the electron to the hole state the transition experiences a red shift. On the other hand interwell transitions usually decrease in intensity and finally disappear when the field grows. The linear shift in coupled quantum wells offers great advantages when a large tunability of the transitions is needed at moderate electric fields.

There has been a great deal of work on the electrooptic properties of coupled double quantum well systems, both symmetrical [60,61] and asymmetrical [62,63].

In superlattices the strongest excitonic peak corresponds to the transition from the top of the hole miniband to the bottom of the electron miniband (Figure 9.10(a)). When a strong enough electric field is applied (see, e.g. [47] for a review) the resonance between the levels is reduced and the minibands split into a series of equally spaced discrete levels, the so called *Stark ladder* [64–67]. Each level corresponds to an electronic state localized around a different well [68,69] (Figure 9.10(b)). For fields of the order of $E \approx \Delta\mathcal{E}/eD$, where $\Delta\mathcal{E}$ and D are the miniband width and the superlattice period, respectively, the states of the Stark ladder become completely localized in the corresponding well (Figure 9.10(c)). This effect is difficult to observe in bulk crystals because the bandwidths are of the order of a few eV and extremely large electric fields are required. However, in superlattices the bandwidths are much smaller and the needed fields are easily attained. For example, in a typical 40/20 Å $GaAs/Al_{0.35}Ga_{0.65}As$ superlattice the electron miniband has a width of about 65 meV and the electric field necessary for Stark localization is of the order of 115 kV/cm. The heavy-hole miniband has a width of only 6 meV and the heavy holes become localized for fields as small as 10 kV/cm.

The absorption spectrum of a superlattice (or related spectra, like photocurrent spectra, Figure 9.11(a)) reflects the field-induced localization. At zero electric field there is an exciton peak for the transition from the top of the hole miniband to the bottom of the electron miniband, corresponding to transitions between delocalized states (Figure 9.10(a)). At intermediate fields (Figure 9.10(b)) the spectrum splits into a series of peaks corresponding to interwell transitions between the different states of the Stark ladder [70,71] (Figure 9.11(b)). At the fields where Stark localization occurs (Figure 9.10(c)) the interwell transitions disappear and only a peak corresponding to the intrawell transitions remain. Since the single well level is close to the centre of the miniband, the absorption edge of the superlattice experiences a blue shift from zero to high fields. The blue shift, which opposes the usual red shift found in thick single quantum wells, is roughly equal to half the electron miniband width and has been the basis of some electrooptic devices (see Section 9.9).

9.8. Electrooptics of quantum wires and quantum dots

We have seen that quantum wells present new electrooptic effects not present in bulk materials. Similarly one could expect that in quantum wires and quantum dots, where the confinement is even larger, new or more pronounced effects should appear. However, to date most of the expectations are based on theoretical calculations due to the lack of experimental results.

Miller and coworkers [72] have calculated the electroabsorption properties of rectangular semiconductor boxes surrounded by infinite potential barriers, neglecting excitonic effects. By changing the sides of the boxes and their ratio they can simulate any degree of confinement, from the bulk to a quantum dot, passing through the quantum well and the quantum wire regimes. For quantum wires and quantum dots the Stark shifts induced by an electric field applied along a given side are the same as the shifts in a quantum well of the same thickness. However, since the density of states is much more concentrated in energy, the change in the absorption coefficient is much larger for the same electric field. Therefore quantum wire and quantum dot devices would require less energy for operation than their quantum well counterparts. Stark shifts larger than in a quantum well have been calculated in quantum wires with finite barriers [73]. This could further enhance electrooptic effects in highly confined structures. Coupled quantum wire arrays have been proposed for electrooptic devices with different response depending on the polarization of the incident light [74], based on the optical anisotropy for light beams perpendicular to the plane of the array.

Most of the experimental work in highly confined systems has been done for semiconductor-doped glasses, which contain microcrystallites of CdS, CdSe, or similar semiconductors with dimensions down to a few angstroms. For small particle size a quantum dot behaviour should be expected. The electroabsorption properties have been observed as a function of the particle size and of the electric field [75,76]. For the smallest particles a typical Stark effect is observed. For strong fields the dependence of the absorption coefficient with electric field changes from quadratic to linear [77].

9.9. Electrooptic applications

In this section we consider some applications of electrooptic effects in semiconductor quantum wells and superlattices. These effects were discovered in the 1980s and there are already some devices which could probe very important for optoelectronic integrated circuits and for optical computing and communications.

The main advantages of quantum well devices over bulk devices are the possibility of fine tuning the characteristics, the appearance of completely new phenomena due to the quantum confinement, and the large quadratic electrooptic effect. This makes possible the use of very small devices with reduced capacitance and low power requirements. Moreover, they could be integrated with conventional microelectronic elements and optical devices such as semiconductor laser diodes. This will be very interesting in the near future for the interface between optic fibre communications to carry the information and conventional electronics to process this information. They look also promising for the production of photonic devices that would be the basis for optical computing, that is, the optical treatment of information, which could be faster than present technologies and more suitable for alternative approaches like parallel computing and neural networks [78].

Major disadvantages of quantum-well electrooptic devices are their narrow wavelength response and their temperature sensitivity, both consequences of the abruptness of the absorption spectrum near the absorption edge.

The electric-field-induced change in the absorption coefficient is the basis for electroabsorption or intensity modulators which are used to control the intensity of an input light beam by the applied bias voltage.

To obtain a high contrast ratio the photon energy of the light beam must be just below the absorption edge at zero electric field (see Figure 9.5(a)). When the field is increased the exciton peak of the absorption spectrum red shifts and the device switches from transparent to highly absorbing (Section 9.6.1), reducing the intensity of the output light. The Stark effect in quantum wells is so strong that one can get effective modulations of two or three in a single pass through one micrometre of the active material.

Figure 9.12 shows a typical multiple quantum well electroabsorption modulator in the perpendicular configuration, where the input light beam enters perpendicular to the layer planes. The quantum wells are in the intrinsic region of a p–i–n diode in order to apply an electric field in reverse bias. The number of wells must be large to have a reasonable absorption in the absorbing (off) state. A disadvantage of this design is that the layers surrounding the active region must be transparent for the working wavelength. When the substrate is opaque, like in GaAs/(Al,Ga)As systems, it must be etched away at least in the central part. This is avoided in surface reflection modulators [79] where the light beam is reflected by a multi-layer dielectric stack mirror placed between the substrate and the quantum wells.

Multiple passes of the light beam are obtained in vertical Fabry–Perot modulators where the beam undergoes multiple reflections between two multilayer mirrors grown above and below the multiple quantum well [80,81]. Contrast ratios over 100 have been reported [82].

In waveguide modulators [83] the light propagates parallel to the quantum wells which are at the centre of a slab waveguide (Figure 9.2). In this case the interaction length is much larger and the modulation is limited only by the device length. Therefore one does not need many wells and the operating voltage can be smaller.

Electrooptic modulators with double coupled quantum wells have also been fabricated [60]. Recently, the Stark localization in superlattices has been applied in modulators [84]. In this case the photon energy is just above the absorption edge at zero electric field and the

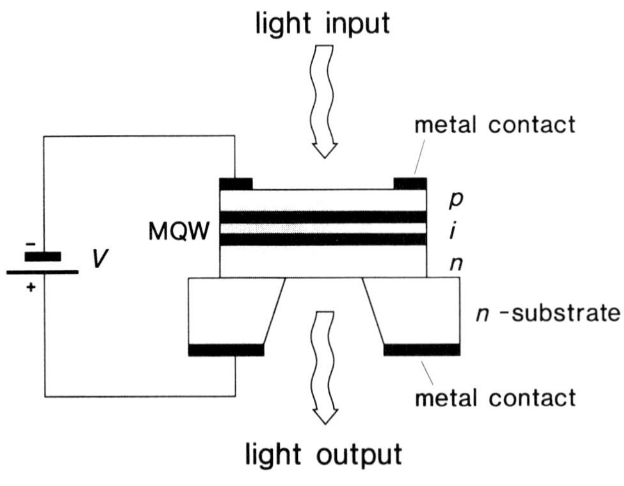

Figure 9.12 ***Multiple quantum well (MQW) electrooptic modulator. The substrate has been etched away at the centre to permit the transmission of light.***

device is normally absorbing. At higher fields the edge blue-shifts and the device becomes transparent.

Quantum well modulators have been used to build more complex devices like a photonic memory switch [85], a Q-switched multiple quantum well laser diode [86], or spatial light modulator arrays [87,88].

The change in refractive index is the basis for waveguide quantum well phase modulators [15], which modify the phase shift of an incident beam. Other electrorefractive devices are waveguide directional couplers [13,89,90].

The shift of the optical transition energies with the electric field has been applied to tune the emission wavelength of quantum well lasers [91] and the frequency modulation of laser diodes [18]. These are only some applications of the effect of electric fields on the density of states of rectangular quantum wells. There are also some important uses of electrooptic effects in other shape quantum wells, like the wavelength tuning of infrared detectors in asymmetric quantum wells [92,93], and electrooptic effects on the population of free carriers. Among the latter one can mention modulators based on the Burstein shift [94,95], and devices based on *n–i–p–i* superlattice devices (see the review by Döhler [96]).

9.10. The self-electrooptic effect device (SEED)

A quantum well electrooptic device that has drawn considerable attention is the *self-electrooptic-effect device* (SEED), which is a bistable optical device in its simplest configuration (for a review see [97]).

The SEED (Figure 9.13) consists of a reversed-biased *p–i–n* diode and a resistor (typically of a few megohms) connected in series. A multiple quantum well is in the intrinsic layer of the diode. The photon energy of the input light beam coincides with the heavy-hole exciton peak of the absorption spectrum at zero bias (Figure 9.5(a)). For certain values of V and R, the power of the output beam presents some bistability as a function of the power of the input beam (Figure 9.14). When the latter is very small there is almost no photocurrent through the circuit and all the source voltage appears on the intrinsic region

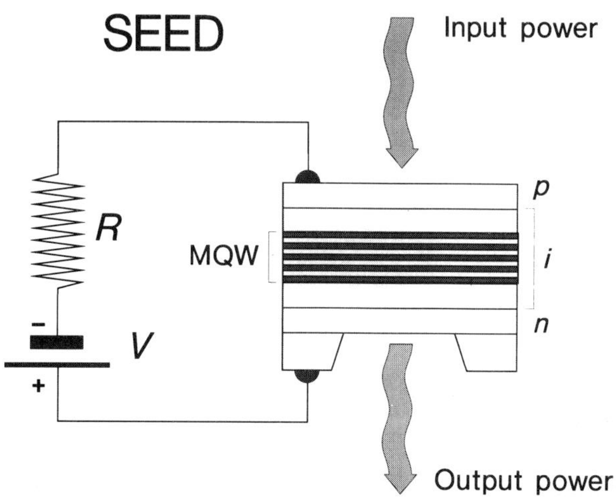

Figure 9.13 ***Multiple quantum well self-electrooptic-effect device (SEED).***

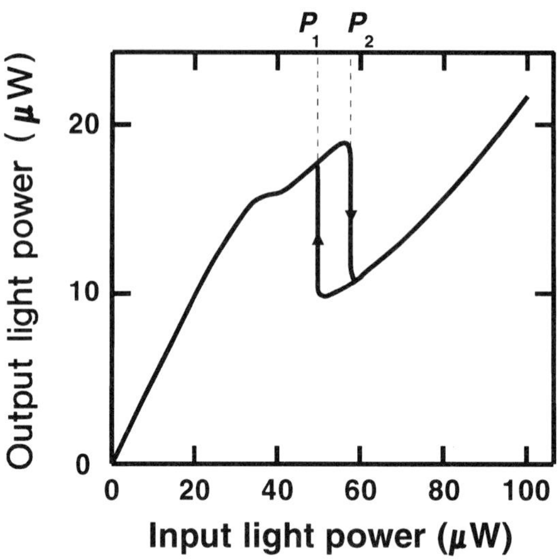

Figure 9.14 *Typical characteristic of a SEED.*

of the diode. The exciton peak is then red-shifted with respect to the operating photon energy and the absorption is low. As the input power increases some photocurrent is produced which induces a voltage drop in the resistor, reducing the voltage applied on the diode. Therefore the exciton peak moves closer to the operating wavelength (Figure 9.5(a)), triggering a high absorption state for some input power P_2. When the input power

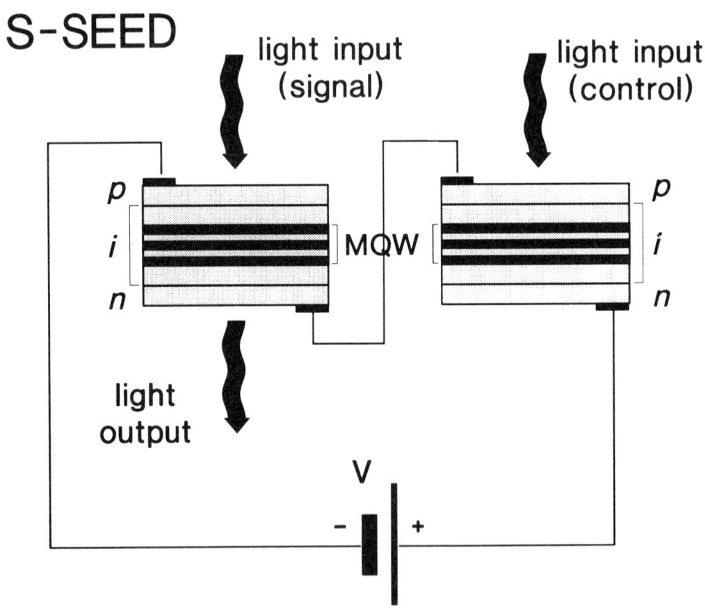

Figure 9.15 *Sketch of a multiple quantum well symmetrical SEED (S-SEED). In a real device the two elements are integrated together.*

decreases the jump from high to low absorption states occurs at a lower power P_1, producing a hysteresis loop and a bistable behaviour. The device can switch between the two states with small changes in the input power. A detailed modelling of a SEED has been provided recently [97a].

Although the first SEED consisted of a GaAs-(Al,Ga)As multiple quantum well [98] other materials and the Stark localization in superlattices [84,99] have been employed since then.

A more practical device is the symmetrical SEED [100] (S-SEED), which consists of two SEEDs connected in series to the voltage source (Figure 9.15). It operates with two light beams, one for each diode. It has been used to build optoelectronic devices that control light with light, such as logic gates [101], dynamic memories and parallel processors [102].

9.11. Concluding remarks

In this chapter we have introduced the electrooptic properties of semiconductor quantum wells and superlattices, which offer the possibility of tailored specifications or bandgap engineering and novel effects due to the quantum confinement. Finally we have reviewed some applications of these properties in devices that, although not yet fully commercial, look very promising for optoelectronics and photonics. We firmly believe that what we have seen in this chapter is only a sample of the versatility and possibilities that lie ahead with materials artificially structured down to atomic dimensions.

References

1. G. Allan, G. Bastard, N. Boccara, M. Lannoo and M. Voos (eds), *Heterojunctions and Semiconductor Superlattices*, Springer, Berlin (1986).
2. E. E. Mendez and K. von Klitzing (eds), *Physics and Applications of Quantum Wells and Superlattices*, Plenum Press, New York (1987).
3. G. Bastard, *Wave Mechanics Applied to Semiconductor Heterostructures*, Editions de Physique, Les Ulis, France (1988).
4. M. Jaros, *Physics and Applications of Semiconductor Microstructures*, Clarendon Press, Oxford (1989).
5. C. Weisbuch and B. Vinter, *Quantum Semiconductor Structures*, Academic Press, New York (1991).
6. L. L. Chang and L. Esaki, *Phys. Today* **45**, 36 (1992).
7. C. M. Sotomayor-Torres and S. P. Beaumont (eds), *Science and Engineering of 1- and 0-Dimensional Semiconductor Systems*, Plenum, New York (1990).
8. M. Reed (ed.), *Nanostructured Systems*, Academic Press, Boston (1991).

8a. J. E. Zucker, *Laser Focus World*, March, 101 (1993).

9. H. Sakata and S. Takeuchi, *Appl. Phys. Lett.* **59**, 3081 (1991).
10. M. Glick, F. K. Reinhart, G. Weimann and W. Schlapp, *Appl. Phys. Lett.* **48**, 989 (1986).
11. T. H. Wood, R. W. Tkach and A. R. Chraplyvy, *Appl. Phys. Lett.* **50**, 798 (1987).
12. A. Jennings, C. D. W. Wilkinson and J. S. Roberts, *Semicond. Sci. Technol.* **7**, 60 (1992).

12a. M. J. Bloemer and K. Myhei, *J. Appl. Phys.* **74**, 4849 (1993).

13. J. E. Zucker, I. Bar-Joseph, B. I. Miller, U. Koren and D. S. Chemla, *Appl. Phys. Lett.* **54**, 10 (1989).
14. B. Pezeshki, S. M. Lord, T. B. Boykin and J. S. Harris, Jr. *Appl. Phys. Lett.* **60**, 2779 (1992).
15. U. Koren, T. L. Koch, H. Presting and B. I. Miller *Appl. Phys. Lett.* **50**, 368 (1987).

16. T. K. Woodward, T. Sizer, II and T. H. Chiu, *Appl. Phys. Lett.* **58**, 1366 (1991).
17. E. Bigan, M. Allovon, M. Carre and A. Carenco, *Electron. Lett.* **26**, 355 (1990).
18. T. Wolf, K. Drögemüller, B. Borchert, H. Westermeier, E. Veuhoff and H. Baumeister, *Appl. Phys. Lett.* **60**, 2472 (1992).
19. T. Miyazawa, S. Tarucha, Y. Ohmori, Y. Suzuki and H. Okamoto, *J. Appl. Phys. Jpn.* **25**, L200 (1986).
20. M. Jupina, E. Garmire, T. C. Hasenberg and A. Kost, *Appl. Phys. Lett.* **60**, 686 (1992).
21. Q. Fu, A. V. Nurmikko, L. A. Kolodziejski, R. L. Gunshor and J. Wu, *Appl. Phys. Lett.* **51**, 578 (1987).
22. D. Lee, J. E. Zucker, M. D. Divino, R. F. Austin, R. D. Feldman, K. L. Jones and A. M. Johnson, *Appl. Phys. Lett.* **59**, 1867 (1991).
23. H. C. Casey Jr and M. B. Panish, *Heterostructure Lasers*, Academic Press, New York (1978).
24. S. M. Sze, *Semiconductor Devices. Physics and Technology*, John Wiley, New York (1985).
25. A. C. Gossard, *IEEE J. Quantum Electron.* **QE-22**, 1649 (1986).
26. M. A. Herman and H. Sitter, *Molecular Beam Epitaxy*, Springer-Verlag, Heidelberg (1989).
27. T. Sakamoto, H. Funabashi, K. Ohta, T. Nakagawa, N. J. Kawai and T. Kojima, *Superlattices Microstruct.* **1**, 347 (1985).
28. W. Richter, *Festkörperprobleme* **26**, 335 (1986).
29. G. B. Strigfellow, *Organometallic Vapor Phase Epitaxy*, Academic Press, Boston (1989).
30. B. L. Sharma, in *Semiconductors and Semimetals*, Academic Press (1981), Vol. 15, Ch. 1.
31. F. J. Leonberger and J. P. Donnelly, in *Guided-Wave Optoelectronics* (ed. T. Tamir), 2nd edn, Springer, Berlin (1990).
32. S. D. Mukherjee and D. W. Woodard, in *Gallium Arsenide* (eds M. J. Howes and D. V. Morgan), John Wiley, New York (1985), Ch. 4.
33. S. Tiwari and D. J. Frank, *Appl. Phys. Lett.* **60**, 630 (1992), and references therein.
34. S. Adachi, *J. Appl. Phys.* **58**, R1 (1985).
35. C. Bosio, J. L. Staehli, M. Guzzi, G. Burri and R. A. Logan, *Phys. Rev. B* **38**, 3263 (1988).
36. G. Duggan, *J. Vac. Sci. Technol.* **B3**, 1224 (1985).
37. R. C. Miller, D. A. Kleinman and A. C. Gossard, *Phys. Rev. B* **29**, 7085 (1984).
38. R. Dingle, *Festkörperprobleme* **15**, 21 (1975).
39. J. I. Pankove, *Optical Processes in Semiconductors*, Dover Publications, New York (1975).
40. R. T. Collins, L. Viña, W. I. Wang, L. L. Chang, L. Esaki, K. von Klitzing and K. Ploog, *Phys. Rev. B* **36**, 1531 (1987).
41. R. L. Greene, K. K. Bajaj and D. E. Phelps, *Phys. Rev. B* **29**, 1807 (1984).
42. K. B. Kahen and J. P. Leburton, *Appl. Phys. Lett.* **49**, 734 (1986).
43. S. Schmitt-Rink, D. S. Chemla and D. A. B. Miller, *Adv. Phys.* **38**, 89 (1989).
44. D. A. B. Miller, D. S. Chemla and S. Schmitt-Rink, in *Optical Nonlinearities and Instabilities in Semiconductors* (ed. H. Haug), Academic Press, New York (1988), Ch. 13.
45. E. E. Mendez, G. Bastard, L. L. Chang, L. Esaki, H. Morkoc and R. Fischer, *Phys. Rev. B* **26**, 7101 (1982).
46. L. Viña, E. E. Mendez, W. I. Wang, L. L. Chang and L. Esaki, *J. Phys. C: Solid State Phys.* **20**, 2803 (1987).
47. E. E. Mendez and F. Agulló-Rueda, *J. Lumin.* **44**, 223 (1989).
48. G. Bastard, E. E. Mendez, L. L. Chang and L. Esaki, *Phys. Rev. B* **28**, 3241 (1983).
49. D. Ahn and S.-L. Chuang, *IEEE J. Quantum Electron.* **QE-23**, 2196 (1987).
50. D. A. B. Miller, J. S. Weiner and D. S. Chemla, *IEEE J. Quantum Electron.* **QE-22**, 1816 (1986).
51. D. A. B. Miller, D. S. Chemla and S. Schmitt-Rink, *Phys. Rev. B* **33**, 6976 (1986).
52. F. L. Lederman and J. D. Dow, *Phys. Rev. B* **13**, 1633 (1976).
53. J. A. Brum and G. Bastard, *Phys. Rev. B* **31**, 3893 (1985).
54. D. A. B. Miller, D. S. Chemla, T. C. Damen, A. C. Gossard, W. Wiegmann, T. H. Wood and C. A. Burrus, *Phys. Rev. B* **32**, 1043 (1985).

55. S. Nojima, *Phys. Rev. B* **37**, 9087 (1988).
56. M. Whitehead, G. Parry, K. Woodbridge, P. J. Dobson and G. Duggan, *Appl. Phys. Lett.* **52**, 345 (1988).
57. J. E. Zucker, T. L. Hendrickson and C. A. Burrus, *Appl. Phys. Lett.* **52**, 945 (1988).
58. J. E. Zucker, I. Bar-Joseph, G. Sucha, U. Koren, B. I. Miller and D. S. Chemla, *Electron. Lett.* **24**, 458 (1988).
59. J. S. Weiner, D. A. B. Miller and D. S. Chemla, *Appl. Phys. Lett.* **50**, 842 (1987).
60. M. N. Islam, R. L. Hillman, D. A. B. Miller and D. S. Chemla, *Appl. Phys. Lett.* **50**, 1098 (1987).
61. Y. J. Chen, E. S. Koteles, B. S. Elman and C. A. Armiento, *Phys. Rev. B* **36**, 4562 (1987).
62. J. W. Little, J. K. Whisnant, R. P. Leavitt and R. A. Wilson, *Appl. Phys. Lett.* **51**, 1786 (1987).
63. H. Q. Le, J. J. Zayhowski and W. G. Goodhue, *Appl. Phys. Lett.* **50**, 1518 (1987).
64. E. E. Mendez, F. Agulló-Rueda and J. M. Hong, *Phys. Rev. Lett.* **60**, 2426 (1988).
65. P. Voisin, J. Bleuse, C. Bouche, S. Gaillard, C. Alibert and A. Regreny, *Phys. Phys. Lett.* **61**, 1639 (1988).
66. J. Bleuse, P. Voisin, M. Allovon and M. Quillec, *Appl. Phys. Lett.* **53**, 2632 (1988).
67. F. Agulló-Rueda, E. E. Mendez and J. M. Hong, *Phys. Rev. B* **40**, 1357 (1989).
68. P. W. A. McIlroy, *J. Appl. Phys.* **59**, 3532 (1986).
69. J. Bleuse, G. Bastard and P. Voisin, *Rev. Phys. Lett.* **60**, 220 (1988).
70. J. A. Brum and F. Agulló-Rueda, *Surf. Sci.* **229**, 472 (1990).
70a. R. P. Leavitt and J. W. Little, *Phys. Rev. B* **42**, 1784 (1990).
71. M. M. Dignam and J. E. Sipe, *Phys. Rev. B* **43**, 4084 (1991).
72. D. A. B. Miller, D. S. Chemla and S. Schmitt-Rink, *Appl. Phys. Lett.* **52**, 2154 (1988).
73. C. Juang and C. Y. Chang, *Appl. Phys. Lett.* **58**, 1527 (1991).
74. D. S. Citrin and Y.-C. Chang, *Appl. Phys. Lett.* **59**, 582 (1991).
75. F. Hache, D. Ricard and C. Flytzanis, *Appl. Phys. Lett.* **55**, 1504 (1989).
76. D. Cotter, H. P. Girdlestone and K. Moulding, *Appl. Phys. Lett.* **58**, 1455 (1991).
77. D. Cotter, M. G. Burt and H. P. Girdlestone, *Semicond. Sci. Technol.* **5**, 631 (1990).
78. M. Murdocca, *A Digital Design Methology for Optical Computing*, The MIT Press, Cambridge, MA (1990).
79. G. D. Boyd, D. A. B. Miller, D. S. Chemla, S. L. McCall, A. C. Gossard and J. H. English, *Appl. Phys. Lett.* **50**, 1119 (1987).
80. R. J. Simes, R. H. Yan, R. S. Geels, L. A. Coldren, J. H. English, A. C. Gossard and D. G. Lishan, *Appl. Phys. Lett.* **53**, 637 (1988).
81. R.-H. Yan, R. J. Simes and L. A. Coldren, *IEEE J. Quantum Electron.* **25**, 2272 (1989).
82. M. Whitehead, A. Rivers, G. Parry, J. S. Roberts and C. Button, *Electron. Lett.* **25**, 984 (1989).
83. S. Tarucha, H. Iwamura, T. Saku and H. Okamoto, *Jpn. J. Appl. Phys.* **24**, L442 (1985).
84. I. Bar-Joseph, K. W. Goossen, J. M. Kuo, R. F. Kopf, D. A. B. Miller and D. S. Chemla, *Appl. Phys. Lett.* **55**, 340 (1989).
85. S. Matsuo, C. Amano and T. Kurokawa, *Appl. Phys. Lett.* **60**, 1547 (1992).
86. K. Berthold, A. F. J. Levi, T. Tanbun-Ek, R. A. Logan and S. N. G. Chu, *Appl. Phys. Lett.* **55**, 1940 (1989).
87. T. H. Wood, E. C. Carr, C. A. Burrus, J. E. Henry, A. C. Gossard and J. H. English, *Electron Lett.* **23**, 916 (1987).
88. G. D. Henshall, P. D. Greene, W. O. Bourne, D. R. P. Guy, D. D. Besgrove, N. Apsley, L. L. Taylor and S. J. Bass, *Semicond. Sci. Technol.* **5**, 512 (1990).
89. K. G. Ravikumar, K. Shimomura, T. Kikugawa, A. Izumi, S. Arai, Y. Suematsu and K. Matsubara, *Electron Lett.* **24**, 415 (1988).
90. M. Cada, B. P. Keyworth, J. M. Glinski, C. Rolland, A. J. SpringThorpe, K. O. Hill and R. A. Soref, *Appl. Phys. Lett.* **54**, 2509 (1989).
91. L. Y. Liu, E. E. Mendez and H. Meier, *Appl. Phys. Lett.* **60**, 2971 (1992).
92. Y. J. Mii, R. P. G. Karunasiri and K. L. Wang, *Appl. Phys. Lett.* **56**, 1986 (1990).
93. E. Martinet, F. Luc, E. Rosencher, Ph. Bis and S. Delaitre, *Appl. Phys. Lett.* **60**, 895 (1992).

94. D. S. Chemla, I. Bar-Joseph, J. M. Kuo, T. Y. Chang, C. Klingshirn, G. Livescu and D. A. B. Miller, *IEEE J. Quantum Electron.* **QE-24**, 1664 (1988).
95. J. E. Zucker, M. Wegener, K. L. Jones, T. Y. Chang, N. Sauer and D. S. Chemla, *Appl. Phys. Lett.* **56**, 1951 (1990).
96. G. Döhler, *Superlattices Microstruct.* **8**, 49 (1990).
97. D. A. B. Miller, *Opt. Quantum Electron.* **22**, S61 (1990).
97a. P. J. Mares and S. L. Chuang, *J. Appl. Phys.* **74**, 1388 (1993).
98. D. A. B. Miller, D. S. Chemla, T. C. Damen, A. C. Gossard, W. Wiegmann, T. H. Wood and C. A. Burrus, *Appl. Phys. Lett.* **45**, 13 (1984).
99. K. Kawashima, K. Fujiwara, T. Yamamoto and K. Kobayashi, *Appl. Phys. Lett.* **60**, 1679 (1992).
100. A. L. Lentine, H. S. Hinton, D. A. B. Miller, J. E. Henry, J. E. Cunningham and L. M. F. Chirovsky, *IEEE J. Quantum Electron.* **25**, 1928 (1989).
101. C. R. Giles, T. Li, T. H. Wood, C. A. Burrus and D. A. B. Miller, *Electron. Lett.* **24**, 848 (1988).
102. N. C. Craft and M. E. Prise, *Laser Focus Wld*, May, 191 (1990).

Chapter Ten

The Photorefractive Effect

10.1. Introduction

Electrooptic photoconductive materials present a very interesting phenomenon, known as the photorefractive effect, with a variety of potential applications in photonics and optoelectronics. The effect was discovered by Ashkin *et al.* [1] in 1966, as degradation of a laser beam traversing a plate of $LiNbO_3$ whose large faces contained the trigonal c-axis of the crystal (Figure 10.1). If the beam is linearly polarized along a direction perpendicular to the c-axis (*ordinary ray*), no degradation occurs. On the other hand, for polarization along the c-axis (*extraordinary ray*), the beam expands along this axis and gives rise to an elongated light spot on an observation screen. The phenomenon involves a light-induced modification of the refractive index. The recorded index pattern can be measured by a compensator technique. A narrow slit moves along the c-direction, keeping parallel to the large faces of the sample. The phase of the beam traversing the slit, and thus the refractive index change, can be recorded as a function of slit displacement. An index profile, as illustrated in Figure 10.2, was obtained for $LiNbO_3$ excited with a He–Ne laser beam (power density: $50\,W/cm^2$). Δn was found to be negative at the illuminated spot and reached a value of 5×10^{-5}. The authors explained the degradation effect as a self-diffraction of the beam by the refractive index pattern generated during propagation. A more thorough discussion of the physical model will be presented in a later section.

Since the first report of the effect, a large number of experiments have been performed in a variety of materials, including inorganic insulators, semiconductors and semiconductor superlattices and very recently organic compounds. Some detailed reviews and monographs on the effect are available to the interested reader [2–9].

At room temperature, the induced photorefractive damage has a persistence ranging from microseconds or less for semiconductors to months or more for $LiNbO_3$. However, the effect is reversible in the sense that can be erased by intense uniform illumination.

In principle, one can consider the photorefractive response as a third-order nonlinear effect relating the index change to the excitation light intensity. As shown in Figure 10.2, the light-induced change in refractive index Δn is negative, leading to defocusing of the beam during propagation (see Figure 10.1), whereas many ordinary (Kerr) nonlinearities cause $\Delta n > 0$ and lead to beam focusing. Moreover, at variance with other third-order phenomena, the photorefractive effect is sensitive not to the local light intensity but to its spatial variation (*nonlocal response*). It is mediated by charge transport and its magnitude is

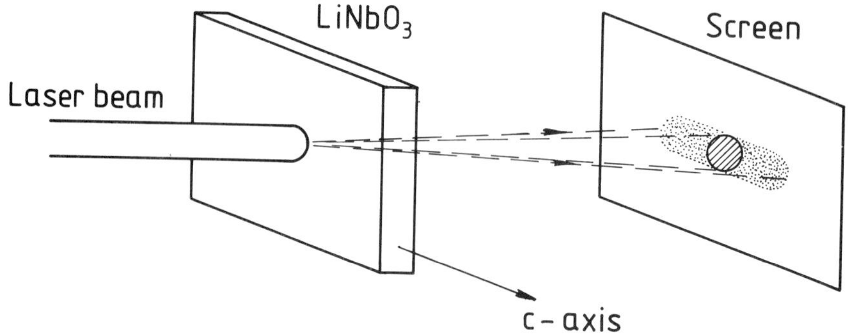

Figure 10.1 *Degradation of a laser beam propagating through a $LiNbO_3$ plate (photorefractive effect).*

usually determined by the Pockels electrooptic effect which is second-order. This latter feature accounts for the very high sensitivity of the effect, which can be observed for illumination levels down to mW/cm^2. It is this property which is mainly responsible for the exciting possibilities of technological application.

10.2. Holographic recording and erasure

A common experimental method to investigate and apply the photorefractive effect is by means of the two-wave-mixing set-up shown in Figure 10.3, which constitutes a

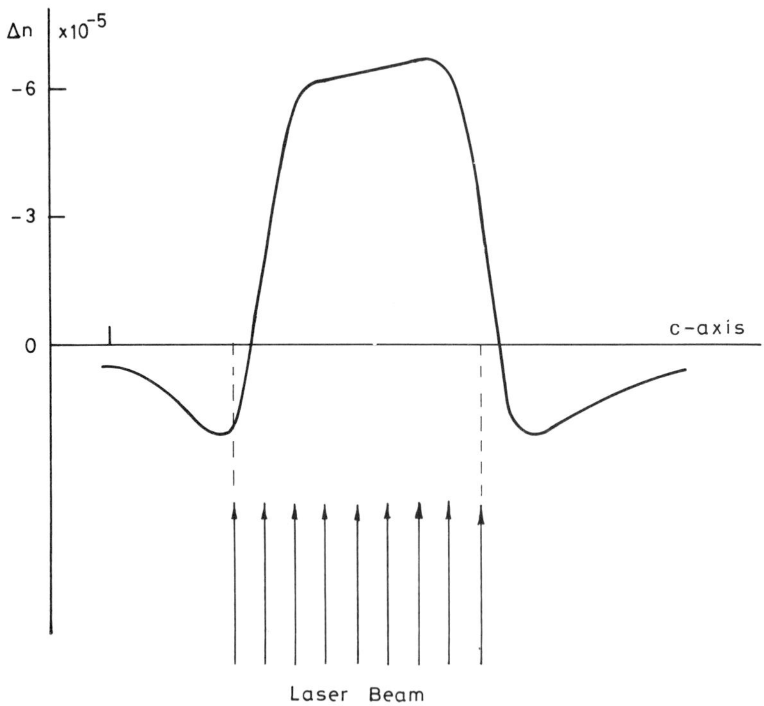

Figure 10.2 *Refractive index profile along the c-axis for the experiment of Figure 10.1 [1].*

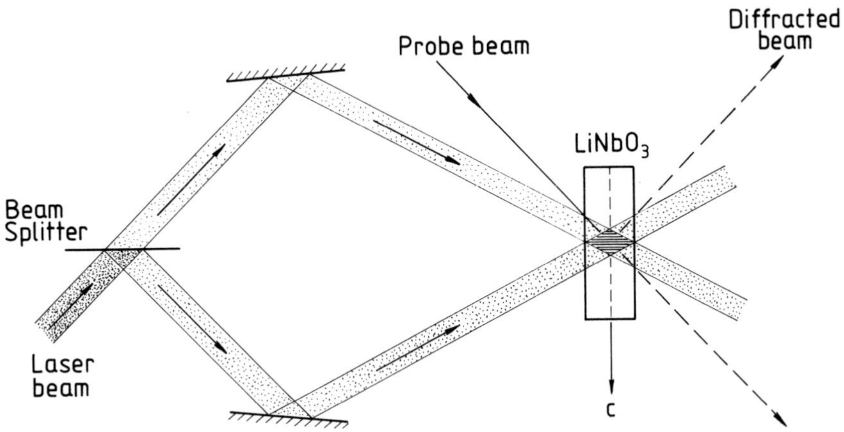

Figure 10.3 ***Schematic set-up for a holographic recording and reading experiment on a photorefractive crystal.***

well-known technique in holography and nonlinear optics [10]. Two coherent laser beams with wavevectors $\mathbf{k}_1$ and $\mathbf{k}_2$ are made to interfere inside the crystal and induce a light intensity grating with period $\Lambda = 2\pi/K_g$, $\mathbf{K}_g = \mathbf{k}_1 - \mathbf{k}_2$ being the grating wavevector. Then, the photorefractive effect gives rise to an associated refractive index pattern with the same period, but generally not the same phase (see Section 10.6), that acts as a *volume phase-grating*.

From an analogy with the holographic imaging technique, one of the recording beams in Figure 10.3 may be considered as the reference beam, whereas the other one is the signal beam carrying the information. The recorded index grating is a hologram of the object. After turning off the signal beam, the reference beam, at the Bragg condition, acts as a probe to read the hologram and reconstruct the signal wave. In fact, any beam, even with different frequency, obeying the Bragg condition will be able to read the hologram. A more thorough discussion of this reading process is given in Section 10.3. In summary, the photorefractive effect provides a method to perform the recording and reconstruction (reading) stages of holography in real time, without the intermediate chemical processing of the hologram needed for photographic films or dichromated gelatin.

As already mentioned in Section 10.1, an intense uniform light beam can be used to erase the grating. The erasure rate depends on light intensity as well as on the particular material and physicochemical state. The erasability of the photorefractive grating is an additional advantage, offering the possibility of *in-situ* reconfigurable holograms for optical computing and information processing applications.

Thermal excitation can also erase the grating. Thermal decay rates at room temperature offer a very wide range of values dependent on material, structural perfection and doping. Heating of the sample causes homogeneous liberation of carriers that enhances the erasure rate. Very long room-temperature storage times (from seconds to months) have been reported for good insulating crystals such as $LiNbO_3$ and $BaTiO_3$. At the other end of the scale, the dark relaxation time for semiconductor materials is much shorter and can be widely controlled by properly adjusting the doping. For GaAs and InP, the time can be varied from milliseconds to even picoseconds.

10.3. Reading the hologram: diffraction efficiency

The diffraction of a beam obeying the Bragg condition provides the read-out of the recorded index grating. From the wavevectors $\mathbf{k}_i$ and $\mathbf{k}_d$, respectively corresponding to the incident and diffracted waves, the grating wavevector $\mathbf{K}_g$ (i.e. the orientation and period of the grating) is immediately determined through

$$\mathbf{K}_g = \mathbf{k}_d - \mathbf{k}_i \qquad \text{(Bragg condition)} \tag{10.1}$$

However, in order to determine the amplitude, the intensity of the diffracted beam has to be measure and compared with theoretical predictions. Kogelnik [11] has studied the diffraction of plane wave, near the Bragg condition, by a sinusoidal grating in an isotropic medium. Other useful references are [12,13]. The main results will be summarized here.

Let us consider the geometry of Figure 10.4, where the diffracted beam grows on moving from the entrance ($z = 0$) to the exit ($z = l$) face (*transmission grating*). A nonslanted grating (fringes perpendicular to the plate faces) has been assumed for simplicity. The index pattern can be written:

$$n = \bar{n} + n_1 \cos K_g x = \bar{n} + \tfrac{1}{2} n_1 (e^{iK_g x} + e^{-iK_g x}) \tag{10.2}$$

the grating vector $\mathbf{K}_g$ being parallel to the X direction. The corresponding dielectric constant pattern is

$$\epsilon = \bar{\epsilon} + \tfrac{1}{2}\epsilon_1 (e^{ik_g x} + e^{-ik_g x}) \tag{10.3}$$

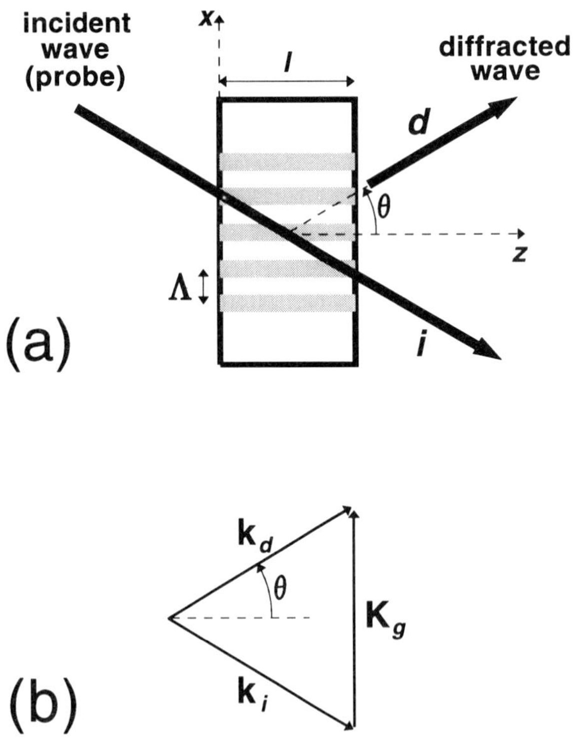

Figure 10.4 *Bragg diffraction by a volume phase grating: (a) schematic diagram; (b) wavevector diagram.*

From $n^2 = \epsilon$, one obtains

$$\bar{n}^2 = \bar{\epsilon}, \qquad 2\bar{n}n_1 = \epsilon_1 \tag{10.4}$$

where $\bar{\epsilon}$ and ϵ_1 are assumed to be real, i.e. the medium is *lossless*.

The scalar Helmholtz equation governing the propagation of a monochromatic wave field, $E_y(z, x)\, e^{i\omega t}$, with linear polarization along Y (perpendicular to the incidence plane) is,

$$\nabla^2 E_y + \frac{\omega^2}{c^2} \epsilon E_y = 0 \tag{10.5}$$

with ϵ given by (10.3). In order to solve (10.5), Kogelnik used a coupled wave formalism, i.e. the proposed solution was written as a superposition of two plane waves, corresponding to the incident and diffracted optical fields. Their complex amplitudes A_i and A_d are allowed to depend on the penetration distance z into the material, i.e.

$$E_y = A_i(z)\, e^{i\mathbf{k}_i \cdot r} + A_d(z)\, e^{i\mathbf{k}_d \mathbf{r}} \tag{10.6}$$

with $\mathbf{k}_i$ and $\mathbf{k}_d$ forming the same angle θ with z (inside the material). Higher diffraction orders have been ignored in (10.6) since a thick sample, compared with the grating period $\Lambda = 2\pi/K_g$, is assumed (*volume hologram*).

Substituting (10.6) into (10.5), using (10.3) and ignoring additional exponential terms arising from the ϵE products, the two following second-order differential equations for the amplitudes are obtained,

$$\frac{d^2 A_i}{dz^2} - 2ik_i \cos\theta \frac{dA_i}{dz} + \frac{\omega^2}{c^2} n_1 \bar{n} A_d = 0 \tag{10.7a}$$

$$\frac{d^2 A_d}{dz^2} - 2ik_d \cos\theta \frac{dA_d}{dz} + \frac{\omega^2}{c^2} n_1 \bar{n} A_i = 0 \tag{10.7b}$$

where $k = k_i = k_d = 2\pi/\lambda$ refers to propagation in free space.

It is reasonable to neglect the second derivatives in (10.7), as already discussed in Chapter 4 on nonlinear optics (*slowly varying amplitude approximation*). Then, one finally arrives at the following two coupled equations,

$$\frac{dA_i}{dz} + i\kappa A_d = 0 \tag{10.8a}$$

$$\frac{dA_d}{dz} + i\kappa A_i = 0 \tag{10.8b}$$

where

$$\kappa = \frac{n_1 \omega}{2c \cos\theta} \tag{10.9}$$

is a coupling constant. If $\kappa = 0$, i.e. $n_1 = 0$, the optical field amplitudes remain constant along z.

The general solution of (10.9) is

$$A_i(z) = R_1\, e^{\gamma_1 z} + R_2\, e^{\gamma_2 z} \tag{10.10a}$$

$$A_d(z) = S_1\, e^{\gamma_1 z} + S_2\, e^{\gamma_2 z} \tag{10.10b}$$

with

$$\gamma_{1,2} = \pm i\kappa \tag{10.10c}$$

To find particular solutions, the appropriate boundary conditions have to be specified. For a transmission hologram, as depicted in Figure 10.4,

$$A_{\mathrm{i}}(0) = A_0, \qquad A_{\mathrm{d}}(0) = 0 \tag{10.11}$$

Then, after some simple algebra, the amplitudes of the incident and diffracted waves result,

$$A_{\mathrm{i}}(l) = \left(\frac{A_0}{2}\right)\{\mathrm{e}^{i\kappa l} + \mathrm{e}^{-i\kappa l}\} = A_0 \cos \kappa l \tag{10.12a}$$

$$A_{\mathrm{d}}(l) = \left(\frac{A_0}{2}\right)\{\mathrm{e}^{-i\kappa l} - \mathrm{e}^{i\kappa l}\} = -iA_0 \sin \kappa l \tag{10.12b}$$

l being the sample thickness. Obviously $A_{\mathrm{i}}A_{\mathrm{i}}^* + A_{\mathrm{d}}A_{\mathrm{d}}^* = A_0^2$ as required by energy conservation.

The diffraction efficiency η is defined as

$$\eta = \frac{I_{\mathrm{d}}}{I_{\mathrm{i}}} = \frac{|A_{\mathrm{d}}A_{\mathrm{d}}^*|}{|A_{\mathrm{i}}A_{\mathrm{i}}^*|} \tag{10.13}$$

that for the solution described by equations (10.12) reduces to

$$\eta = \sin^2 \kappa l = \sin^2\left(\frac{\pi n_1 l}{\lambda \cos \theta}\right) \tag{10.14}$$

According to (10.14), the diffraction efficiency experiences periodic variations between 0 and 1 with slab thickness l, as a consequence of the successive stages of energy conversion from the incident to the diffracted wave and *vice versa*. Values $\eta \approx 1$ can be obtained for visible light incident on a plate with $l = 1$ mm and an index modulation amplitude $n_1 = 10^{-4}$. For weak coupling or a relatively thin sample, equation (1.14) reduces to

$$\eta \approx \left(\frac{\pi^2 l^2}{\lambda^2 \cos^2 \theta}\right) n_1^2 \tag{10.15}$$

establishing a proportional relationship between η and the square of n_1. The minimum η that can be measured is limited by light scattering from material imperfections and background noise. For good crystalline samples one may reach down to $\eta \approx 10^{-8}$.

A more general analysis including wave absorption, parallel polarization, off-Bragg geometry and slanted gratings is given in references [11–13]. It should be mentioned here that the diffraction process has a very strong wavelength and angular selectivity. This is relevant to evaluate the capacity of a photorefractive sample to store different holograms by using either *angular* or *wavelength multiplexing* (see Chapter 11).

Thumb rule values for the full half-widths are

$$\Delta\theta_{1/2} \approx \frac{\Lambda}{l} \tag{10.16a}$$

$$\frac{\Delta\lambda_{1/2}}{\lambda} \approx \frac{\Lambda}{l} \cot \theta \tag{10.16b}$$

so that for $\lambda = 0.5\,\mu$m, $l = 1$ mm and $\theta = 14.5°$ ($\Lambda = 1\,\mu$m), they yield $\Delta\theta_{1/2} = 10^{-3}$, $\Delta\lambda_{1/2}/\lambda \approx 4 \times 10^{-3}$.

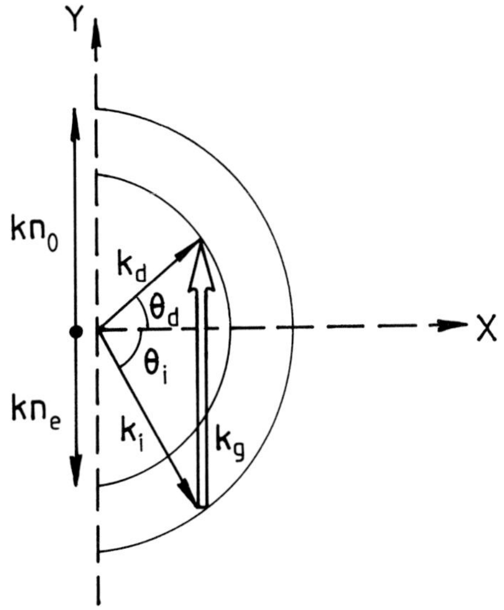

Figure 10.5 *Wavevector diagram for anisotropic diffraction.*

Anisotropic diffraction

For an anisotropic medium the mathematics describing the diffraction of light by the grating becomes more complex. However, the results obtained here are still valid, if one assumes that the incident and diffracted waves are polarized along a principal direction, and n and ϵ refer to that polarization state. On the other hand, the possibility exists that the polarization states of the incident and diffracted beams be different (e.g. one ordinary and the other extraordinary for a uniaxial crystal). Such *anisotropic diffraction* is discussed in reference [4] and provides interesting applications. The Bragg wavevector diagram for this effect in a uniaxial crystal with optic axis along Z is shown in Figure 10.5. In this example the grating wavevector is parallel to Y. Incident and diffracted waves are assumed to propagate in the XY plane and present ordinary and extraordinary polarization, respectively. The extremes of the corresponding wavevectors lie on the circumferences with radii ω/cn_0 (for the incident wave) and ω/cn_e (for the diffracted wave) and obey the Bragg rule, $\mathbf{k}_i + \mathbf{K}_g = \mathbf{k}_d$ (*momentum conservation*). This geometrical construction determines both the angle of incidence, θ_i, and of diffraction, θ_d, with regard to the X-axis. For this anisotropic diffraction to occur, it is necessary that the applied field provides the off-diagonal terms in the dielectric tensor through electrooptic coupling.

10.4. Microscopic mechanisms for the photorefractive effect

The physical mechanism now generally accepted for the photorefractive effect can be discussed with reference to Figure 10.6. Other models mentioned in the literature, such as those proposed by Chen [14], Johnston [15], Amodei [16] and Glass *et al.* [17], correspond to partial approximations to the problem and are mostly of historical interest.

The electrooptic material has both donor and acceptor centres, which may correspond to two valence states of the same impurity (or defect), such as Fe^{2+} (donor) and Fe^{3+}

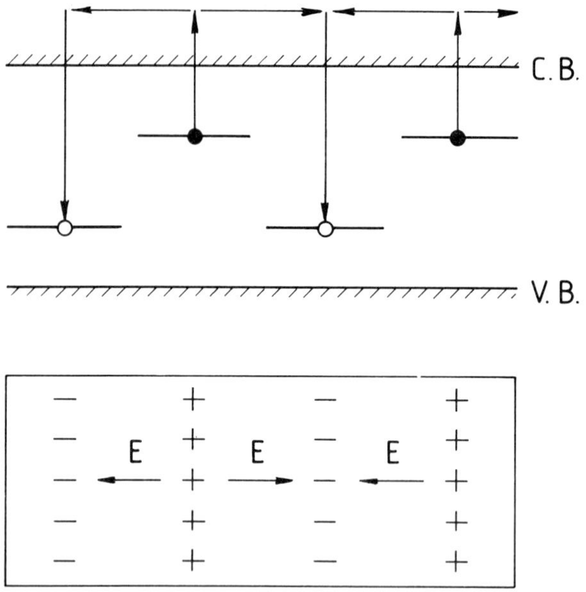

Figure 10.6 *Schematic diagram illustrating the microscopic mechanism for the photorefractive effect.*

(acceptor). Nonuniform illumination with light of suitable wavelength ionizes the donors and generates free carriers (either electrons or holes, or both). They move through the conduction (and/or valence) band and are finally trapped at the acceptors. As a result, a charge redistribution between the illuminated and dark areas is induced. The associated electric field causes a refractive index pattern via the electrooptic coupling. Although most experiments rely on the Pockels effect, the active role of the Kerr effect has also been demonstrated [18].

For the most common Pockels case, the local change in refractive index can be written as in equation (3.11), i.e.

$$\Delta\left(\frac{1}{n^2}\right)_{ij} = r_{ijk}E_k \tag{10.17}$$

where **E** is now the space-charge electric field. For a grating wavevector along a principal axis and light in a polarization *eigenstate*, expression (10.17) reduces to

$$\Delta\left(\frac{1}{n^2}\right) = rE \tag{10.18}$$

r being the appropriate electrooptic coefficient and E the modulus of the field.

Clamped or unclamped coefficients

The question of whether free or clamped values for the coefficients r should be used in (10.17) has been addressed in the literature and it is still a matter of discussion. In a general case, neither the clamped nor the unclamped coefficients should be used. They must be calculated for each particular crystal symmetry and orientation, taking into account the rotation of the optical axes caused by shear deformations [19].

10.4.1. Charge-transport processes

A key step in the photorefractive process is the carrier transport from the bright to the dark areas. Contributions to it come from: (a) carrier diffusion due to nonhomogeneous population, (b) electric-field induced drift and (c) photovoltaic effect. The relative importance of each transport mechanism depends on material as well as recording conditions. As an example photovoltaic drift, which implies an asymmetrical light excitation of electrons, is very important for $LiNbO_3$ or $LiTaO_3$ but is not operative in other oxides such as BSO and semiconductors.

Assuming electron transport, the photovoltaic contribution to the current density along the photovoltaic axis is written

$$j_{ph} = \chi_{ph} I N_D = \chi_{ph} h\nu \Phi N_D \tag{10.19}$$

where χ_{ph} is a photovoltaic coefficient, N_D the donor concentration, and Φ the photon flux ($\Phi = I/h\nu$, I being the light intensity). In a more rigorous analysis (not used in this book), χ_{ph} is a third-rank tensor [20]. The photovoltaic current density (10.19) can be attributed to an equivalent applied field E_{ph} (the equivalence is not strict and one should be careful when using it [21]) given by

$$E_{ph} = \frac{\chi_{ph} N_D I}{q\mu \mathcal{N}} \tag{10.20}$$

where $\mathcal{N}$ is the carrier concentration, μ the carrier mobility and q the electronic charge (absolute value). Since $\mathcal{N}$ results from a balance between photoionization and trapping process, one can write

$$\mathcal{N} = \frac{\sigma_i \Phi N_D}{\gamma_t N_A} \tag{10.21}$$

N_A being the acceptor concentration, σ_i the photo-ionization cross-section and γ_t a trapping coefficient. Substitution of (10.21) into (10.20), leads to

$$E_{ph} = \frac{\chi_{ph} \gamma_t N_A h\nu}{q\mu\sigma_i} \tag{10.22}$$

For the diffusion and electric-field induced drift, carrier mobility μ is the relevant parameter. The electron current density arising from the latter process is

$$j_E = q\mu \mathcal{N} E \tag{10.23}$$

and that resulting from carrier diffusion

$$j_D = q\mathcal{D}_C \nabla \mathcal{N} = k_B T \mu \nabla \mathcal{N} \tag{10.24}$$

where $\mathcal{D}$ is the diffusion coefficient and use has been made of the Einstein relationship $\mathcal{D}_C = (k_B T/q)\mu$, k_B being the Boltzmann constant. In (10.23) and (10.24), μ is the electron mobility.

For hole transport, (10.23) is still valid using the corresponding hole mobility, whereas a minus sign has to be added to the right side of (10.24).

In most materials, both types of carrier contribute to the current densities and to the photorefractive response, although usually one of them is dominant for a given excitation wavelength and physico-chemical conditions (doping, prior treatments etc.).

10.4.2. Transport lengths

Instead of the fundamental transport parameters $(\chi, \mathscr{D}_C, \mu)$, it is sometimes convenient to use characteristic transport lengths: L_{ph} (photovoltaic), L_D (diffusion) and L_E (drift). For each case, they measure the average distance travelled by carriers during their life-time τ (before being trapped). Expressions for these transport lengths are as follows:

$$L_{ph} = \chi_{ph} \frac{h\nu}{q s_i} \tag{10.25a}$$

$$L_D = (\mathscr{D}_C \tau)^{1/2} \tag{10.25b}$$

$$L_E = \mu \tau E \tag{10.25c}$$

L_D and L_E are related through the expression

$$\frac{L_D^2}{L_E} = \frac{k_B T}{qE} \tag{10.26}$$

so they convey essentially the same information.

Table 10.1 summarizes representative transport parameters for some useful photorefractive materials. Note the much shorter transport length L_D (and consequently $\mu\tau$ and L_E) of $LiNbO_3$ and $BaTiO_3$ in comparison with BSO and GaAs.

10.4.3. Carrier dynamics

Carrier transport may result from two different microscopic mechanisms. For some materials and temperature ranges, carriers move in a coherent or band-like way with large mobilities. This is the case for semiconductors such as GaAs and InP, where mobilities as high as $10^4\,cm^2/(V\,s)$ (at room temperature) are reached in pure samples, as well as for many insulating materials. In other cases, carriers are self-trapped in the lattice as *small polarons* [22,23], and move by thermally assisted hopping, as electrons in $LiNbO_3$ and some other oxides. As a consequence, mobilities are typically lower than $1\,cm^2/(V\,s)$ and increase with temperature at variance with band-like motion.

Intrinsic mobilities can be markedly reduced by existing shallow traps that temporarily capture carriers and prevent their motion. Therefore, reported data offer a very large scatter of values, mostly due to differences in material growth and preparation. This is one of the sources accounting for the lack of reproducible photorefractive responses. At variance with μ, values for coefficients s and γ_t are not generally available and are often estimated from simple geometrical arguments.

Table 10.1 *Transport parameters.*

Material	μ_e ($cm^2/(V\,s)$)	μ_h ($cm^2/(V\,s)$)	L_D (μm)	L_{ph} (μm)	$\mu\tau$ (cm^2/V)
$LiNbO_3$	0.1		5×10^{-3}	5×10^{-4}	10^{-11}
$BaTiO_3$	0.5				10^{-10}
BSO	0.03		3.5		4×10^{-7}
GaAs	8.000	400	0.5	0.5	10^{-7} *

* This value refers to electrons.

10.5. Rate equations

It will be assumed in this section that the light interference pattern on the material is not influenced by the development of the grating, i.e. grating formation and wave propagation can be handled independently (coupling is neglected). Then, the growth of the index grating can be described by a set of material equations first put forward by Kuktharev and coworkers [24]. This theoretical scheme, based on standard semiconductor physics, assumes coherent carrier motion and summarizes the present state of the art [25].

Let us consider a plate of photorefractive material as in Figure 10.4, with infinite dimensions along x and y directions. For simplicity, we will consider that only one type of carrier is operative (*monopolar transport*) as well as that a single photorefractive centre is involved. The centre is supposed to be in two stable charge states acting as donors and acceptors for the free carriers. Assuming electron transport, the rate equations governing the time evolution of the free carrier, donor and acceptor concentrations, can be written

$$\frac{\partial \mathcal{N}}{\partial t} = (s_{\mathrm{i}}\Phi + s_{\mathrm{th}})N_{\mathrm{D}} - \gamma_{\mathrm{t}}\mathcal{N}N_{\mathrm{A}} + \frac{1}{q}\frac{\partial j}{\partial x} \tag{10.27a}$$

$$\frac{-\partial N_{\mathrm{D}}}{\partial t} = \frac{\partial N_{\mathrm{A}}}{\partial t} = s_{\mathrm{i}}\Phi N_{\mathrm{D}} - \gamma_{\mathrm{t}}\mathcal{N}N_{\mathrm{A}} \tag{10.27b}$$

s_{th} being a thermal ionization rate. It will be neglected from here on since for normal operating conditions, $s_{\mathrm{th}} \ll s_{\mathrm{i}}\Phi$. One should then add the continuity equation

$$\frac{\partial \rho}{\partial t} = -\frac{\partial j}{\partial x} \tag{10.27c}$$

and the Poisson equation

$$\frac{\partial E}{\partial x} = \frac{\rho}{\epsilon_0 \epsilon} \tag{10.27d}$$

where the charge ρ and the current density j are given by

$$\rho = -q(\mathcal{N} + N_{\mathrm{D}} - N_{\mathrm{D}}(0)) \tag{10.28a}$$

and

$$j = q\mu\mathcal{N}E + \mathcal{D}_{\mathrm{C}}q\left(\frac{\partial \mathcal{N}}{\partial x}\right) + \chi_{\mathrm{ph}}IN_{\mathrm{D}} \tag{10.28b}$$

including all transport mechanisms. In this latter equation E stands for the sum of both, the applied and the space-charge fields.

Feinberg approach

A different model has been put forward by Feinberg *et al.* [26], where carriers are assumed to experience successive jumps among trap sites, with a rate dependent on jump distance. However, although it uses a different language and a different set of physical parameters, it can be shown [27] to be essentially equivalent to the Kukhtarev *et al.* model and will not be discussed further here.

10.5.1. Main approximations

Equations (10.27) are nonlinear and a completely general solution is not presently available, so that numerical methods and computer simulations have often been used [28–32]. On the other hand, different approximation schemes, leading to linearization of the equations, have been essayed to obtain analytical solutions.

Here, we shall briefly discuss the solution of equations (10.27) for two relevant cases: (a) steady-state situation under sinusoidal light excitation and (b) erasure kinetics under homogeneous light flux. An analytical solution for both cases has been obtained under the assumption of a low modulation of the light intensity and trap patterns as well as a relatively small free carrier concentration, $\mathcal{N} \ll N_D, N_A$. This second assumption permits us to write the charge density as $\rho = q\{N_A - N_A(0)\}$ and it applies to continuous wave experiments where light density powers are low or moderate. The first hypothesis allows linearization of the equations, and so the expected solutions for $\mathcal{N}$, N_D and N_A (and ρ and E) only contain zero- and first-order terms in a Fourier expansion.

One key point in the mathematical procedure is to assume that the light-induced $\mathcal{N}$ is always in equilibrium with the instantaneous N_D and N_A i.e. $\partial\mathcal{N}/\partial t = 0$ (*pseudoequilibrium condition*). Then, equation (10.27a) can be written

$$\sigma_i \Phi N_D - \gamma_t \mathcal{N} N_A = \frac{1}{q}\frac{\partial j}{\partial x} \tag{10.27e}$$

10.5.2. Steady-state recording (low modulation depths, $m \ll 1$)

During recording, the excitation photon flux obtained from the interference of two light waves of complex amplitudes A_1 and A_2 is

$$\Phi = \Phi_0(1 + m\cos K_q x) \tag{10.29a}$$

where the *modulation depth m* of the interference pattern is given by

$$m = \frac{A_1 A_2^* + A_1^* A_2}{|A_0|^2} = \frac{2|A_1||A_2|}{|A_0|^2} \tag{10.29b}$$

with $|A_0|^2 = |A_1|^2 + |A_2|^2$. The condition $m \ll 1$ is assumed to assure the validity of a linear approximation (*linear regime*).

In the steady state, time derivatives are zero and equations (10.27), with the approximations of Section 10.5.1, simplify to

$$\sigma_i \Phi N_D - \gamma_t \mathcal{N} N_A = 0 \tag{10.30a}$$

$$\frac{\partial j}{\partial x} = 0 \tag{10.30b}$$

$$\frac{\partial E}{\partial x} = -\frac{q}{\epsilon_0 \epsilon}\{N_D - N_D(0)\} \tag{10.30c}$$

The solution for the space-charge field E, in the linear regime, takes the form

$$E = E_S m \cos(K_g x + \phi) \tag{10.31a}$$

ϕ being the dephasage with regard to the light interference pattern. After some lengthy but straightforward algebra, one arrives at the following expressions for E_S and ϕ [20],

$$E_S = E_q\left(\frac{E_D^2 + (E_{ph} + E_a)^2}{(E_q + E_D)^2 + [E_{ph}N_r/N_D(0) + E_a]^2}\right)^{1/2} \tag{10.31b}$$

and

$$\tan\phi = \frac{(E_q + E_D)E_D + [E_{ph}^2 N_r/N_D(0)] + E_{ph}E_a[1 + N_r/N_D(0)] + E_a^2}{E_q(E_{ph} + E_a) + E_{ph}E_D N_r/N_D(0)} \tag{10.31c}$$

where E_a and E_{ph} are, respectively, the *applied* and *photovoltaic fields*. $E_D = K_g \mathscr{D}/\mu = 2\pi kT/q\Lambda$ is the so-called *diffusion field*, for which diffusion and drift currents become equal. It depends on grating spacing, but not on material parameters ($E_D \sim 1\,\mathrm{kV/cm}$ for $\Lambda = 1\,\mu\mathrm{m}$, at room temperature). $N_r = (1/N_D) + (1/N_A)$ is a *reduced* trap concentration and $E_q = qN_r/\epsilon\epsilon_0 K_g$, a *limiting space-charge field*, determined by the available concentration N_r.

Equations (10.31) are very general since all transport processes have been included. They show that the amplitude of the space-charge field is proportional to the modulation depth m, but is independent of the average intensity of the light pattern. E_S, which is a space-charge field *normalized* to $m = 1$, depends exclusively on the material parameters and grating spacing. On the other hand, the dephasage ϕ does not depend on either the average or the modulated light intensity.

In many practical cases the limiting field E_q is very high in comparison with E_D, E_{ph} and E_a. Then, assuming $N_r/N_D(0) < 1$, equations (10.31) simplify to

$$E_S \approx \{E_D^2 + (E_{ph} + E_a)^2\}^{1/2} \tag{10.32a}$$

$$\tan\phi = \frac{E_D}{E_{ph} + E_a} \tag{10.32b}$$

Expression (10.32a) yields $E_S = E_D$, $E_S = E_{ph}$, and $E_S = E_a$ when the diffusion, photovoltaic or drift mechanisms are respectively dominant. Therefore, a clear physical meaning is attached to those transport fields.

Depending on material and physical parameters, a span of steady-state field values is found, which typically covers the range 10^3–10^5 kV/cm. On the other hand, (10.32b) indicates that a dephasage $\phi \approx \pi/2$ between light and index gratings develops when diffusion is dominant ($E_D \gg E_{ph}$, E_a), whereas $\phi \sim 0$ if $E_D \ll E_a$, E_{ph}.

The occurrence of a dephasage between the index and light patterns is a peculiar feature of the photorefractive nonlinearity and has deep implications for the coupling between the interfering optical waves (see Section 10.7).

π/2-shifted and unshifted components

The space-charge field pattern (10.31a) is equivalent to the superposition of two gratings, one in-phase with the light and having an amplitude $mE_S \cos\phi$, and another one shifted $\pi/2$ and with amplitude $mE_S \sin\phi$. They can also be considered as the *real* and *imaginary* parts, respectively, of a complex space-charge field grating $mE_S\,e^{i\phi}$.

10.5.3. Kinetics of optical erasure

Initial conditions for the donor and acceptor concentrations are those corresponding to a recorded grating, i.e.

$$N_D = N_D^0 + M\cos K_g x = N_D^0(1 + M_D \cos K_g x) \tag{10.33a}$$

$$N_A = N_A^0 - M\cos K_g x = N_A^0(1 - M_A \cos K_g x) \tag{10.33b}$$

where

$$M_{\mathrm{D}} = \frac{M}{N_{\mathrm{D}}^0}, \qquad M_{\mathrm{A}} = \frac{M}{N_{\mathrm{A}}^0} \tag{10.34}$$

with M_{D}, $M_{\mathrm{A}} \ll 1$ (linear approximation). The time-dependent solution of the rate equations (10.27) under a uniform light flux Φ_0 (or intensity $I_0 = h\nu\Phi_0$) can be written in the form (10.33) by allowing time-dependent amplitudes M_{D} or M_{A} and the occurrence of a dephasage $\phi(t)$, [33], i.e.

$$N_{\mathrm{D}}(x,t) = N_{\mathrm{D}}^0\{1 + M_{\mathrm{D}}(t)\cos[K_{\mathrm{g}}x + \phi(t)]\} \tag{10.35a}$$

$$N_{\mathrm{A}}(x,t) = N_{\mathrm{A}}^0\{1 - M_{\mathrm{A}}(t)\cos[K_{\mathrm{g}}x + \phi(t)]\} \tag{10.35b}$$

leading to a charge density,

$$\rho(x,t) = -M(t)q\cos[K_{\mathrm{g}}x + \phi(t)] \tag{10.36a}$$

and a space charge field,

$$E(x,t) = -M(t)\left(\frac{q}{\epsilon_0\epsilon K_{\mathrm{g}}}\right)\sin[K_{\mathrm{g}}x + \phi(t)] \tag{10.36b}$$

Now, at variance with the steady-state case, differences arise in the mathematical procedures [33], depending on whether transport lengths L_{D}, L_{E} and L_{ph} are lower than or comparable with the grating period Λ. The simplest case corresponds to small transport lengths, which is equivalent to neglect carrier transport with regard to trapping in (10.27a). Under this assumption the pseudoequilibrium carrier concentration is

$$\mathcal{N}(x,t) = \bar{\mathcal{N}} + \mathcal{N}_1\cos(K_{\mathrm{g}}x + \phi) \tag{10.37a}$$

with the average value,

$$\bar{\mathcal{N}} = \frac{s_{\mathrm{i}}\Phi_0 N_{\mathrm{D}}}{\gamma_{\mathrm{t}} N_{\mathrm{A}}} \tag{10.37b}$$

and the modulation amplitude,

$$\mathcal{N}_1 = \mathcal{N}M\left(\frac{1}{N_{\mathrm{D}}^0} + \frac{1}{N_{\mathrm{A}}^0}\right) = \frac{\mathcal{N}M}{N_{\mathrm{r}}} \tag{10.37c}$$

Substituting (10.37a) and (10.35a,b) into (10.27e), one obtains for the time-dependent grating parameters,

$$M(t) = M(0)\,\mathrm{e}^{-(a+bk_{\mathrm{g}}^2)t} \tag{10.38a}$$

$$\phi(t) = vK_{\mathrm{g}}t \tag{10.38b}$$

with

$$a = \frac{q\mu\bar{\mathcal{N}}}{\epsilon_0\epsilon} = \frac{\sigma_{\mathrm{i}}}{\epsilon\epsilon_0} \tag{10.38c}$$

$$b = \mathcal{D}_{\mathrm{C}}\bar{\mathcal{N}}\left(\frac{1}{N_{\mathrm{D}}^0} + \frac{1}{N_{\mathrm{A}}^0}\right) = \frac{\mathcal{D}\bar{\mathcal{N}}}{N_{\mathrm{r}}} \tag{10.38d}$$

$$v = \frac{\mu\bar{\mathcal{N}}E_{\mathrm{a}}}{N_{\mathrm{r}}} + \frac{\chi_{\mathrm{ph}}I_0}{q} \tag{10.38e}$$

where σ_{i} is the light-induced conductivity (*photoconductivity*).

In summary, the grating amplitude decays exponentially, with a time constant

$$t = \frac{1}{a + bK_g^2} \tag{10.39}$$

known as the *dielectric relaxation time*. It has a clear physical meaning, namely the time needed by the photocurrent to neutralize the space-charge grating. The decay rate is proportional to $\mathcal{N}$, and consequently to the light intensity. At the same time, the equal-phase fringes move in the $+C$ direction at a constant speed v determined by the applied and photovoltaic fields. In practice, decay curves are often nonexponentials, very likely due to the contribution of several active centres and traps (see below).

One should note that the erasure rate for small K_g (wide gratings) is independent of grating spacing. Then, the *erasure time* is $t = 1/a = \epsilon_0\epsilon/q\mu\bar{\mathcal{N}}$, i.e. inversely proportional to the light intensity. More meaningful than t_E is the energy required to erase a given grating, $W_E = I_0 t$. Illustrative values are 10^{-1} J/cm^2 for $LiNbO_3$ and 10^{-4} J/cm^2 for BSO (see Section 10.11).

For arbitrary gratings the above formulae predict a linear increase of the decay rate on the square of the grating wavevector, that has been experimentally confirmed in a reduced range of K_g [34].

The above model can be improved if an arbitrary transport length for the carriers is assumed. Although the decay curves are still exponentials, the dependence on the grating wavevector is more complex. Depending on physical parameters, the erasure rate increases or decreases with increasing K_g in accordance with experiment [35,36].

Dark erasure

The kinetics of thermal (*dark*) erasure can be described with the same formalism used for that induced optically [37]. In the single-centre model used in this section, one has only to keep the thermal ionization rate $s_{th}N_D$, and ignore the photoionization rate $s_i\Phi N_D$ in equation (10.27a). Then, equations (10.38) are valid, as long as one takes $\mathcal{N} = s_{th}N_D/\gamma_t N_A$. In particular, one should expect an exponential dark decay with a time constant given by (10.39) after substituting s_{th} for s_i (*dark relaxation time*). This is not often the case [38,39], as remarked for the optical erasure case. However, the thermal ionization rate of the donors could be obtained, in principle, from the grating decay curves measured at various temperatures. It may provide information on the depth of the donor energy levels below the conduction band minimum.

Some complicating features

The theoretical analysis presented so far has assumed a single active centre. The situation is more complex if several centres participate in the process [40–43]. These centres may be either optically or thermally active or both. Then, the recording and erasure curves may become nonmonotonic and present peculiar features due to the competition among the various space-charge gratings. At some stages of the recording, growth rates for the field amplitude can be negative, implying some grating anihilation processes. Moreover, the recording curve can be, in general, light-intensity dependent (in a scale of deposited energy). On the other hand, erasure curves may present stages of increasing grating amplitude.

Also, in many experimental situations (e.g. for BSO), both electrons and holes contribute to the photorefractive dynamics (*bipolar transport*) [44–47]. Here, the competition of the two opposite charge carriers may lead to a decrease in the grating recording efficiency and to similar effects as those described for multiple active centres.

10.6. Fixing of holographic gratings

Optical or light-induced erasure is detrimental for some applications since it leads to loss of the stored information during multiple readouts. There are several possibilities to avoid the unwanted erasure of the hologram. First, one may read the hologram with light of wavelength outside the efficiency range of the photorefractive response. For example in $LiNbO_3$ infrared light may be used with very little (but not zero) erasure. Other interesting possibility is offered by two-photon excitation at frequencies ω_1 and ω_2 $(\omega_1 \neq \omega_2)$ via a virtual or a real intermediate level (Figure 10.7). The reading process with light of frequency ω_1 (or ω_2) does not erase the grating since no free carriers are generated for that frequency.

Fixing or stabilization of the photoinduced gratings in ferroelectric materials by means of polarization reversal in regions of high space-charge field has been accomplished in $BaTiO_3$ [48], SBN [49] and KTN [50]. The fixed grating can be erased by applying an external field larger than the coercive field of the crystal.

Another useful way to fix a hologram is by transforming the electronic-charge distribution into a correlated ionic-charge pattern. For $LiNbO_3:Fe$ this process has been extensively studied and can be described schematically as follows. During or after hologram recording the sample is heated to temperatures in the range 100–200 °C. There is evidence that during this treatment, protons present in the sample (typically in concentrations $10^{18}\,cm^{-3}$ or more) move to form another grating that neutralizes the previously written electronic grating (*fixing stage*). After cooling to room temperature, a homogeneous illumination erases the electronic grating and brings out a proton replica (*developing stage*).

If the electronic pattern is assumed to be thermally stable, the index modulation of the grating should decay to zero during the fixing stage. The time-constant τ_F is given by the dielectric relaxation time for proton motion, i.e.

$$\tau_F = \frac{\epsilon\epsilon_0}{q\mu_p H} \tag{10.40}$$

$\mu_p = \mathscr{D}_p kT/q$ being the proton mobility and H the average proton concentration.

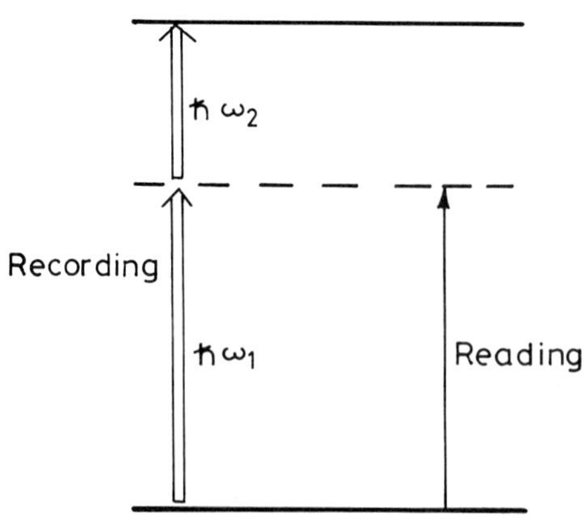

Figure 10.7 *Two-photon recording to avoid erasure during one-photon read-out.*

During the developing stage, the electronic charge grating is erased, leaving the previously formed proton grating. This final level of ionic charge modulation is stable *vs* illumination. It is generally much lower than the initial electronic value.

Detailed theoretical models to account for the kinetics of the fixing process have been put forward by several authors [51,52]. In particular, the behaviour of $LiNbO_3$ has been successfully simulated [52], using the following diffusion parameters for protons: $\mathcal{D}_0 = 0.085\,\text{cm}^2/\text{s}$ (*pre-exponential factor*) and $\mathcal{E}_H = 1.1\,\text{eV}$ (*activation energy*). They had been obtained from independent diffusion experiments. It is expected that careful experiments and comparison with theory should yield interesting information on proton configuration and dynamics, as recently reported for $LiNbO_3$ [53].

Thermal fixing in other oxides (BSO and BTO) has also been observed and investigated but physical mechanisms involved are less well understood [54–57].

10.7. Wave-coupling effects

During the holographic recording process, the two interfering waves are diffracted by the generated index grating and give rise to very interesting energy coupling phenomena. To describe these self-diffraction phenomena it is no longer possible to ignore the wave equation when dealing with the solution of the material rate equations. In other words, material and wave equations are intimately coupled and have to be solved together. Here, the steady situation will be exclusively considered, and again, light absorption and crystal anisotropy will be neglected.

A light interference pattern as (10.39) will be assumed at the entrance ($z = 0$) face. At a depth z in the material, the steady index pattern will be of the general form obtained in Section 10.5, i.e.

$$n = \bar{n} + n_1(z)\cos\{K_g x + \varphi(z) + \phi\} \tag{10.41a}$$

or in complex form

$$\hat{n} = \bar{n} + \hat{n}_1(z)\,e^{iK_g x} \tag{10.41b}$$

with $\hat{n} = n_1\,e^{i(\phi+\varphi)}$. Allowance has been made for the grating amplitude n_1 to depend on depth z. Moreover a shift in the light interference pattern on z is also included through the phase term $\varphi(z)$, whereas the dephasage ϕ between the light and index gratings should be kept constant (see 10.31c). These z-dependencies describe the coupling between light waves and index pattern.

The corresponding dielectric constant grating is

$$\epsilon = \bar{\epsilon} + \epsilon_1(z)\cos\{K_g x + \varphi(z) + \phi\} \tag{10.41c}$$

or

$$\hat{\epsilon} = \bar{\epsilon} + \hat{\epsilon}_1\,e^{iK_g x} \tag{10.41d}$$

with $\bar{\epsilon} = \bar{n}^2$ and $\epsilon_1 = 2\bar{n}n_1$.

The method to solve the wave equation with the x- and z-dependent dielectric constant is a simple generalization of the coupled wave formalism used in Section 10.3. Neglecting higher diffraction orders (*Bragg regime*), the solution is the sum of two interfering plane waves coupled by the Bragg condition. Assuming perpendicular polarization and designating the two complex wave amplitudes as $E_{1+}(z, t)$ and $E_{1-}(z, t)$, one may write for the total complex field amplitude,

$$E = A_{1+}(z)\,\mathrm{e}^{-ik_+ z} + A_{1-}(z)\,\mathrm{e}^{-ik_- z} \tag{10.42}$$

with $A_{1\pm}(z) = |A_{1\pm}|\,\mathrm{e}^{i\Psi_\pm(z)}$, and $k_+ = k_- = 2\pi\bar{n}/\lambda$. By following exactly the same procedure outlined in Section 10.3, (*slowly varying amplitude approximation*), one obtains for the transmission geometry of Figure 10.4,

$$\frac{\mathrm{d}A_{1+}}{\mathrm{d}z} + i\kappa^* A_{1-} = 0 \tag{10.43a}$$

$$\frac{\mathrm{d}A_{1-}}{\mathrm{d}z} + i\kappa A_{1+} = 0 \tag{10.43b}$$

with

$$\kappa = \frac{\hat{n}_1\omega}{2c\cos\theta} \tag{10.44a}$$

$$\hat{n}_1 = \bar{n}^3 r E_s\,\mathrm{e}^{i\phi}\,\frac{A_{1+}^* A_{1-}}{|A_0|^2} \tag{10.44b}$$

E_s is the *normalized* amplitude and ϕ the phase (with regard to the light) of the space charge grating field (see equations (10.31)). The phase φ in the index pattern $\hat{n}_1$ is given by the argument of $A_{1+}^* A_{1-}$. Now, substituting (10.44) into (10.43), one obtains the coupled equations

$$\frac{\mathrm{d}A_{1+}}{\mathrm{d}z} + i\Gamma\,\mathrm{e}^{-i\phi}\,\frac{A_{1+}|A_{1-}|^2}{2|A_0|^2} = 0 \tag{10.45a}$$

$$\frac{\mathrm{d}A_{1-}}{\mathrm{d}z} + i\Gamma\,\mathrm{e}^{i\phi}\,\frac{A_{1-}|A_{1+}|^2}{2|A_0|^2} = 0 \tag{10.45b}$$

with

$$\Gamma = \frac{2\pi\bar{n}^3 r E_s}{\lambda\cos\theta} \tag{10.46}$$

Multiplying the first equation by A_{1+}^* and adding the resulting equation to its complex conjugate, it comes out

$$\frac{\mathrm{d}I_{1+}}{\mathrm{d}z} + \Gamma\,\frac{I_{1+}I_{1-}}{I_0}\sin\phi = 0 \tag{10.47a}$$

By doing the same with equation (10.45b), one arrives at

$$\frac{\mathrm{d}I_{1-}}{\mathrm{d}z} - \Gamma\,\frac{I_{1+}I_{1-}}{I_0}\sin\phi = 0 \tag{10.47b}$$

The solution of (10.47) is

$$I_{1+}(z) = \frac{I_{1+}(0) I_0\,\mathrm{e}^{-\Gamma z\sin\phi}}{I_{1-}(0) + I_{1+}(0)\,\mathrm{e}^{-\Gamma z\sin\phi}} \tag{10.48a}$$

$$I_{1-}(z) = \frac{I_{1-}(0) I_0\,\mathrm{e}^{\Gamma z\sin\phi}}{I_{1+}(0) + I_{1-}(0)\,\mathrm{e}^{\Gamma z\sin\phi}} \tag{10.48b}$$

These equations show energy transfer between the two beams during propagation.

10.7.1. Beam amplification

For the particular case where $I_{1-} \ll I_{1+}$, equations (10.48) lead to

$$I_{1+}(z) \approx I_{1+}(0) \approx I_0 \tag{10.49a}$$

$$I_{1-}(z) \approx I_{1-}(0)\, \mathrm{e}^{\Gamma z \sin\phi} \tag{10.49b}$$

indicating that the weak beam experiences an exponential growth (*beam amplification*), whereas the strong one becomes only slightly depleted. The *logarithmic gain* per unit length is $\Gamma \sin\phi$, i.e. it is associated with the occurrence of a dephasage $\phi \neq 0$ between the light and index gratings. Alternatively, since $\Gamma \sin\phi \sim E_S \sin\phi$, one can say that the gain is caused by the component of the grating which is shifted by $\pi/2$ with regard to the light. If absorption by the material is taken into account, the above calculated gain is reduced in the value of the absorption coefficient α, i.e. the effective gain is $\Gamma' = \Gamma \sin\phi - \alpha$.

Gain coefficients Γ are large (>10/cm) for $LiNbO_3$, $BaTiO_3$ or $KNbO_3$, and small (<1/cm) for BSO and GaAs. However, a value $\Gamma = 16.3$/cm has been recently reported for GaAs when recording is performed with wavelengths near the band edge [58].

The gain in a material such as BSO can be markedly enhanced by applying alternating electric fields [59] or by oscillating the light interference fringes during recording (e.g. by means of a piezoelectric mirror placed in the path of one of the beams) whereas a DC field is applied [60–62]. Gains as high as 20 or more have been reported for BSO. The increase in Γ can be understood as a resonance effect enhancing the $\pi/2$-shifted component of the complex amplitude of the field.

10.7.2. Fringe bending

Taking into account that

$$\frac{\mathrm{d}A_{1\pm}}{\mathrm{d}z} = i|A_{1\pm}|\,\mathrm{e}^{i\psi_\pm}\frac{\mathrm{d}\psi_\pm}{\mathrm{d}z} + (\mathrm{e}^{i\psi_\pm})\frac{\mathrm{d}|A_{1\pm}|}{\mathrm{d}z} \tag{10.50}$$

one derives from the imaginary components of (10.45)

$$\frac{\mathrm{d}\psi_+}{\mathrm{d}z} + \Gamma\cos\phi\frac{|A_{1-}|^2}{2|A_0|^2} = 0 \tag{10.51a}$$

$$\frac{\mathrm{d}\psi_-}{\mathrm{d}z} + \Gamma\cos\phi\frac{|A_{1+}|^2}{2|A_0|^2} = 0 \tag{10.51b}$$

Subtraction of these equations yields

$$\frac{\mathrm{d}\psi}{\mathrm{d}z} - \Gamma\,\frac{I_{1+} - I_{1-}}{2I_0}\cos\phi = 0 \tag{10.52}$$

where $\psi = \psi_+ - \psi_-$. Equation (10.52) describes the variation with depth z of the light interference pattern. Solution of this equation for the simple case $I_{1-} \ll I_{1+}$ is

$$\Psi = \frac{\Gamma z \cos\phi}{2} \tag{10.53}$$

indicating that the light fringes (and consequently the index fringes) become bent inside the sample.

Role of optical activity

Optical activity occurring in some materials (e.g. sillenites) strongly affects light propagation and photorefractive coupling. A strong effort has been made to study polarization behaviour inside the material and establish optimal conditions for image readout and beam coupling [63,64]. In general, optical activity causes a reduction in wave coupling and diffraction efficiency.

Light-induced scattering

Light scattering by crystal inhomogeneities (*stray light*) may be strongly enhanced by beam interference and coupling effects [65]. In fact, interference between coherent scattered and incident light writes a space-charge grating and causes energy transfer from the incident to the scattered wave (*beam coupling*). The dependence of the gain coefficient on the physical and geometrical parameters gives rise to strongly favoured directions in the scattered light, leading to very anisotropic patterns.

10.8. Short-pulse experiments

Photorefractive experiments with short light pulses are quite useful, not only because of their technological relevance for signal processing and optical computing, but also because they can provide significant information on trap levels and transport parameters. As an example, holographic gratings in BSO-family crystals have been recorded by two-wave mixing of nanosecond [66,67] and picosecond [68] pulses. Comparison with pulsed photoconductivity experiments permits a more reliable analysis of the results. A major conclusion is that carrier mobilities are much larger than those obtained under quasi-continuous (CW) light excitation. The difference has been generally attributed to the presence of shallow traps which limit the apparent mobility in the CW experiments. On the other hand, the high peak intensities used in the short-pulse experiments lead to substantial trap filling and even saturation, enhancing the mobility.

The rigorous analysis of the holographic experiments under pulsed excitation requires the appropriate solution of the Kukhtarev *et al.* equations [10–27]. This has been carried out by Valley [69] for a single active centre and two extreme situations respectively corresponding to the pulse length $\mathscr{T}$ being much larger or smaller than the recombination time τ_R for the carriers. For $\mathscr{T} \gg \tau_R$, the situation can be assimilated to that found in CW experiments. In this case, the limit value for the rise time of the holographic grating formation is τ_R. For pulse duration $\mathscr{T} \ll \tau_R$ the situation is quite different and the photorefractive response depends on the value of the diffusion time τ_D relative to τ_R. τ_D is defined as

$$\tau_D = \frac{q\Lambda^2}{4\pi\mu k_B T} = \frac{\Lambda^2}{4\pi^2 \mathscr{D}_C} \tag{10.54}$$

i.e. the time required for the carrier to reach a distance $\Lambda/2\pi$. For $\tau_D \gg \tau_R$ or $\Lambda \gg L_D$, no grating can be recorded in the material. If $\Lambda < L_D$, recording is possible, the grating amplitude being dependent on the relative value of those distances. On the other hand, the response time is influenced by the trapping time τ_T at existing shallow traps.

10.9. Four-wave mixing experiments: phase conjugation

In addition to two-wave mixing, four-wave mixing experiments can also be performed on photorefractive materials, such as other nonlinear media (e.g. local Kerr materials) [70–73]. Particularly interesting is the so-called *degenerate four-wave mixing* (DFWM) discussed in Chapter 4 for a general nonlinear medium. All waves have the same frequency ω and their wavevectors are arranged as indicated in Figure 10.8. The sample is illuminated by two pump beams 1 and 2, propagating in opposite directions ($\mathbf{k}_1 = -\mathbf{k}_2$) and by a signal beam $\mathbf{k}_3$. The $\mathbf{k}$-matching condition ($\Sigma\mathbf{k} = 0$) ensures the generation of a fourth beam 4, with $\mathbf{k}_4 = -\mathbf{k}_3$. This beam is *phase-conjugated* or *time-reversed* with respect to the signal beam. In other words, the photorefractive medium behaves as a special mirror whose peculiar reflection law, schematically illustrated in Figure 4.9, has very interesting applications to be discussed in the next chapter.

The mixing process can be understood as a real-time holography, that generates the conjugate of the signal wave, with the recording and reconstruction stages being simultaneously achieved [74]. The recording implies the formation of four index gratings caused by the interference of all pairs of waves: 1–4 and 2–3 (grating **1**), 1–3 and 2–4 (grating **2**), 1–2 (grating **3**) and 3–4 (grating **4**). The simultaneous reconstruction stage results from the diffraction of the waves by the recorded grating and, in particular, the conjugate of the signal wave arises from the diffraction of wave 1 by grating **1**.

The theoretical analysis is based essentially on the Kogelnik coupled-wave approach [10] with four, instead of two, interacting waves. One starts with the complex waves

$$E_j(\mathbf{r}\cdot t) = A_j(\mathbf{r})\,e^{i(\mathbf{k}_j\cdot\mathbf{r}-\omega t)}, \quad (j = 1,\ldots,4) \tag{10.55}$$

causing a refractive index pattern made up of the four gratings **1** to **4**. The general expression for the complex $\tilde{n}$ is

$$\tilde{n} = \bar{n} + n_1\,e^{i\phi_1}\frac{A_1^*A_4 + A_2A_3^*}{2|A_0|^2}\,e^{i\mathbf{k}_1\cdot\mathbf{r}} + \ldots \tag{10.56}$$

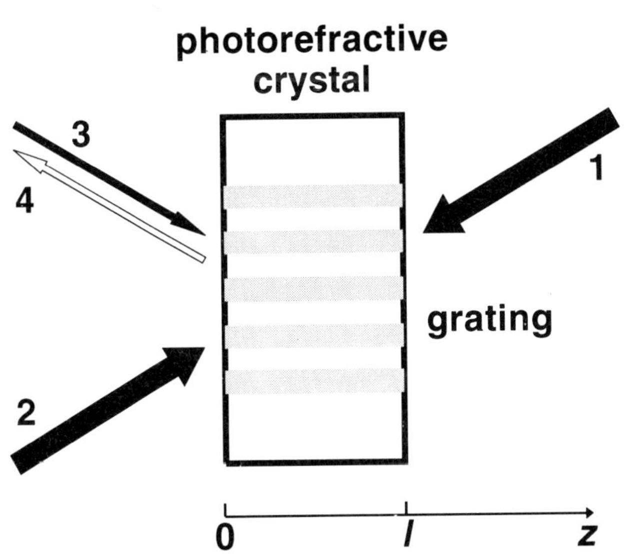

Figure 10.8 *Beam layout in a degenerate four wave-mixing experiment.* ***k*** *conservation is assured.*

where only grating **1** has been explicitly written. The factor between brackets is the modulation depth of the light pattern due to interference of waves 1 and 4 and 2 and 3. Similar expressions apply to gratings **2**, **3** and **4**. $|A_0|^2 = \Sigma_j |A_j|^2$, $\bar{n}$ is the average index of the medium, and the subscripts on n, ϕ and $\mathbf{k}$ refer to each particular grating. Expressions for n and ϕ are obtained by solving the material rate equations as indicated in Section 10.5.

To obtain the steady-state solution, the index distribution (10.56) has to be substituted into the Helmholtz scalar equation to yield four coupled differential equations for the z-dependence of the complex wave amplitudes. The general solution is cumbersome and given in reference [73]. A very simplified situation, while keeping the key features of the effect, occurs under the following assumptions:

(1) lossless medium;
(2) predominance of grating **1** (*transmission grating geometry*);
(3) undepleted approximation, i.e. I_1, $I_2 \gg I_3$, I_4, so that I_1 and I_2 (as well as the corresponding amplitudes) remain constant.

From condition (2) one infers that the z-dependence of wave 4 arises from the diffraction of wave 2 by grating **1**. Similarly, the evolution of wave 3 with z is caused by the diffraction of wave 1. Then, the coupled equations can be immediately obtained from those derived by Kogelnik (10.8) for two-wave coupling by a grating. For the transmission geometry of Figure 10.8, they can be written [73]

$$\frac{\mathrm{d}A_1(z)}{\mathrm{d}z} = 0 \tag{10.57a}$$

$$\frac{\mathrm{d}A_2(z)}{\mathrm{d}z} = 0 \tag{10.57b}$$

$$\frac{\mathrm{d}A_3^*(z)}{\mathrm{d}z} = i\kappa \frac{A_1^* A_4 A_2^* + A_1 A_1^* A_3^*}{|A_0|^2} \tag{10.57c}$$

$$\frac{\mathrm{d}A_4(z)}{\mathrm{d}z} = i\kappa \frac{A_1 A_2 A_3^* + A_2 A_2^* A_4}{|A_0|^2} \tag{10.57d}$$

with $\kappa = (\omega n_1 \,\mathrm{e}^{i\phi_1})/2c\cos\theta_1$ being the coupling constant applicable to diffraction by grating **1**.

The boundary conditions are

$$A_3(0) = A(0) \tag{10.58a}$$

$$A_4(l) = 0 \tag{10.58b}$$

The solution is as follows [73],

$$A_1(z) = A_1(0) \tag{10.59a}$$

$$A_2(z) = A_2(l) \tag{10.59b}$$

$$A_4(z) = -\frac{A_2(l)A_3^*(0)}{A_1^*(0)} \frac{\mathrm{e}^{-i\kappa l} - \mathrm{e}^{-i\kappa z}}{r + \mathrm{e}^{-\kappa l}} \tag{10.59c}$$

$$A_3^*(z) = -A_3^*(0)\,\mathrm{e}^{i\kappa z} \frac{r + \mathrm{e}^{-i\kappa(z-l)}}{r + \mathrm{e}^{-i\kappa l}} \tag{10.59d}$$

where $r = I_2/I_1$ is the pump ratio.

The phase-conjugated reflectivity is

$$\rho = \frac{A_4(0)}{A_3^*(0)} = -\frac{A_2(l)}{A_1^*(0)}\,\frac{1-\mathrm{e}^{-i\kappa l}}{r+\mathrm{e}^{-i\kappa l}} = -\left(\frac{A_1A_2}{A_1^*A_2^*}\right)^{1/2}\frac{\sinh(i\kappa l/2}{\cosh((i\kappa l/2)+\frac{1}{2}\ln r)} \tag{10.60}$$

For the particular case with $r = 1$ (equal intensity pumps), expression (10.60) simplifies to

$$\rho = \tanh\left(\frac{i\kappa l}{2}\right) \tag{10.61}$$

and the intensity reflectance

$$\mathscr{R} = \frac{I_4(0)}{I_3(0)} = \tanh^2\left(\frac{i\kappa l}{2}\right) \tag{10.62}$$

From these equations it is found that conjugated reflectivities $\mathscr{R} > 1$ can be achieved for $\kappa l > \pi/2$. Under these conditions the conjugated mirror has *gain*. This possibility can be also inferred from the more general expressions (10.60). In fact, optimum reflectivities result for pumping beams having unequal intensities. Detailed curves for the dependence of ρ and $\mathscr{R}$ with r and κl are given in reference [73]. It is possible to obtain $\mathscr{R} = \infty$, and so get a *parametric oscillator*.

Conjugated reflectivities higher than 1 have indeed been achieved in several photorefractive materials, e.g. $BaTiO_3$, SBN, and BSO.

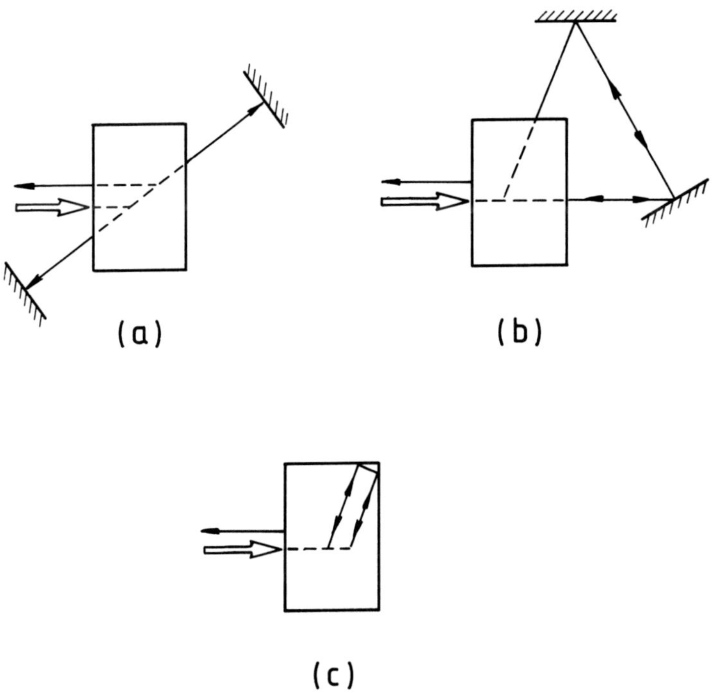

Figure 10.9 *Self-pumping configurations: (a) linear; (b) ring; (c) CAT. (From [75,76].)*

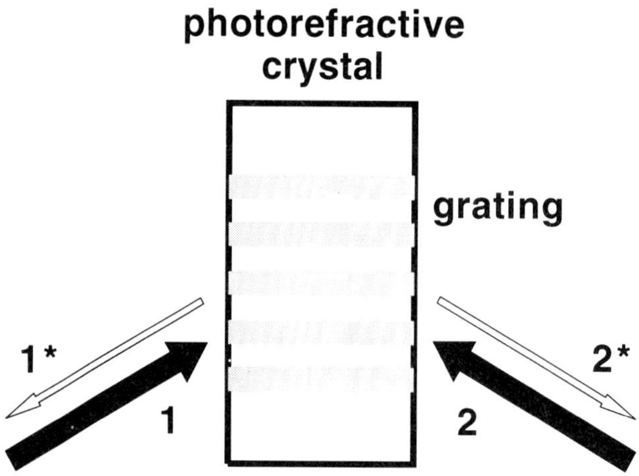

Figure 10.10 *Beam layout in a double-phase conjugated mirror [77]. 1* and 2* are the phase-conjugated replica of input beams 1 and 2 respectively.*

10.9.1. Self-pumping

In self-pumped phase conjugators the pump waves are generated from a single incident (*signal*) beam, through amplification of stray light (*beam fanning*). In some proposed configurations (Figure 10.9(a), (b)), the photorefractive crystal is part of an optical cavity defined by external mirrors [75]. The pump beams are two counterpropagating beams, aligned along the resonator axis, which become amplified by photorefractive coupling. A particularly interesting device is the CAT self-pumped conjugator [76] demonstrated by Feinberg on $BaTiO_3$ (Figure 10.9(c)). It is an arrangement without external feedback, i.e. pump waves are originated from internal reflection at the crystal faces. Conjugated reflectivity of 30% was achieved, as well as good image fidelity. A key requirement for self-pumped operation is high beam-coupling gain of the material. These devices provide *thresholding* and are used as nonlinear elements in optical implementations of *associative memories* (see Chapter 11).

Double-phase conjugated mirrors

For such an operation (Figure 10.10), two separate beams 1 and 2 are incident on a photorefractive crystal. Through self-diffraction of each of the beams, the phase-conjugated replica of the other is generated [75,77], i.e. 1* and 2*. In this way, the information carried by one beam is imposed on the other's conjugated replica, leading to a *bidirectional spatial light modulator*. In accordance with the above operating principle, the two external beams do not need to be mutually coherent and may originate from different lasers. They could even have different (but close enough) wavelengths.

10.10. Photorefractive materials and active centres

It has been already mentioned that a large number of materials present photorefraction. Although most work on microscopic mechanisms has focused on ferroelectric oxides, such as $LiNbO_3$ and $BaTiO_3$, the research effort is being progressively extended to other inorganic and even organic materials.

10.10.1. Inorganic crystals and ceramics

The case of $LiNbO_3$ is illustrative and is now described in some detail. Several transition metal impurities, such as Mn, Cu and Fe are active [78], as shown by the data in Figure 10.11. Iron, which is a typical defect in all oxides, is more efficient. It enters into the $LiNbO_3$ lattice substituting for Li [79] and presents two charge states, Fe^{2+} and Fe^{3+}, whose relative concentrations depend on growth conditions and total doping. The $[Fe^{2+}]/[Fe^{3+}]$ ratio can be modified by suitable thermal treatments in oxidizing or reducing atmospheres, as well as by ionizing radiation [80]. Fe^{2+} presents a rather broad and intense optical absorption band peaked at 3 eV, associated to a charge-transfer transition to Nb^{5+}, together with a small d-d crystal field band at 1.1 eV. Fe^{3+} shows only some low-intensity crystal field transitions, which are not appreciable for usual concentrations in the range up to 0.1%. When illuminated at the charge transfer band, Fe^{2+} acts as an electron donor centre, whereas Fe^{3+} is the corresponding acceptor. In principle, one should expect a correlation between the absorption band of the Fe donor centre and the spectral efficiency of the photorefractive response. The two curves shown in Figure 10.12 indeed appear in the same spectral region, but have a different shape, with the absorption band extending more towards the infrared [81]. This situation, not yet completely clarified, seems to indicate that the absorption spectra may include additional bands in the 1.5–2.5 eV region corresponding to either inactive or less efficient centres for photorefraction. It should be mentioned, as a complicating feature, that the two curves markedly depend on crystal preparation (growth conditions, doping and thermal treatments).

For $LiTaO_3$ [82] (isomorphous to $LiNbO_3$) and $KNbO_3$ [2], different valence states of iron are also considered to be active donor and acceptor centres, although more information is needed, particularly for $KNbO_3$. The situation is also unsettled for $BaTiO_3$

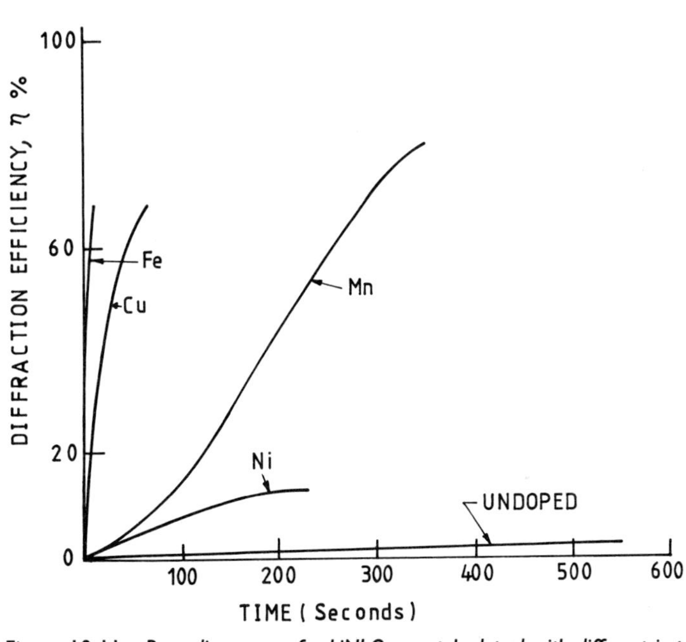

Figure 10.11 *Recording curves for $LiNbO_3$ crystals doped with different impurities [78].*

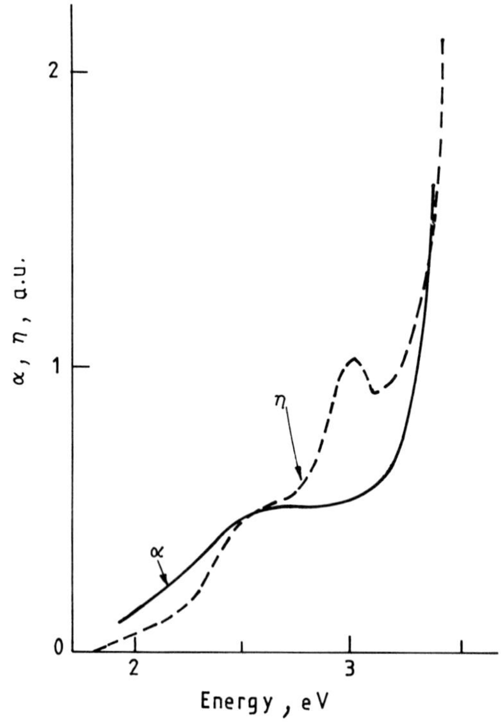

Figure 10.12 *Comparison between the optical absorption and the spectral photorefractive efficiency for Fe–$LiNbO_3$ [81].*

[83], in spite of the intense effort devoted to the study of defects, both intrinsic and extrinsic, in this material. It is generally believed that iron, present in all as-grown samples, is responsible for photorefractive action. However, more than two valence states (at least Fe^{3+} and Fe^{4+}) have been claimed and, moreover, association with oxygen vacancies has been demonstrated. As a consequence of all these complicating features, definite identification of donor and acceptor centres has not yet been achieved.

It is known that BSO crystals grow in a nonstoichiometric composition, i.e. an excess of 9% Bi atoms is present in the lattice [84]. The silicon deficiency was soon correlated with a shoulder in the optical absorption lying close to the main edge, and it has been associated with the occurrence of an intrinsic defect, either a silicon vacancy or a Bi antisite. It is this spectral region which is active for photorefraction. A detailed comparison between the absorption and photorefractive efficiency spectra has been recently made [84]. The situation appears quite complex and still far from clear. However, the data have, indeed, suggested that more than one centre is involved. In conclusion, intrinsic defects arising from the excess of Bi play an important role in optical absorption, photoconductivity and photorefraction, although a definite microscopic picture is still missing. Anyway, by using holographic methods, the simultaneous contribution of electrons and holes to the photoconductivity and photorefractive response has been ascertained.

In addition to single crystals, electrooptic transparent ceramics have shown photorefractive behaviour [85]. Particular attention has been paid to PLZT. Chemical and structural composition provide interesting parameters for the optimization of the effect.

10.10.2. Semiconductor crystals and nanostructures

In the case of semiconductor crystals, information is expanding at a fast rate. It has been recently reviewed by Glass and Strait [86] and Fabre *et al.* [87]. Semiconductors offer the possibility of photorefraction in the infrared, which has particular relevance for communication technologies. Work has been performed on III–V (GaAs, InP and GaP) and II–VI (CdTe, ZnS, ZnSe, CdSe, CdS) compounds. Most data refer to GaAs and InP because of their well-established optoelectronic applications. Photorefractive action has been demonstrated in GaAs doped with Fe, V, Cr as well as containing intrinsic EL2 centres. Photorefractivity has been also investigated in InP doped with Fe. It is clear that many more studies of the correlation between particular defects and photorefractive behaviour are needed for a better understanding of the process.

Semiconductor multiple quantum wells (MQW) and superlattices are attractive materials due to the high value of the electrooptic coefficients in transverse configuration, i.e. electric field perpendicular to the layer surfaces. This is associated to the quantum confinement of exciton wavefunctions (QCSE) described in Chapter 9. A few experiments have already been performed on semiinsulating GaAs/AlGaAs [88,89] and ZnCdTe/ZnTe [90] MQW to substantiate the expected advantages. Parallel [88] as well as perpendicular [89,90] configurations for the applied field have been used. Diffraction efficiencies as high as 3% [89] and 1% [90] have been reported for active-layer thicknesses of 2 μm and 1.56 μm, respectively. Response times around and below 1 μs have been measured. The theoretical understanding of these devices is still very preliminary, although some simple models, accounting for a few gross features have been proposed [88,91]. Much work is still needed before the full potential of those quantum structures can be definitely ascertained.

An extremely fast (ps) transient response has been obtained in a graded superlattice [92], i.e. with band energy profiles as illustrated in Figure 10.13. Electron–hole pairs are created by photoexcitation at the large gap sites. The rapid down-motion of the electrons in the

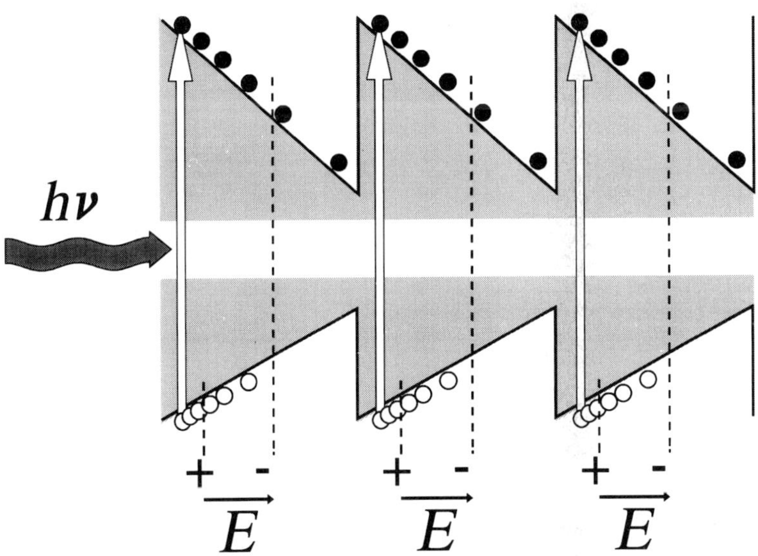

Figure 10.13 *Fast transient photorefractive (photovoltaic) response in a graded semiconductor superlattice [92].*

conduction band, in contrast with holes which are essentially immobile, creates an electric field and consequently a transient refractive index change. Hole motion and recombination finally restore the initial situation. Although the effect involves carrier transport, it may be properly considered as a kind of internal photovoltaic process.

10.10.3. Organic materials

Finally, it should be mentioned that photorefractive behaviour has been recently observed in an organic crystals [93] and in some doped polymeric materials [94,95]. This is a very promising field that has just started. In particular, polymers are ideally suited for waveguide devices and offer clear advantages for fabrication of large area samples. In the work reported by the IBM group [94], space-charge fields of 26 kV/cm have been obtained for applied fields of 126 kV/cm. Unfortunately, diffraction efficiencies are low ($\sim 10^{-3}$), because of the thin samples used (350 μm) and the small electrooptic coefficients of the polymer.

10.11. Performance parameters

As a conclusion to this chapter, the following relevant performance parameters are listed in Table 10.2, for a few useful materials:

(1) *Electrooptic figure of merit*, $n^3 r$, determining maximum index change.
(2) *Photorefractive sensitivity*, S, or more often its reciprocal, S^{-1}, defined as the energy density required to reach a $\eta = 1\%$ diffraction efficiency (for a given crystal thickness). In fact, this energy density, i.e. S^{-1} is the parameter listed in Table 10.2. Other sensitivities are also defined and commonly used [2].
(3) *Response time* for grating recording and erasure (for a given light intensity).
(4) *Storage time* referring to the time necessary for the erasure of a grating in the dark.

Since those parameters vary from sample to sample and are very dependent on material conditions (growth method, doping, etc.), only rough representative values are given in the Table 10.2.

For other performance parameters, such as resolution and dynamic range, the interested reader should consult the appropriate literature [2,4,5,7].

Table 10.2 *Performance parameters for photorefractive materials.*

Material	$n^3 r$ (pm/V)	S^{-1} (mJ/cm^2)	Response time (s)	Storage time (s)
$LiNbO_3$	320	1000	1	10^6
$BaTiO_3$		60	1	10^3
BSO	80	3	10^{-3}	10^{-2}
GaAs	60	30	10^{-5}	10^{-4}

Sensitivities refer to 1 mm thick crystal and optimum wavelength.
Response times correspond to 100 mW/cm^2 light intensity.
Storage times are for room temperature.

References

1. A. Ashkin, G. D. Boyd, J. M. Dziedzik, R. G. Smith, A. A. Ballman and K. Nassau, *Appl. Phys. Lett.* **9**, 72 (1966).
2. P. Günter, *Phys. Rep.* **93**, 199 (1982).
3. T. J. Hall, R. Jaura, L. M. Connors and P. D. Foote, *Prog. Quantum Electron.* **10**, 77 (1985).
4. H. M. Smith (ed.), *Holographic Recording Materials*, Springer-Verlag, Berlin (1977).
5. P. Günter (ed.), *Electrooptic and Photorefractive Materials*, Springer-Verlag, Berlin (1987).
6. J. Feinberg, *Phys. Today* Oct., 46 (1988).
7. P. Günter and J. P. Huignard (eds), *Photorefractive Materials and Their Applications*, Springer-Verlag, Berlin, Vol. 1 (1988) and Vol. 2 (1989).
8. D. M. Pepper, J. Feinberg and N. V. Kukhtarev, *Scient. Am.* Oct., 34 (1990).
9. M. P. Petrov, S. I. Stepanov and A. V. Khomenko, *Photorefractive Crystals in Coherent Optical Systems*, Springer Series in Optical Sciences, Springer-Verlag, Berlin (1991).
10. P. Yeh, *IEEE J. Quantum Electron.* **25**, 484 (1989).
11. H. Kogelnik, *Bell. Syst. Tech. J.* **48**, 2909 (1969).
12. L. Solymar and D. J. Cooke, *Volume Holography and Volume Gratings*, Academic Press, London (1981).
13. J. M. Heaton, P. A. Mills, E. G. S. Paige, L. Solymar and T. Wilson, *Optica Acta* **31**, 885 (1984).
14. F. S. Chen, *J. Appl. Phys.* **40**, 3389 (1969).
15. W. D. Johnston, *J. Appl. Phys.* **41**, 3279 (1970).
16. J. J. Amodei, *Appl. Phys. Lett.* **18**, 22 (1971).
17. A. M. Glass, D. Von der Linde and T. J. Negram, *Appl. Phys. Lett.* **25**, 233 (1974).
18. F. Micheron, C. Mayeux, A. Hermosin and J. Nicolas, *J. Am. Ceram. Soc.* **57**, 306 (1974).
19. P. Günter and M. Zgonik, *Optics Lett.* **16**, 1826 (1991).
20. H. G. Festl, P. Hertel, E. Kratzig and R. von Baltz, *Phys. Status Solidi* (b) **113**, 157 (1982).
21. M. Carrascosa and F. Agulló-López, *IEEE J. Quantum Electron.* **27**, 509 (1991).
22. T. Holstein, *Ann. Phys. (NY)* **8**, 343 (1959).
23. D. Emin, *J. Solid State Chem.* **12**, 246 (1975).
24. N. V. Kukhtarev, V. B. Markov, S. G. Odulov, M. S. Soskin and L. Vinetskii, *Ferroelectrics* **22**, 949 (1979).
25. J. P. Valley and J. F. Lam, in *Photorefractive Materials and Their Applications* (eds P. Günter and J. P. Huignard), Vol. 1, Springer-Verlag, Berlin (1988).
26. J. Feinberg, D. Helman, A. R. Tanguay Jr and R. W. Hellwarth, *J. Appl. Phys.* **51**, 1279 (1980).
27. R. Jaura, T. P. Hall and P. D. Foote, *Opt. Engng* **25**, 1068 (1986).
28. D. A. Temple and C. Ward, *J. Opt. Soc. Am.* B**5**, 1800 (1988).
29. L. B. Au and L. Solymar, *J. Opt. Soc. Am.* **A7**, 1554 (1990).
30. A. Bledowski, J. Otten and K. H. Ringhofer, *Optics Lett.* **16**, 672 (1991).
31. Y. H. Lee and R. W. Hellwarth, *J. Appl, Phys.* **71**, 916 (1992).
32. G. A. Brost, *J. Opt. Soc. Am.* **B9**, 1454 (1992).
33. M. Carrascosa and F. Agulló-López, *IEEE J. Quantum Electron.* **QE-22**, 1369 (1986).
34. J. K. Tyminski and R. C. Powell, *J. Opt. Soc. Am.* **B2**, 440 (1985).
35. R. A. Mullen and R. W. Hellwarth, *J. Appl. Phys.* **58**, 40 (1985).
36. T. Hirao and T. Saweda, *Optics Commun.* **82**, 83 (1991).
37. F. P. Strohkendl, *J. Appl. Phys.* **65**, 3773 (1989).
38. L. Arizmendi and R. C. Powell, *J. Appl. Phys.* **62**, 896 (1987).
39. L. Arizmendi, P. D. Townsend, M. Carrascosa, J. Baquedano and J. M. Cabrera, *J. Phys.: Condens. Matter* **3**, 5399 (1991).
40. G. Valley, *Appl. Optics* **22**, 3160 (1983).
41. M. Carrascosa and F. Agulló-López, *Appl. Optics* **27**, 2851 (1988).
42. P. Tayebati and D. Mahgerefteh, *J. Opt. Soc. Am.* B**8**, 1053 (1991).

43. F. Jariego and F. Agulló-López, *Appl. Optics* **30**, 4615 (1991).
44. F. P. Strokendl, J. M. Jonathan and R. W. Hellwarth, *Optics Lett.* **11**, 312 (1986).
45. G. C. Valley, *J. Appl. Phys.* **59**, 3363 (1986).
46. M. C. Bashaw, T. P. Ma, R. C. Barker, S. Mroczkowski and R. R. Dube, *J. Opt. Soc. Am.* **12B**, 2329 (1990).
47. S. Zhivkova and M. Miteva, *J. Appl. Phys.* **68**, 3099 (1990).
48. F. Micheron and G. Bismuth, *Appl. Phys. Lett.* **20**, 79 (1972).
49. J. B Thaxter and M. Kestigan, *Appl. Optics* **13**, 913 (1974).
50. V. Leyva, A. Agranat and A. Yariv, *Optics Lett.* **16**, 554 (1991).
51. P. Hertel, K. H. Ringhofer and R. Sommerfeldt, *Phys. Status Solidi* **A104**, 855 (1987).
52. M. Carrascosa and F. Agulló-López, *J. Opt. Soc. Am.* B**7**, 2317 (1990).
53. R. Müller, L. Arizmendi, M. Carrascosa and J. M. Cabrera, *Appl. Phys. Lett.* **60**, 3212 (1992).
54. J. P. Herriau and J. P. Huignard, *Appl. Phys. Lett.* **49**, 1140 (1986).
55. A. E. Attard, *J. Appl. Phys.* **66**, 3211 (1989).
56. L. Arizmendi, *J. Appl. Phys.* **65**, 423 (1989).
57. S. W. McCahon, D. Rytz, G. C. Valley, M. V. Klein and B. A. Wechsler, *Appl. Optics* **28**, 1967 (1989).
58. A. Partovi, A. Kost, E. M. Garmire, G. C. Valley and M. B. Klein, *Appl. Phys. Lett.*, **55**, 1087 (1989).
59. S. I. Stepanov and M. P. Petrov, *Optics Commun.* **53**, 292 (1985).
60. J. P. Huignard and A. Marrakchi, *Optics Commun.* **38**, 249 (1981).
61. S. I. Stepanov, K. Kolikov and M. Petrov, *Optics Commun.* **44**, 19 (1982).
62. Ph. Refregier, L. Solymar, H. Rajbenbach and J. P. Huignard, *J. Appl. Phys.* **58**, 45 (1985).
63. N. V. Kukhtarev, G. E. Dougalenko and V. N. Starkov, *Appl. Phys.* **A33**, 227 (1984).
64. A. Marrakchi, R. V. Johnson and A. R. Tanguay Jr, *J. Opt. Soc. Am.* B**3**, 321 (1986).
65. S. G. Odulov and M. S. Soskin, in *Photorefractive Materials and Their Applications*, Vol. 2 (eds P. Günter and J. P. Huignard), Springer-Verlag, Berlin (1989).
66. G. Lesaux, G. Roosen and A. Brun, *Optics Commun.* **56**, 374 (1986).
67. G. Lesaux and A. Brun, *IEEE J. Quantum Electron.* **QE-23**, 1680 (1987).
68. J. M. C. Jonathan, G. Roosen and Ph. Roussignol, *J. Phys.* **49**, C2-267 (1988).
69. G. C. Valley, *IEEE J. Quantum Electron.* **QE-19**, 1637 (1983).
70. D. M. Pepper, *Opt. Eng.* **21**, 156 (1982).
71. R. A. Fischer (ed.), *Optical Phase Conjugation*, Academic Press, New York (1983).
72. M. B. Klein, R. K. Jain and G. C. Valley, *J. Phys.* **44**, C2-149 (1983).
73. M. Cronin-Golomb, B. Fischer, J. O. White and A. Yariv, *IEEE J. Quantum. Electron.* **QE-20**, 12 (1984).
74. A. Yariv, *Optics Commun.* **25**, 23 (1978).
75. B. Fischer, S. Sternklar and S. Weiss, *IEEE J. Quantum Electron.* **QE-25**, 550 (1989).
76. J. Feinberg, *Optics Lett.* **7**, 486 (1982).
77. S. Weiss, S. Sternklar and B. Fischer, *Optics Lett.* **12**, 114 (1987).
78. W. Phillips, J. J. Amodei and D. L. Staebler, *RCA Rev.* **33**, 94 (1972).
79. L. Rebouta, M. F. da Silva, J. C. Soares, M. Hage-Ali, J. P. Stoquert, P. Siffert, J. A. Sanz-Garcia, E. Dieguez and F. Agulló-López, *Europhys. Lett.* **14**, 557 (1991).
80. L. Arizmendi, J. M. Cabrera and F. Agulló-López, *J. Phys. C* **17**, 515 (1984).
81. J. Baquedano, M. Carrascosa, L. Arizmendi and J. M. Cabrera, *J. Opt. Soc. Am.* B**4**, 309 (1987).
82. P. A. Augustov, K. K. Shvarts and E. Kratzig, *Phys. Status Solidi* (a) **87**, K73 (1985).
83. M. B. Klein and R. N. Schwartz, *J. Opt. Soc. Am.* **B2**, 293 (1986).
84. L. Arizmendi, J. M. Cabrera and F. Agulló-López, *Int. J. Optoelectron.* **7**, 149 (1992).
85. A. Krumins and A. Sternberg, in *Electrooptic and Photorefractive Materials* (ed. P. Günter), Springer-Verlag, Berlin (1987).
86. A. M. Glass and J. Strait, in *Photorefractive Materials and Their Applications* (eds P. Günter and J. P. Huignard), Vol. 1, Springer-Verlag, Berlin (1988).

87. J. C. Fabre, E. Brotons, P. V. Halter and G. Roosen, *Int. J. Optoelectron.* **4**, 459 (1989).
88. D. D. Nolte, D. H. Olson, G. E. Doran, W. A. Knox and A. M. Glass, *J. Opt. Soc. Am.* B7, 2217 (1990).
89. A. Partovi, A. M. Glass, D. H. Olson, G. J. Zydzik, H. M. O'Bryan, T. H. Chiu and W. H. Knox, *Appl. Phys. Lett.* **62**, 464 (1993).
90. A. Partovi, A. M. Glass, D. H. Olson, G. J. Zydzik, K. T. Short, R. D. Feldman and R. F. Austin, *Optics Lett.* **17**, 655 (1992).
91. M. Carrascosa, F. Agulló-Rueda and F. Agulló-López, *Appl. Phys.* **A55**, 25 (1992).
92. F. Capasso, S. Luryi, W. T. Tsang, G. C. Bethea and B. F. Levine, *Phys. Rev. Lett.* **51**, 2318 (1983).
93. K. Sutter and P. Günter, *J. Opt. Soc. Am.* **B7**, 2274 (1990).
94. S. Ducharme, J. C. Scott, R. J. Tweig and W. E. Moerner, *Phys. Rev. Lett.* **66**, 1864 (1991).
95. Y. Cui, Y. Zhang, P. Prasad, J. C. Schildkraut and D. O. Williams, *Appl. Phys. Lett.* **61**, 2132 (1992).

Chapter Eleven

Photorefractive Applications

11.1. Introduction

This chapter presents a short review of relevant applications of photorefractive materials. Up to the present, not one photorefractive device has been marketed, in many cases because of the lack of sufficiently large, homogeneous or reproducible crystals. Nonetheless, a large variety of exciting laboratory demonstrations look very promising, and others are expected. The following applications will be discussed here: information storage, real-time holographic recording, beam deflectors, image amplifiers, optical interconnects, phase conjugation, associative memories and neural networks. The determination of material parameters by photorefractive measurements, which can be considered as an interesting side application, will also be briefly discussed. The interested reader will find a more complete description of these and other applications in references [1] and [2]. Major advantages of photorefractive materials with respect to other nonlinear optical media are based on two specific features of photorefraction. On one hand, they show much higher sensitivities and diffraction efficiencies than ordinary nonlinear optical media, i.e. required light intensities are much lower and low power lasers can be used. On the other hand, the nonlocal response of photorefractive media gives rise to a mismatch between light and refractive index patterns and this is responsible for the energy transfer between beams (*amplification effect*). Both features are due to the peculiar use that photorefractive materials make of the electrooptic effect via charge transport along a distance much greater than the lattice parameter, a mechanism not present in ordinary nonlinear optical materials. Charge transport makes also feasible "quasi-permanent" changes of the refractive index that can be either optically or thermally erased (reversibility), another specific feature of these materials.

11.2. Information storage

Since the first experiments with $LiTaO_3$ and $LiNbO_3$, the photorefractive effect was recognized as a means of information storage by recording phase volume holograms. The dark storage time of these holograms at room temperature is simply given by the dark dielectric relaxation time (see also Section 10.5.3) $\tau = \epsilon_0\epsilon/\sigma_d$, where σ_d is the dark conductivity that can be either ionic or electronic. Typical examples of τ values are given in Table 10.2. By using the thermal fixing process described in Section 10.6, up to approximately 10^5 years (virtually infinite) either dark or light storage times at room temperature have been proved in $LiNbO_3$.

High density optical memories were soon demonstrated [3] by selective superimposition of fixed holograms in the same crystal, taking advantage of the pronounced angular and wavelength selectivity of volume holograms (equation (10.16)). Up to 500 holograms, exhibiting a minimum diffraction efficiency above 2.5% at the final stage, were recorded in a 1 cm-thick $LiNbO_3$–Fe sample with a 488 nm wavelength, giving a total capacity of 0.5 Gbits/cm^3. Hologram superimposition was carried out by rotating the crystal by 0.1° after each recording and using a sample with a highly asymmetrical write/erase cycle. Recording and fixing were performed simultaneously at high temperature (160 °C) in order to increase fixing efficiency and reduce build-up of light scattering effects during storage. No degradation of diffraction efficiency or image quality have been observed [3,4]. Recently, up to 750 holograms with diffraction efficiencies above 0.5% have been reported [5].

Any specified information block can also be selectively erased by using the method of coherent image subtraction demonstrated by Huignard *et al.* [6]. In addition, Petrov *et al.* [7] have successfully used a bias electric field applied to a $LiNbO_3$ sample to control the Bragg condition for image reconstruction of a selected hologram among three holograms written in the sample under different voltages. Advantage is taken of the large electrooptic coefficient of $LiNbO_3$ and the maximum efficiency is reached at the same voltage used for the recording.

For nonfixed photorefractive volume holograms, a major limitation is the degradation they suffer during multiple readouts. Other factors limiting the information capacity are related with crosstalk between different holograms in multiple hologram recording, the maximum value that can be reached for Δn, as well as the concentration of active donor/acceptor centres [8–10]. To avoid hologram degradation during readout, aside from thermal fixing, several other methods have been developed and have already been discussed in Section 10.6. Further details can be found in the review papers [11] and [12].

These limitations, together with the concurrent development of semiconductor technologies, made that photorefractive holographic memories had no practical impact. However, the subject has become important again in connection with the new field of all-optical associative memories, as well as in applications which utilize dynamic holographic recording (*holographic interferometry*, *pattern recognition*, *photolithography*, *image convolution* and *correlation*, *parallel optical logic* and *reconfigurable optical interconnections*). These subjects are discussed below under separate sections. An evident advantage of photorefractive materials for these applications is that they need neither development of the recording medium, nor ultrahigh speed, since high information rates are already achieved by the high degree of parallelism of the input image.

11.3. Real-time holographic interferometry

Holographic interferometry is recognized as an efficient tool in remote nondestructive testing, with wide use in industry, science, medicine, metrology and other areas. The most complicated wavefronts, including those scattered from diffusive objects, can be tested with holographic recording techniques. One needs to compare two or more wavefronts, one of them coming from the hologram reconstruction. Typical measurement accuracy is at the submicrometre level ($\sim 0.1\ \mu m$), determined by the light wavelength ($\lambda \sim 0.5\ \mu m$) and the fringe width ($\sim \lambda/4$).

Classical materials for holographic recording (silver halide plates, photoresist, etc.) require three basic steps: recording, developing and readout of the hologram. With photorefractive materials all these steps occur simultaneously, and hologram destruction also takes place at the same time due to the space-charge field relaxation under uniform illumination with the reference beam. These properties make photorefractive materials particularly suitable for the holographic observation of a changing object in real time, since no processing of the recording medium is needed. In addition, no noticeable degradation of the photorefractive material is observed either in continuous or high-cycle repetition rate modes. It should be remembered, however, that in order to keep linearity in the recording, the reference beam must be about 10 times more intense than the signal beam, i.e. $m \ll 1$ in equation (10.29a).

Successful experiments on holographic interferometry using photorefractive media were first reported by Huignard and Micheron in 1976 at Thomson CSF [13], after the discovery of the photorefractive crystal $Bi_{12}SiO_{20}$ (BSO). This crystal continues to be one of the most convenient for holographic interferometry. Its sensitivity (see Section 10.11) in the blue-green spectral region is comparable to that of high-resolution photographic plates, and ranges from $S \approx 10^3\,cm^2/J$ for a spatial frequency of $\Lambda^{-1} \approx 1000\,mm^{-1}$ to $S \approx 10^4\,cm^2/J$ for $\Lambda^{-1} \approx 100\,mm^{-1}$, with a diffraction efficiency above 10% when an external electrical field is applied. Several methods have been reported for holographic interferometry, and applied to many practical examples using BSO and two other crystals of its family, BGO and BTO.

One of the methods consists of recording a single hologram at some initial time t_1. Then, the object is illuminated with the original beam and viewed through the hologram, whilst the hologram is simultaneously illuminated with the same reference beam. The real object wave scattered from the object at time t_2, and the reconstructed wave representing the object at time t_1, will interfere to produce an image of the object covered with fringes. The fringes provide information on the changes of the object during the interval $t_2 - t_1$. A continuous observation of the object evolutions can be carried out (*real-time holographic interferometry*), provided the hologram and the illumination geometry are kept unchanged. With photorefractive crystals, however, the hologram is continuously erased and rewritten. That limits the maximum possible observation time and requires special precautions to be taken. The method has been demonstrated in a BTO crystal [14] under an a.c. electric field with an amplitude of about 10 kV/cm.

In the double-exposure technique, two holograms of the object are stored in the photorefractive medium, delayed by a certain time interval Δt by using a pulsed laser. Reconstruction of this superimposition hologram by the original reference wave results in

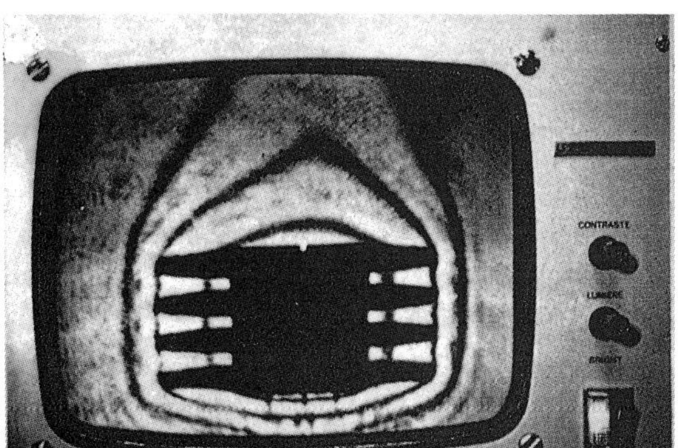

Figure 11.1 *Example of double-exposure holographic interferometry: refractive index gradient in air thermally induced by a transistor radiator [15].*

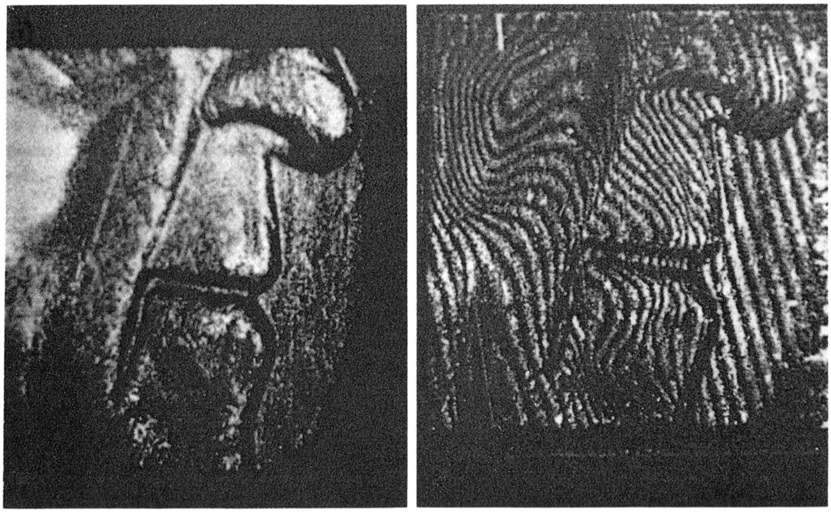

(a) (b)

Figure 11.2 *Coin surface relief determined in [17] by two-wavelength holographic interferometry, using a BSO crystal with $\lambda_1 = 520.83$ nm and $\lambda_2 = 530.87$ nm; the relief depth is 13.88 μm. (a) Image of part of the coin. (b) Real-time contour lines of (a). (Reproduced with permission.)*

an image of the object covered by interference fringes. The reason is the coherent subtraction of the two reconstructed wave fields. Bright fringes correspond to object regions displaced at distances multiple of $\lambda/2$, whereas dark fringes correspond to regions displaced by $\lambda/4 + \nu\lambda/2$ (where ν is an integer). Figure 11.1 shows an application of this technique to the analysis of the refractive index gradient induced by the thermal radiator of a transitor [15]. In this application a BSO crystal was used with an applied d.c. electric field of 6 kV/cm, a wavelength of 488 nm and a light intensity of 13 mW/cm^2. Since recording of the second hologram partially erased the first one, the recording time given to the first one was twice that of the second. Also, the reference beam erased both holograms during readout and a vidicon memory tube was used to observe the image. A full cycle was completed in less than 0.1 s. A four-wave mixing configuration has also been used for the double-exposure method [16].

A variant of the double-exposure technique uses simultaneous recording with two slightly different wavelengths, i.e. $|\lambda_2 - \lambda_1| \ll \lambda_{1,2}$ and a four-wave mixing set-up. By reading out with either wavelength, interference fringes will show a phase relief of the object due to different longitudinal magnification. Figure 11.2 shows the application of this technique to the determination of the relief of a coin surface using a BSO crystal and a krypton laser [17].

11.4. Beam modulators, deflectors and interference filters

Transient or steady photorefractive gratings can also be used as active electrooptic elements that deflect a beam at the Bragg condition. One method implies the application of an external d.c. electric field E_a during readout of a hologram in order to modify the Bragg condition and entirely "switch out" the diffracted beam. At the same time, the Bragg condition of another hologram that could not be reached for $E_a = 0$ can be reached after applying the field, and the diffracted beam from this second hologram "switched on". Repeating the process with a number of previously recorded holograms may produce fast optical interconnection with a fibre matrix. This technique obviously requires stable and nondegrading volume holograms during readout, and this can be achieved by fixing methods as described in Section 10.6. The amplitude $E_{0.5}$ of the characteristic electric field required to decrease by a factor 2 the hologram diffracted intensity, is of the same order of magnitude as the driving field used with conventional electrooptic modulators in a similar geometry [18]. This switching mechanism was first suggested and experimentally demonstrated with a $LiNbO_3$ crystal in reference [19] and later with different geometries in references [20–22]. The technique allows switching of complicated wavefronts and can be used to control electrically the retrieval of complicated images [22,23].

Another technique for laser beam deflection [24–26] is based on the diffraction by a volume phase grating with a variable spacing Λ. Deflection of a near-infrared beam is achieved over a large scan angle of 10°. The grating spacing is varied by changing the wavelength λ of the recording light, while keeping fixed the crossing angle θ of the writing beams, according with the Bragg condition $\Lambda = \lambda/(2\sin\theta)$. The operation speed of such a deflector is limited by the characteristic time τ required to record a hologram. The deflector bandwidth can be further increased with an additional tilt of the grating. One of the writing beams is diffracted at a high-efficiency holographic grating before left to interfere with the other beam [24,26]. Up to 200 resolvable positions have been achieved with a photorefractive deflector using this method [26]. Experimental data were the following: 2.7 mm thick $Bi_{12}SiO_{20}$ crystal, with a grating spacing $\Lambda \approx 1\,\mu m$, and recording intensity of $10\,mW/cm^2$ providing a switching time $\tau \approx 0.1$ s and a diffraction efficiency of $\eta \approx 1\%$. To vary the recording wavelength a dye laser was used ($530\,nm \lesssim \lambda \lesssim 570\,nm$). With a similar geometry but using anisotropic Bragg diffraction in a biaxial $KNbO_3$ crystal, Voit *et al.* [27] have demonstrated deflection of a He–Ne laser beam within an angle of about 5.7°. They used different wavelengths ($457.9\,nm < \lambda < 514.5\,nm$) from an argon-ion laser, with an average intensity of $0.5\,W/cm^2$. A different method [26] to increase the deflection angle involves simultaneous recording of two (or more) holograms of different wavevectors to obtain a combined hologram.

Optical amplification via two- or four-wave mixing in a photorefractive crystal can also be employed to deflect light coming from selected area of a spatial light modulator (SLM) [28]. It has been demonstrated with a $BaTiO_3$ crystal and an array of 4×4 piezomirrors working as a SLM. By exciting with a ramp generator all piezomirrors except the one corresponding to the chosen direction, efficient energy transfer in the photorefractive optical amplifier was only possible for the beam coming at that direction.

All methods described above, except the one that switches off the Bragg condition by the application of an electric field (described in the first paragraph of this section), suffer from the same important limitation: the speed of operation is determined by the characteristic response time ($1\,ms < \tau < 10\,s$) of typical photorefractive materials.

Very narrow-band interference filters have emerged as the first commercial application of photorefractive materials. They are based on holographic Bragg reflectors made up with sinusoidal thick photorefractive gratings. Laboratory essays were first performed with transient gratings [28a], although practical interest was not relevant. Recently, however, efficient Bragg reflectors have been achieved with photorefractive fixed gratings [28b,28c], taking advantage of substantial improvements in the diffraction efficiency of fixed gratings. Bandwidths below 10 pm, with reflectivities about 50% and angular fields of views about 4°, have permitted these filters to compete favorably with anisotropic Lyot-Ohman and Sölc filters. Main advantages are: fewer optical elements, lower weight and lower cost.

11.5. Optical amplifiers

Optical amplifiers are defined as devices that generate a coherent light beam at the output of greater intensity than that of the input beam. Familiar quantum optical amplifiers are based on an active medium where a population inversion is produced by external pumping, and stimulated emission processes result in the desired amplification. As discussed in Section 10.7, coupling of interacting beams during photorefractive hologram recording can be utilized as a novel type of optical amplifier. At variance with the quantum amplifier, the holographic amplifier is based on the classical phenomena of interference and diffraction arising from the light-induced refractive index change. The resulting amplification is nonlinear, and, under appropriate conditions (see equation (10.49)), it can be exponential with gains within the 10–100/cm range. These gains enable one to obtain, in photorefractive crystals a few millimetres thick, hundred- to thousandfold amplifications of signal beams that many times exceed those attainable with commonly used laser media such as ruby or neodymium-doped materials. As with ordinary amplifiers, coherently pumped oscillators have also been developed with holographic amplifiers. Usually the photorefractive response time is not a limitation here, since complex light wavefronts carrying a huge amount of information can be amplified in a few milliseconds.

11.5.1. Image amplification

Figure 11.3 shows schematically the optical set-up based on nearly degenerate two-wave mixing with a BSO crystal for amplification of a signal wavefront carrying an image taken from a three-dimensional object (loudspeaker). Amplification proceeds through the energy transfer from the pump beam to the signal beam occurring in the photorefractive crystal. The pump beam is stirred with the help of a piezo-mirror, and the fringe velocity and the angle between reference and signal beams are adjusted in order to maximize the gain. With a beam intensity ratio of approximately 10^3, the transmitted image of the loudspeaker is amplified by a factor of 10 [29] (Figure 11.4).

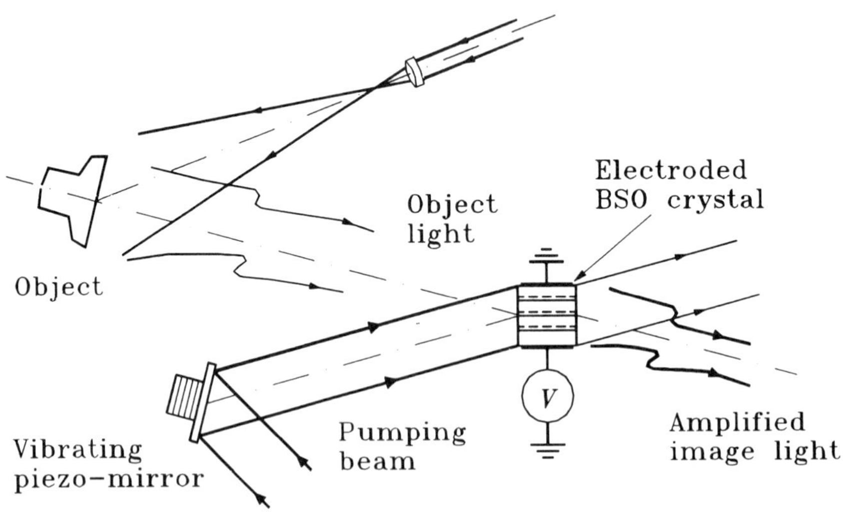

Figure 11.3 *Optical set-up for image amplification by nearly degenerate two-wave mixing in a BSO crystal. (From [29].)*

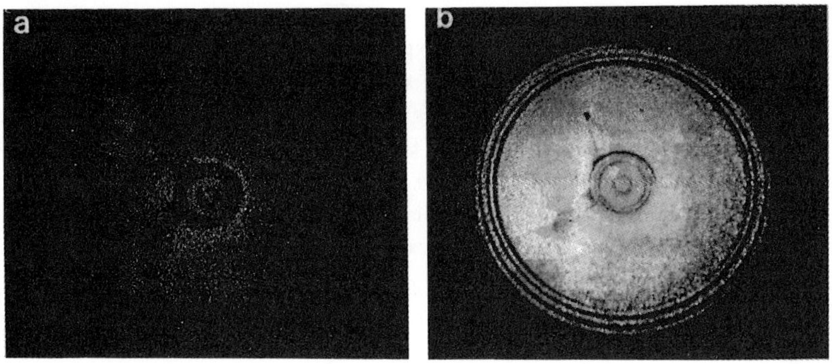

Figure 11.4 *(a) Image of a loudspeaker transmitted through the crystal. (b) Amplified image due to energy transfer. (From [29].)*

Efficient image amplification has also been demonstrated with other photorefractive crystals such as $LiNbO_3$, $BaTiO_3$ and $KNbO_3$, where the phase-shift between volume hologram and light pattern is achieved by diffusion recording (grating period $\Lambda^{-1} \approx 1000/\text{mm}$) without the application of any external field. A similar kind of amplification can be achieved with a four-wave mixing setup and this will be described in the section devoted to phase conjugation. Major limitations of coherent image amplifiers based on photorefractive crystals arise from crystal inhomogeneities, that limit both the resolution and the minimum intensity of the image to be amplified. Local fluctuations of the crystal dielectric constant cause scattered waves which are efficiently amplified by two-wave mixing with the pump beam. A compromise has to be reached between the gain (not too high, to avoid high nonlinearities) and the signal-to-noise ratio.

11.5.2. Laser beam clean-up

Laser beam clean-up is referred to as the improvement of the spatial characteristics of optical beams in real time. This process can be achieved by two-wave mixing in photorefractive crystals: the amplitude of the aberrated pump beam is transferred to a plane-wave signal beam of low intensity, without transferring the phase. The process can be

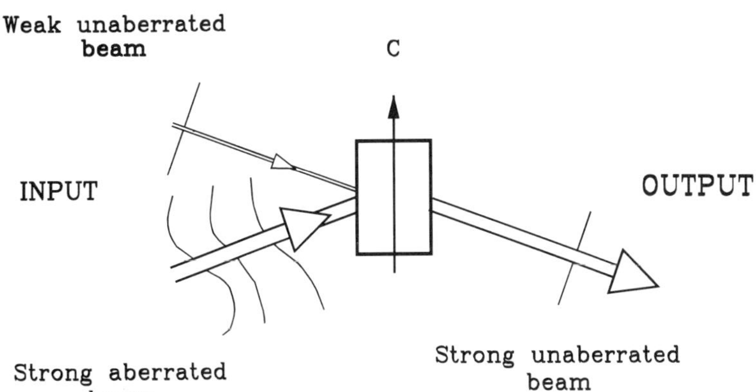

Figure 11.5 *Illustration of the laser beam clean-up using a two-wave mixing set-up.*

seen as the dynamic holographic reconstruction of the reference beam when reading out with a complex object wavefront (Figure 11.5). Experimental demonstrations have been achieved in SBN and $BaTiO_3$ crystals [30], where diffraction-limited output wavefronts have been obtained from diffraction-limited input signal beams. This proves that the phase of the weak signal is unaffected by the distortions of the pump beam. A different set-up [31] uses a single mode photorefractive ring oscillator pumped by the aberrated beam to generate a self-induced clean beam. The photorefractive response time limits the speed at which time-varying phase distortions of the beam can be corrected.

11.6. Optical phase conjugation

Although optical phase conjugation through degenerate four-wave mixing was first observed in liquid CS_2 and later in many other nonlinear media, the high power needed (in the megawatt range) has been a major drawback for practical use of this interesting phenomenon. On the other hand, as discussed in Section 10.9, photorefractive materials can act as efficient CW phase conjugators with weak light beams (in the milliwatt range), in addition to a remarkable optical amplification effect from which associate processes such as oscillation, thresholding and others are available. These facts have triggered an explosion of exciting applications of photorefractive phase conjugators, and some of the most relevant will be described here. Applications rely either on the phase of the conjugated wave or on the image content of the generated wave, and, as with other photorefractive applications, major drawbacks of present phase conjugators come from material limitations: slow response time, low light sensitivity, expensive materials or crystal growing difficulties. In spite of the lack of commercial success due to these limitations, many of the demonstrated applications are very promising.

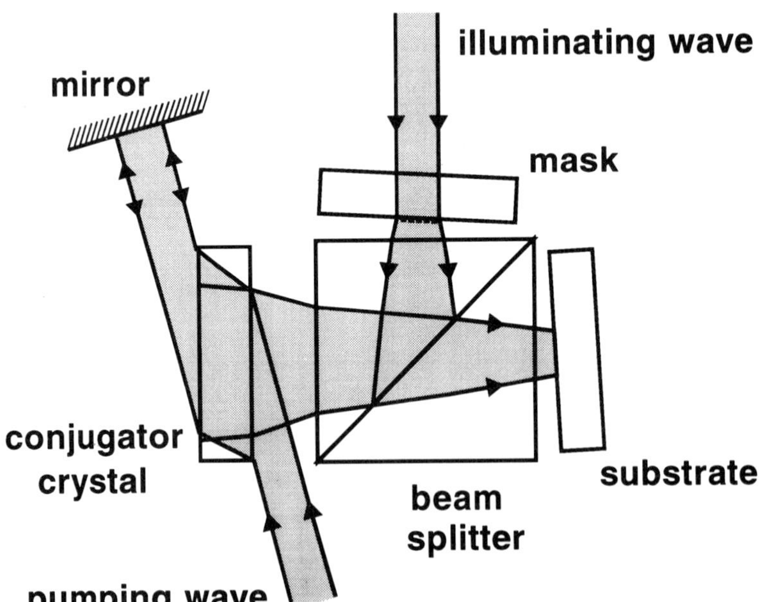

Figure 11.6 *Scheme of a lensless photolithographic set-up using a photorefractive phase conjugator. (From [32].)*

11.6.1. Lensless and speckle-free imaging

Photorefractive phase conjugators have been successfully applied to lensless imaging, an important example being mask-copying in photolithography. In the set-up sketched in Figure 11.6 [32], the signal beam coming from the illuminated mask (with features of the order of 10^{-6} m) is reflected at a high quality beam splitter and addressed to a γ-irradiated $LiNbO_3$ phase conjugator. The retroreflected conjugate wave then forms an image on the substrate surface, coated with a light-sensitive material. As shown in Section 11.6.2, most of the optical aberrations are corrected in the projection process because of the phase conjugation. Higher spatial resolutions (submicrometre features free of speckle noise) over image fields larger than with conventional lens lithography have been obtained [32]. Further improvements of this experiment (elimination of pumping beam and faster response time) have been achieved with a $BaTiO_3$ self-pumped phase conjugator [33].

A major problem when using coherent light for image reconstruction is the degradation of the signal-to-noise ratio produced by the complex interference pattern (speckle) of scattered light from dust particles, inhomogeneities or surface defects in the optical components of the system. A substantial speckle-noise reduction using a BSO photorefractive conjugator has been demonstrated by Huignard *et al.* [34]. It is based on the integration time in the detection plane of N coherent images having uncorrelated speckle noise pattern, what improves the signal-to-noise ratio by $\sqrt{N}$. This approaches the signal-to-noise ratio of coherent systems to that of incoherent ones, while retaining all the possibilities of coherent light for processing optical fields.

11.6.2. Imaging through a phase disturbing medium

Correction of the phase distortion in imaging experiments has also been demonstrated with photorefractive phase conjugators. Figure 11.7 shows the application of a $BaTiO_3$ phase conjugator, working in the diffusion recording mode, to the transmission of images through multimode fibres [35]. The distortion produced by the modal dispersion of the fibre in the first pass is compensated by the same modal dispersion in the second pass of the phase conjugate wavefront. Step-index fibres 1.75 and 0.75 m in length with up to 10^4 modes were

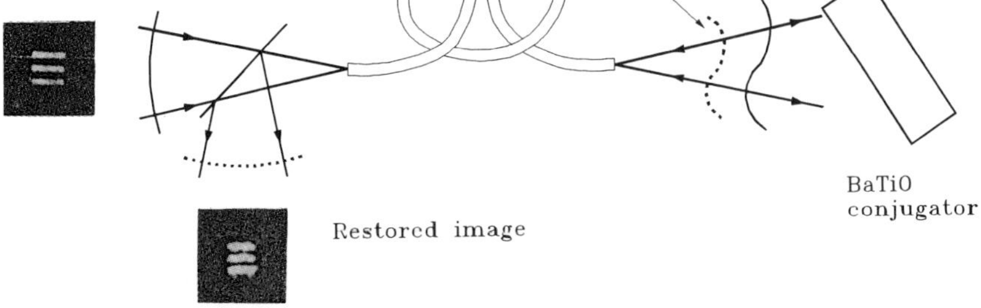

Figure 11.7 *Image restoration after propagation through a multimode optical fibre with a $BaTiO_3$ phase conjugator. (From [35].)*

used. Unfortunately, the difficulty of fabricating two identical fibres at present rules out the applicability of such a scheme to the correction of the modal dispersion of one fibre using propagation in a second one, so that the practical use is still limited. A case of practical interest is when using the multimode fibre as an arm in the Michelson interferometer in order to reproduce the gaussian wave with high fidelity at the fibre input [36], simplifying construction and adjustment of high-sensitivity fibre-optic interferometers.

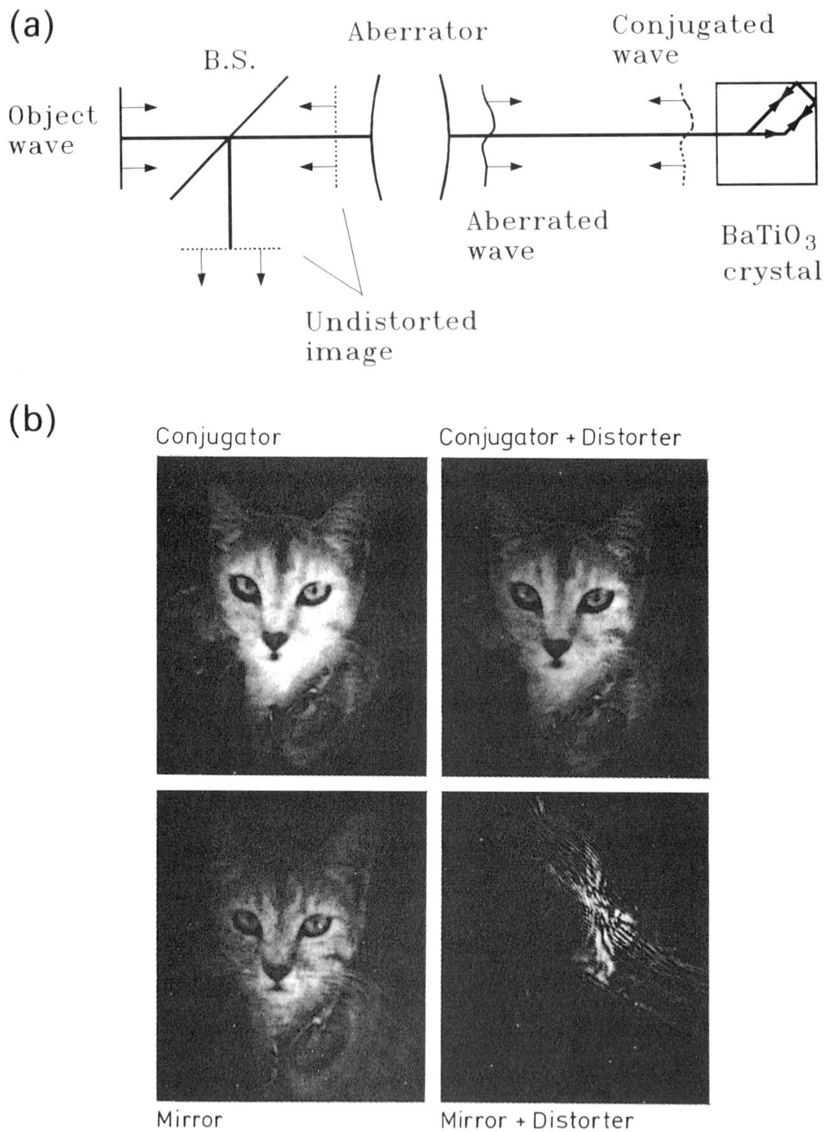

Figure 11.8 *(a) Correction of the phase distortion with a self-pumped $BaTiO_3$ phase conjugator using total reflections at the corner of the crystal instead of external mirrors. (b) Bottom row: an ordinary mirror cannot correct for the distorter (clear glue smeared on a glass slide); top row: the conjugate mirror can restore an image that has passed through the distorter. (From [37].)*

In the example of Figure 11.8, a self-pumped $BaTiO_3$ phase conjugator is used to correct the phase distortion introduced in the image of Feinberg's cat by a distorter made of clear glue smeared on a glass slide [37]. This special geometry uses total reflections at the crystal corner and the self-generated pump waves have the freedom to choose the spatial pattern and orientation that optimize the phase-conjugate signal providing an excellent quality of the reproduction. Conjugator reflectivities as high as 60% have been reported in spite of Fresnel reflection losses.

Similar wavefront corrections by phase conjugation have been demonstrated in a number of other photorefractive materials such as BSO, BGO, $LiNbO_3$, $LiTaO_3$, $KNbO_3$ and SBN. At present, reflectivity values up to 40 at low incident intensity are commonly obtained.

11.6.3. Phase distortion correction in laser cavities

As elements of laser resonator cavities, photorefractive phase conjugators have been proved useful in order to correct aberrations from the laser medium, mirrors or other intracavity components. The first experimental demonstration of such a device was achieved by Cronin-Golomb *et al.* [38], and the set-up and results are illustrated in Figure 11.9. The distorter consists of a piece of etched glass. Lasing with the phase conjugate mirror is initiated with the aid of an ordinary mirror, since the coherence of the fluorescence is insufficient to allow the formation of the required photorefractive grating in the crystal. Once the grating has been established, the auxiliary mirror can be removed. In the normal laser cavity the distorter drastically lowered the spatial mode quality and the output power decreased from 2 W down to 1 mW. On the contrary, with the phase conjugator the mode shape was restored and the power output boosted back to 500 mW. Later on, several other configurations apart from self-pumping, including self-starting oscillation, have been used.

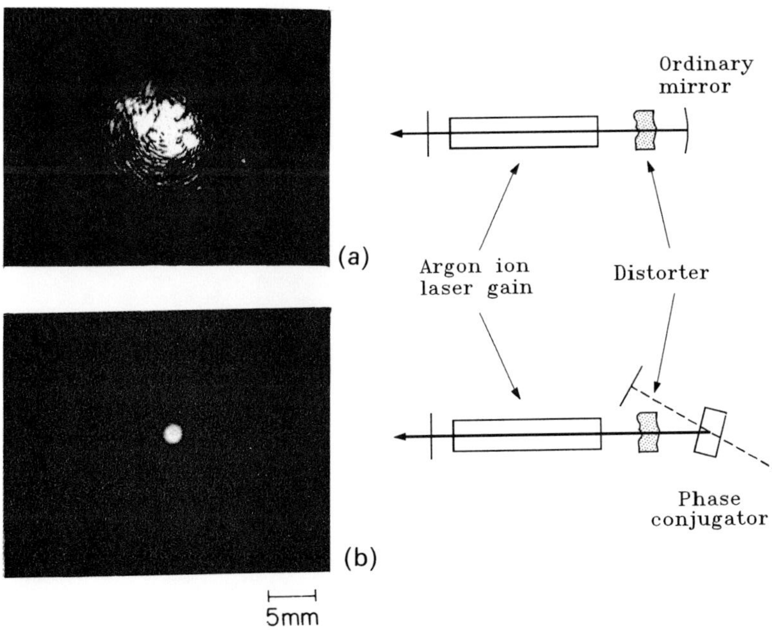

Figure 11.9 *Correction of the phase disturbance introduced by a distorter inside a laser cavity. The pictures show the beam quality when using (a) an ordinary mirror and (b) a $BaTiO_3$ phase conjugator.*

11.6.4. Self-sweeping of the frequency of a dye laser

Self-pumped $BaTiO_3$ phase conjugate mirrors, with the configuration of total reflections at the crystal corners (see Figure 11.8), have been used in dye lasers working in either pulsed [39] or CW [40] modes. Apart from correcting astigmatism and other dye cell aberrations, a remarkable increase of the output beam intensity, a spectral linewidth narrowing to a few GHz, and an unexpected wavelength scan with time over the entire tuning range were observed. The direction of the scan (towards the blue or the red end of the spectrum) depends on the precise location and orientation of the crystal. The laser output increase is explained in terms of the loss reduction in the cavity due to the phase conjugation. On the other hand, the line narrowing is attributed to the high spectral selectivity of the volume holograms formed in the photorefractive crystal. The wavelength scanning is caused by the self-pumped phase conjugator itself at the rate of around 1 Hz frequency shift per round trip, which makes it easily observable at 10^9 Hz.

Such a self-scanning dye laser has been used [41] to perform spectroscopic measurements. The wavelength scans are obtained without any moving parts, and the scanning rate can be made constant by controlling the dye laser power. Coupling of a semiconductor diode laser to a self-pumped phase conjugator also has produced rapid self-scanning of the diode laser tuning range [42].

11.6.5. Phase-locking of independent lasers

In this application, two or more independent lasers are phase-locked with the aid of photorefractive phase conjugators. For example, the image transmitted by an optical fibre

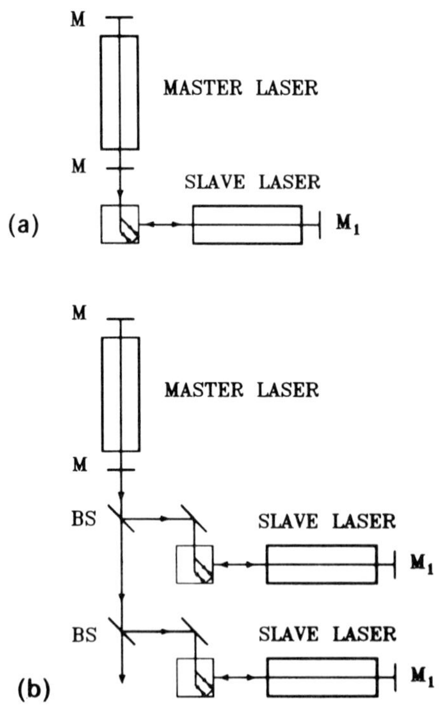

Figure 11.10 *Scheme for phase-locking (a) two independent lasers and (b) several independent lasers.*

can be coherently processed if a local laser is phase-locked to the frequency of the incoming beam. In the configuration sketched in Figure 11.10(a) [43], the beam from the *master* argon-ion laser excites a self-pumped $BaTiO_3$ phase conjugator which generates a conjugate beam, and a *slave* laser, without output mirror, is pointing at the photorefractive crystal. In the slave laser cavity, lasing takes place between the phase conjugator and the remaining mirror. The frequencies of the two lasers stay locked indefinitely to within $\Delta\omega \approx 1/\tau$ (about 1 Hz for $BaTiO_3$), τ being the characteristic time for photorefractive grating recording. Figure 11.10(b) illustrates the set-up for phase-locking several independent lasers. The demonstration of this process with semiconductor lasers [42] (a $BiTiO_3$ phase conjugator working at $\lambda \approx 850$ nm) suggests the applicability of the method to phase-locked arrays of diode lasers.

11.6.6. Phase-conjugate mirrors in interferometers

Self-pumped phase conjugate mirrors can also replace the remote mirror in free-air or fibre-optic interferometers, in order to correct wavefront distortions induced by atmospheric turbulence or modal dispersion respectively [44]. In this case, phase conjugation is only required of the high spatial-frequency phase distortion (the *noise*) and not of the low spatial frequency phase change (the *signal*) caused by propagating to and from the mirror. This is indeed the behaviour of the self-pumped phase conjugator: since it has no external phase reference, it cannot compensate the uniform phase changes of the signal.

11.7. Optical information processing and computing

11.7.1. Optical image processing

Optics is particularly well adapted to process the information contents of an optical image [45,46]. Chiou *et al.* [47] have demonstrated the addition and subtraction of two optical images using a photorefractive phase-conjugating interferometer. Two optical images can be subtracted by shifting the phase of the whole wavefront of one of the beams by 180° and then making it to interfere with the other beam. The amplitude images to be subtracted are placed in the interferometer arms. Real-time subtraction and differentiation of digital images in highly sensitive BSO crystals using reflection type holograms have also been demonstrated in reference [48]. These optical logic operations are performed in data planes superimposed in the same crystal volume, where the corresponding refractive index changes are added or subtracted according with their relative phases. Further work along these lines, including weighted-image addition, has been reported in reference [49]. These results illustrate the capabilities of the photorefractive effect as an efficient way of handling parallel data with a high computation rate.

Coherent optical processing is mostly based on the Fourier-transform properties of lenses. A bidimensional object placed in the front focal plane of a lens, yields, in the back focal plane, the Fourier transform of the object pattern. Then adequate filtering techniques in the Fourier plane can be used to manipulate the information and perform operations like contrast enhancement [45]. All these operations can be advantageously carried out with photorefractive materials, eliminating the need for independent exposure and development stages. Moreover, they allow for selective spatial amplification, instead of filtering, to process the image [50].

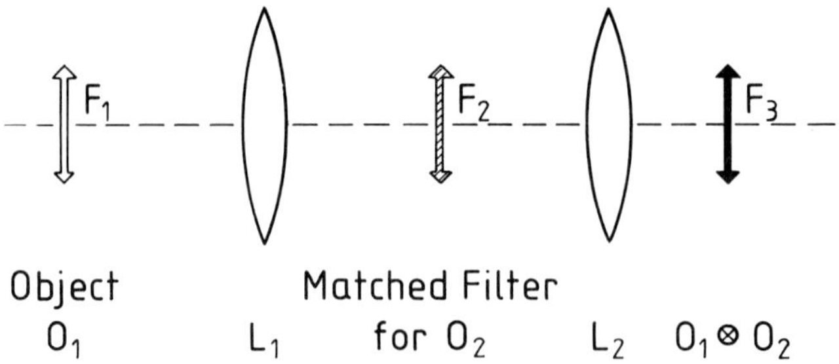

Figure 11.11 *Van der Lugt optical correlator. F_1, F_2 and F_3 are the focal points of the lenses L_1 and L_2.*

A highlight in the field of image processing was the image correlator proposed by Van der Lugt [51]. One places the object O_1 to be recognized at the front focal plane of a lens L_1 and a *matched filter* (Fourier transform) of a reference object O_2 at the common focal plane between lenses L_1 and L_2 (Figure 11.11). Then, the correlation product $O_1 \otimes O_2$ is obtained at the back focal plane of L_2. From this correlation pattern, one may assess whether O_1 and O_2 are like or unlike objects (*pattern recognition*).

Real-time correlation with photorefractive materials has been performed with the basic layout shown in Figure 11.12 [52–55]. It uses a four-wave mixing arrangement, where one of the pump beams (2) carries the object information represented by the complex amplitude A_2, and the other one (1) is a plane wave with uniform amplitude A_1. The signal beam (3) carries the reference pattern A_3. Beams 2 and 3 are focused on the crystal by lenses that yield the Fourier transforms $F(A_2)$ and $F(A_3)$ of A_2 and A_3. The three beams 1, 2, 3 interact inside the photorefractive crystal (e.g. BSO) and generate the conjugated beam (4)

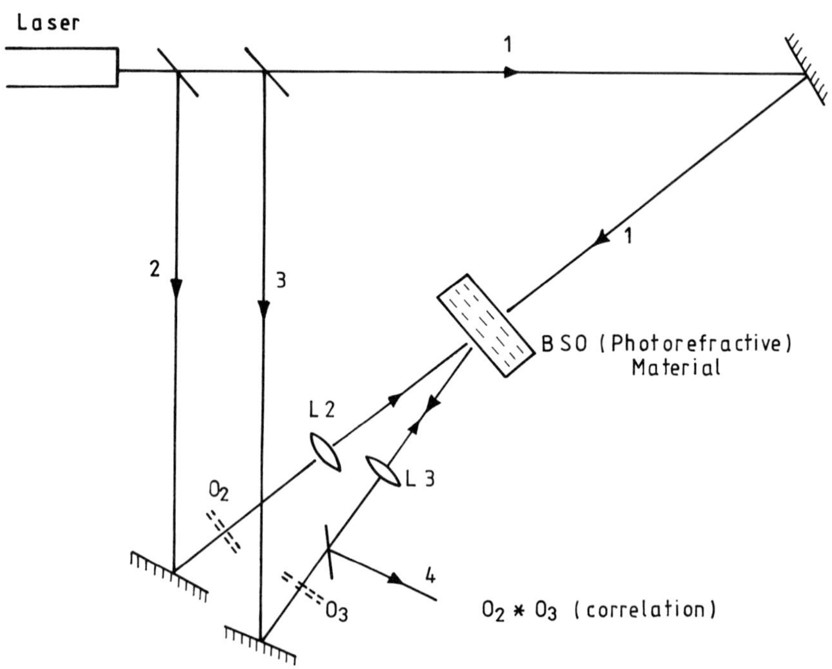

Figure 11.12 *Real-time optical correlator using nonlinear four-wave mixing on a photorefractive (BSO) crystal.*

whose amplitude $F(A_4)$ is

$$F(A_4) \approx F(A_1)F(A_2)F^*(A_3) \approx F(A_1)F^*(A_3) \tag{11.1}$$

After Fourier-transforming $F(A_4)$ through another lens, one obtains a final beam, whose amplitude A_4 is

$$A_4 \approx A_2 \otimes A_3 \tag{11.2}$$

i.e. the correlation product of the object and reference patterns.

Other related schemes have also been used, such as that implemented under the British JOERS Program [56]. For recent industrial developments see references [57,58].

11.7.2. Optical computing: optical interconnects

Essentially, any computer is made up of arrays of interconnected elementary processing and memory units [59]. These computing units consist of logic gates relying on some type of nonlinear process. At present this task is satisfactorily performed by electronic devices. Although optics may also provide such nonlinear elements, they cannot yet compete with their electronic counterparts. On the other hand, optics appears particularly suited to the problem of the interconnections at various levels of the computer architecture. In electronics implementations, interconnects are severely limited by hard wiring, crosstalk and *RC* time constants. Therefore, a strong research effort has been applied since the early 1980s to the development of *optical interconnects*.

An interconnection device links an input to an output vector through a coupling matrix. Several philosophies have been proposed and implemented to perform this task, using optical fibres, integrated optics waveguides or through free space by means of SLMs. One possible and promising strategy uses photorefractive materials to fabricate dynamic holographic optical interconnects. Figure 11.13 illustrates schematically the use of a volume hologram in either transmission or reflection to interconnect given arrays of optical sources or sinks. There is already extensive literature on the subject, which is growing at a rapidly increasing rate. General analyses of the various aspects of the topic are discussed in a number of references [60–65]. To be specific, a few proposals are briefly mentioned here. Reconfigurable interconnects have been implemented by using multiplexed volume gratings in photorefractive BGO [66]. The signal beams have a larger wavelength to assure nondestructive readout. Another approach uses a parallel matrix–vector multiplicator scheme, by means of an SLM in conjunction with a photorefractive crystal [67] (Figure 11.14). The SLM provides reconfigurability, whereas the photorefractive crystal acts as an optical amplifier, to compensate the fan-out losses. Finally, holographic dynamic interconnects have also been proposed using self-pumped phase-conjugation in $BaTiO_3$ [68,69].

11.7.3. Optical neural networks: associative memories

Neural network computers are aimed to simulate the performance of the brain [70]. They consist of elementary processing units or *neurons* linked by weighted *interconnections*. The system may learn through suitable modification of the interconnection weights in accordance with a given learning strategy. The massive parallelism of a neural network architecture is very well adapted to optics, as long as elementary processing time is not a critical parameter. On the other hand, interconnections are a key element of the neural computer. Here, again reconfigurable holographic interconnects are a very promising

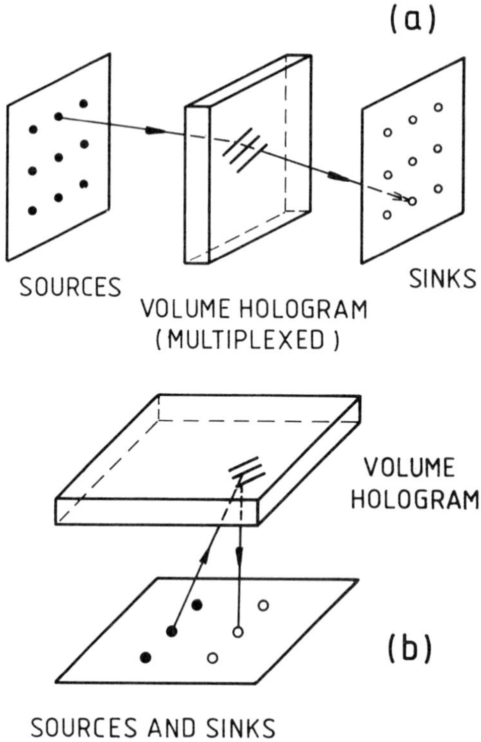

Figure 11.13 *Schematic diagram for a holographic interconnection: (a) transmission hologram; (b) reflection hologram.*

approach [71]. The conceptual similarity between a biological and an optical (holographic) interconnection is shown in Figure 11.15. A particularly simple realization of a reconfigurable optical interconnection between two neural planes has been proposed by Wagner and Psaltis [72]. The light signal from a given "neuron" in the input plane is time-reversed by a phase-conjugated mirror and made to interfere on a photorefractive crystal

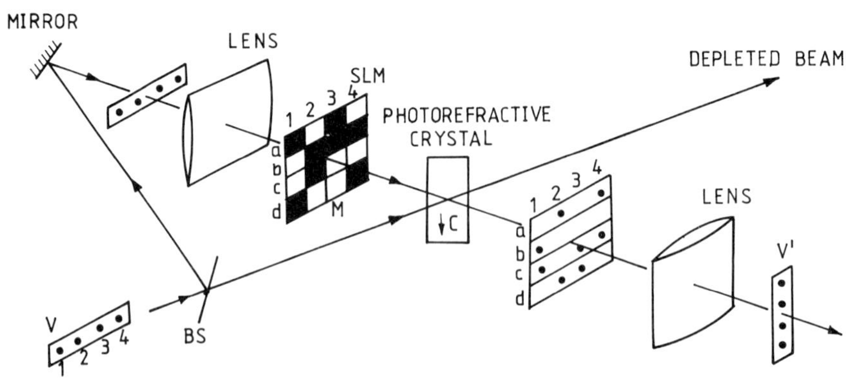

Figure 11.14 *Use of a photorefractive crystal as an amplification stage in a vector-matrix multiplication scheme interconnection sources v_i to detectors v'_j. The matrix is written on a spatial light modulator (SLM).*

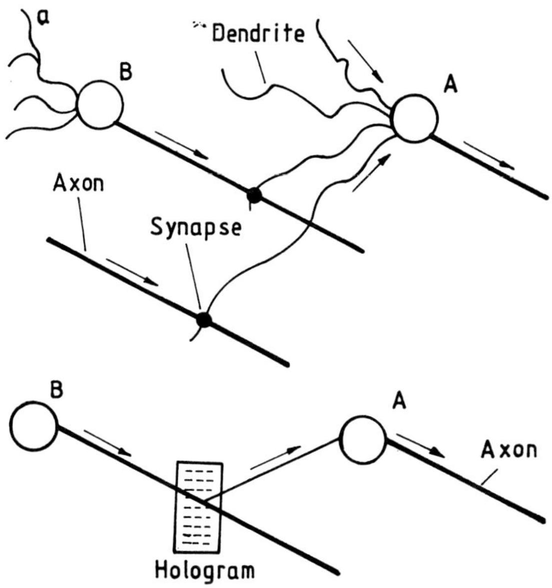

Figure 11.15 *Comparison between a biological and an optical neuron. The optical interconnection is performed by means of a hologram in a photorefractive crystal (synapse).*

with the optical beam coming from another "neuron" in the output plane. The recorded grating assures the connection between the two "neurons". The weight of the connection can be controlled by varying the recording time or the signal intensity. Moreover, this recording process can be adapted to simulate hebbian learning.

A suitable model for neural networks is provided by *associative memories*, where a given stored image has to be retrieved in response to a partial or noisy input. They have been the subject of active research and several optical implementations have already been reported [73]. One such realization is illustrated schematically in Figure 11.16 [74]. A single hologram is simultaneously illuminated by a partial object beam and the approximate reference beam it produces after being phase-conjugated. The associated information is unlocked by this phase-conjugate reference since it retrieves the full image. The key element is a photorefractive $BaTiO_3$ phase conjugator that, at the same time, provides

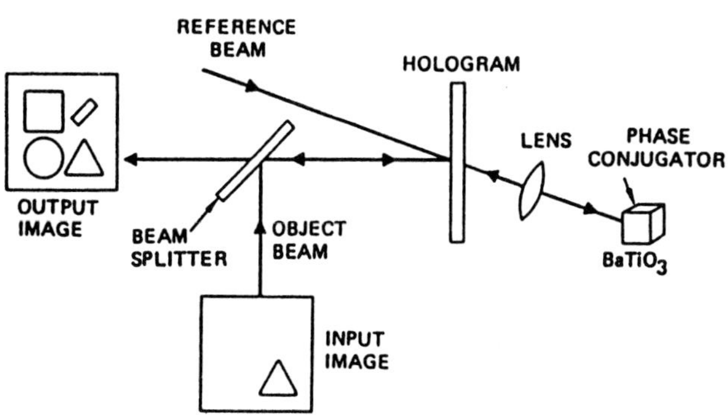

Figure 11.16 *Complete image reconstruction from a partial object input beam. (From [74].)*

retroreflected phase conjugation, gain and thresholding. Due to the combined effect of gain and thresholding, only those signals strongly correlated with the partial object beam are emphasized. The principle has also been demonstrated to work with different objects superimposed in the hologram memory by angular coding of the reference beam directions [75], so that only the image of the partially excited hologram is retrieved from the multiple hologram. The system is particularly relevant in pattern recognition, robotic vision and image processing operations. On the other hand, by using a photorefractive crystal to store the volume hologram, real-time modification of the memory is also possible, making, in principle, the adaptive or learning behaviour of the system accessible.

A related device also demonstrated in reference [76] is the *optical novelty filter* which displays only what is new in a scene. The basic idea is to perform the time derivative of an entire image in parallel by using a photorefractive crystal. A laser beam is spatially modulated in phase with a live scene by using a liquid-crystal TV screen connected to a video camera. Then, the beam is phase conjugated in the photorefractive crystal and retroreflected towards the screen. Those pixels having changed because of changes in the scene will not return light to the source, but deflect it onto a screen, so that only those portions that are moving or changing appear and the stationary background is removed. The device operates correctly when changes occur during the response time of the photorefractive crystal.

11.8. Determination of material parameters

When appropriately designed, photorefractive measurements can be applied to the determination of material parameters. Although these parameters can, in principle, be determined from other optical and/or electrical independent experiments, the high sensitivity of photorefractive measurements may provide a more precise, easier, or complementary way of determining them. From the model presented in Chapter 10, which is generally accepted, the following list of parameters is, in principle, accessible from photorefractive experiments:

ϵ_{ij}	low-frequency dielectric tensor coefficients;
n_{ij}	refractive index tensor coefficients;
r_{ij}	electrooptic tensor coefficients;
σ	cross-section for trap photoionization;
n	free carrier concentration;
N_D	donor concentration;
N_A	acceptor concentration;
μ	free carrier mobility;
γ_t	coefficient for carrier retrapping at acceptors N_A;
$(L_{ph})_{ij}$	photovoltaic tensor coefficients;
$\mathscr{E}_e$	activation energy for electron detrapping.

From proton-induced fixing experiments the following additional parameters can be determined:

H	proton concentration;
$\mathscr{E}_H$	activation energy for proton migration;
D_H	coefficient for proton diffusion.

The determination of some of these parameters may require specific experimental conditions. For example, to obtain thermal activation energies photorefractive measurements have to be performed at various temperatures, and the determination of the spectral dependence of $s_i(\lambda)$ requires measurements at various wavelengths. To obtain the retrapping coefficient γ_t, a combination of CW and short-pulse measurements is required; from CW measurements N_A and the product $\mu\tau$ ($\tau = 1/\gamma_t N_A$ is the carrier lifetime) can be determined, whereas from short-pulse measurements τ is independently obtained. In determining donor/acceptor concentrations, different doping levels or oxidation/reduction states may also help.

It should be emphasized, however, that the experimental determination of these parameters is based on specific models for the photorefractive phenomenon. When the applicability of a model is not guaranteed the results have to be taken with caution. For instance, sometimes it is not easy to ascertain whether a single or double species model is applicable to a specific crystal, and this obviously may drastically affect the final numerical value obtained for a parameter. In the following, some examples of parameters determined from photorefractive experiments are presented.

From anisotropic Bragg diffraction, the local birefringence of $LiNbO_3$ has been measured with a precision of 10^{-4} [77], making possible a fast and precise characterization of the crystal stoichiometry (which is known to affect refractive indices, electrooptic coefficients and other important crystal parameters). Several parameters can be obtained from measurements of the time constants for recording or erasing as a function of the grating period: in $BaTiO_3$ [78,79], the effective trap density $N_D(0)N_A(0)/N$ and the dielectric relaxation time; and in BSO [80], the initial acceptor density $N_A(0)$, the $\mu\tau$ product and the photorefractive absorption coefficient $sN_D(0)$. In $LiNbO_3$ [81], the relative contribution of electron and holes to the photorefractive effect has been determined from both steady-state measurements and independent spectroscopic data (to monitor the Fc^{2+}/Fe^{3+} ratio). Data on the spectral dependence of the nonexponential decay for optical erasure in BSO and BGO [82] have permitted separation of the highly overlapped photorefractive absorption bands of two active centres in these materials, which are also responsible of the optical absorption shoulder observed in these crystals near the absorption edge. From 30 ps pulse experiments [83], the hole trapping time τ_h in $BaTiO_3$ has been estimated to be less than 100 ps. Activation energies for mobility [79] and for detrapping of photorefractive centres [84] have been obtained from measurements of the erase rate as a function of temperature in $BaTiO_3$. Similar measurements have also been made in $LiNbO_3$ [85] and BSO [86]. From fixing experiments also in $LiNbO_3$, the absolute value of the proton concentration [87] and the activation energy for proton migration [88] have been obtained.

References

1. P. Günter and J. P. Huignard (eds), *Photorefractive Materials and Their Applications*, Springer-Verlag, Berlin, Vol. I (1988) and Vol. II (1989).
2. M. P. Petrov, S. I. Stepanov and A. V. Khomenko, *Photorefractive Crystals in Coherent Optical Systems*, Springer Series in Optical Sciences, Springer-Verlag, Berlin (1991).
3. D. L. Staebler, W. J. Burke, W. Phillips and J. J. Amodei, *Appl. Phys. Lett.* **26**(4), 182 (1975).
4. W. J. Burke, D. L. Staebler, W. Phillips and G. A. Alphonse, *Opt. Engng* **17**, 308 (1978).
5. S. Tao, D. Selviah and J. Midwinter, *Optics Lett.*, to appear (1993).

6. J. P. Huignard, J. P. Herriau and F. Micheron, *Appl. Phys. Lett.* **26**, 256 (1975).
7. M. P. Petrov, S. I. Stepanov and A. A. Kamshilin, *Opt. Commun.* **21**, 297 (1977).
8. V. V. Aristov, *Opt. Commun.* **3**, 194 (1971).
9. R. Güther and S. Kusch, *Optica Applicata* **4**(4), 121 (1976).
10. K. Blotekjaer, *Appl. Optics* **18**, 57 (1979).
11. J. P. Huignard, F. Micheron and E. Spitz, in *Optical Properties of Solids. New Developments* (ed. B. O. Seraphin), Ch. 16, North-Holland, Amsterdam (1976).
12. D. L. Staebler, in *Holographic Recording Materials*, Topics in Applied Physics, Vol. 20 (ed. H. M. Smith), Springer-Verlag, Heidelberg (1977), pp. 102–132.
13. J. P. Huignard and F. Micheron, *Appl. Phys. Lett.* **29**, 591 (1976).
14. G. S. Trofimov and S. I. Stepanov, *Pis'ma v Zh. Tech. Fiz.* **11**, 615 (1985). English transl.: *Sov. Tech. Phys. Lett.* **11**, 256 (1985).
15. J. P. Huignard and J. P. Herriau, *Appl. Optics* **16**, 1807 (1977).
16. G. C. Valley and M. B. Klein, *Opt. Engng* **22**, 704 (1983).
17. F. M. Küchel and H. J. Tiziani, *Opt. Commun.* **38**, 17 (1981).
18. See page 202 of reference [2].
19. R. P. Kenan, C. M. Verber and E. Van Wood, *Appl. Phys. Lett.* **24**, 428 (1974)
20. O. Mikami, *Opt. Commun.* **19**, 42 (1976).
21. S. I. Stepanov, A. A. Kamshilin and M. P. Petrov, *Pis'ma v Zh. Tech. Fiz.* **3**, 89 (1977). English transl.: *Sov. Tech. Phys. Lett.* **3**, 36 (1977).
22. M. P. Petrov, S. I. Stepanov and A. A. Kamshilin, *Opt. Commun.* **29**, 44 (1979).
23. M. P. Petrov, S. I. Stepanov and A. A. Kamshilin, *Ferroelectics* **22**, 631 (1978).
24. P. D. Henshaw, *Appl. Optics* **21**, 2323 (1982).
25. G. T. Sincerbox and G. Roosen, *Appl. Optics* **22**, 690 (1983).
26. G. Pauliat, J. P. Herriau, A. Delboulbe, G. Roosen and J. P. Huignard, *J. Opt. Soc. Am.* **B3**, 306 (1986).
27. E. Voit, C. Zaldo and P. Günter, *Optics Lett.* **11**, 309 (1986).
28. D. Rak, I. Ledoux and J. P. Huignard, *Opt. Commun.* **49**, 302 (1984).
28a. P. Mills and E. G. S. Paige, *Electronics Lett.* **21**, 885 (1985).
28b. G.A. Rakuljic and V. Leyva, *Optics Lett.* **18**, 459 (1993).
28c. R. Müller, M. T. Santos, L. Arizmendi and J. M. Cabrera, *J. Phys.: Cond. Matt.* (in press) (1994).
29. J. P. Huignard and A. Marrakchi, *Optics Lett.* **6**, 622 (1981).
30. A. Chiou and P. Yeh, *Optics Lett.* **10**, 621 (1985).
31. S. Kwong and A. Yariv, *Appl. Phys. Lett.* **48**, 564 (1986).
32. M. D. Levenson, K. M. Johnson, V. C. Hanchett and K. Chiang, *J. Opt. Soc. Am.* **71**, 737 (1981).
33. M. C. Gower, in the "Annual Report to the Laser Facility Committee 1985", Laser Division, Rutherford Appleton Laboratory, Chilton, Didcot, Oxon, UK.
34. J. P. Huignard, J. P. Herriau, L. Pichon and A. Marrakchi, *Optics Lett.* **5**, 436 (1980).
35. G. J. Dunning and R. C. Lind, *Optics Lett.* **7**, 558 (1982).
36. B. Fischer and S. Sternklar, *Appl. Phys. Lett.* **46**, 113 (1985).
37. J. Feinberg, *Optics Lett.* **7**, 486 (1982).
38. M. Cronin-Golomb, B. Fischer, J. Nilsen, J. O. White and A. Yariv, *Appl. Phys. Lett.* **41**, 219 (1982).
39. R. A. MacFarlane and D. G. Steel, *Optics Lett.* **8**, 208 (1983).
40. J. Feinberg and G. D. Bacher, *Optics Lett.* **9**, 420 (1984).
41. J. M. Ramsey and W. B. Whitten, *Analyt. Chem.* **56**, 2979 (1984).
42. M. Cronin-Golomb and A. Yariv, *Optics Lett.* **11**, 455 (1986).
43. J. Feinberg and G. D. Bacher, *Appl. Phys. Lett.* **48**, 570 (1986).
44. J. Feinberg, *Optics Lett.* **8**, 569 (1983).
45. *Optical Information Processing. Fundamentals* (ed. S. H. Lee), Springer-Verlag, Berlin (1981).
46. *Optical Processing and Computing* (eds H. Arsenault, T. Szoplik and B. Macuko), Academic Press, Orlando (1989).

47. A. E. T. Chiou and P. Yeh, *Optics Lett.* **11**, 306 (1986). See also: A. E. T. Chiou, P. Yeh, M. Khoshenevisan, *SPIE's O-E Laser '86*, Los Angeles, January 1986, Paper 613-07.
48. Y. H. Ja, *Appl. Phys.* **B36**, 21 (1985).
49. N. A. Vainos, J. A. Khoury and R. W. Eason, *Optics Lett.* **13**, 503 (1988).
50. T. Y. Chang, J. H. Hong and P. Yeh, *Optics Lett.* **15**, 743 (1990).
51. A. B. Van der Lugt, *Proc. IEEE* **54**, 1055 (1966).
52. J. O. White and A. Yariv, *Appl. Phys. Lett.* **37**, 5 (1980).
53. C. R. Petts, M. W. McCall and L. C. Laycock, *Electron. Lett.* **20**, 32 (1984).
54. B. Loisseaux, G. Iliaguer and J. P. Huignard, *Optics Eng.* **24**, 144 (1988).
55. P. D. Foote, T. P. Hall and L. M. Connors, *Optics Laser Technol.* **39**, February (1986).
56. N. Collings, *Optical Pattern Recognition Using Holographic Techniques*, Addison-Wesley (1988).
57. H. Rajbenbach, *Sig. Image Process.* Mar., 1 (1992).
58. L. Cheng, *Opt. Comput. Process.* June, 2, 5 (1992).
59. A. D. McAulay, *Optical Computer Architectures,* John Wiley, New York (1991).
60. S. A. Collins Jr and H. J. Canfield, *J. Opt. Soc. Am.* **A6**, 1568 (1989).
61. H. Lee, X. Gu and D. Psaltis, *J. Appl. Phys.* **65**, 2191 (1989).
62. A. Chiou and P. Yeh, *Appl. Optics* **29**, 1111 (1990).
63. S. Wu, Q. Song, A. Mayers, D. A. Gregory and T. T. S. Yu, *Appl. Optics* **29**, 1118 (1990).
64. A. Marrakchi and W. M. Hubbard, *Optics Lett.* **16**, 417 (1991).
65. K. Y. Tu, H. Lee and T. Tamir, *Appl. Optics* **31**, 1717 (1992).
66. R. McRuer, J. Wilde, L. Hesselink and J. Goodman, *Optics Lett.* **14**, 1174 (1989).
67. P. Yeh, A. E. T. Chiou and J. Horg, *Appl. Optics* **27**, 2093 (1988).
68. S. Weiss, M. Segev, S. Skernklar and B. Fisher, *Appl. Optics* **27**, 3422 (1988).
69. A. Sakamoto, *Jap. J. Appl. Phys.* **29**, L2070 (1990).
70. R. Beale and T. Jackson, *Neural Computing: An Introduction*, Adam Hilger, Bristol (1990).
71. D. Psaltis, D. Brady, X. G. Gu and S. Lin, *Nature* **343**, 325 (1990).
72. K. Wagner and D. Psaltis, *Appl. Optics* **26**, 5061 (1987).
73. Y. Owechko, *IEEE J. Quantum Electron.* **QE-25**, 619 (1989).
74. B. H. Soffer, G. J. Dunning, Y. Owechko and E. Marcom, *Optics Lett.* **11**, 118 (1986).
75. A. Yariv, S. Kwong and K. Kyuma, *Appl. Phys. Lett.* **48**, 114 (1986).
76. D. Z. Anderson, D. M. Lininger and J. Feinberg, *Optics Lett.* **12**, 123 (1987).
77. L. Arizmendi, *J. Appl. Phys.* **64**, 4654 (1988).
78. J. Feinberg, D. Heimen Jr, A. R. Tanguay and R. W. Hellwarth, *J. Appl. Phys.* **51**, 1297 (1980).
79. S. Ducharme and J. Feinberg, *J. Appl. Phys.* **56**, 839 (1984).
80. R. A. Mullen and R. W. Hellwarth, *J. Appl. Phys.* **58**, 40 (1985).
81. R. Orlowski and E. Krätzig, *Solid State Commun.* **27**, 1351 (1978).
82. J. A. Baquedano, L. Contreras, E. Diéguez and J. M. Cabrera, *J. Appl. Phys.* **66**, 5146 (1989).
83. A. L. Smirl, G. C. Valley, R. A. Mullen, K. Bohnert, C. D. Mire and T. F. Boggess, *Optics Lett.* **12** (1987).
84. S. Ducharme and J. Feinberg, *J. Opt. Soc. Am.* **B3**, 283 (1986).
85. L. Arizmendi, P. D. Townsend, M. Carrascosa, J. Baquedano and J. M. Cabrera, *J. Phys.: Cond. Matt.* **3**, 5399 (1991).
86. M. A. Powell and C. R. Petts, *Optics Lett.* **1**, 36 (1986).
87. R. Müller, L. Arizmendi, M. Carrascosa and J. M. Cabrera, *Appl. Phys. Lett.* **60**, 3212 (1992).
88. M. Carrascosa and L. Arizmendi, *J. Appl. Phys.* **73**, 2709 (1993).

Index